ARISTOTLE'S *PARTS OF ANIMALS*

Aristotle's *Parts of Animals* is a foundational text in both the history of philosophy and the history and philosophy of biology. Critically important for understanding his mature philosophical program, the *Parts of Animals* has two chief aims. *PA* Book I is an introduction to the study of animals and plants and provides preliminary considerations for how to investigate all aspects of their nature. *PA* Books II–IV provide the most comprehensive example of the application of Aristotle's philosophical methodology to real-world examples of substances, that is, to animals. In this book, a team of international experts cover topics such as Aristotle's exhortation to study biology; his methodology in the study of natural entities and kinds; the study of mind as part of nature; his analysis and use of concepts such as essence, substance, definition, matter, form, species, analogy, and teleology; and the influence and legacy of the text.

SOPHIA M. CONNELL is Reader in Philosophy at Birkbeck, University of London. She is the author of *Aristotle on Female Animals: A Study of the Generation of Animals* (Cambridge, 2016) and *Aristotle on Women: Physiology, Psychology and Politics* (Cambridge, 2021) and is the editor of *The Cambridge Companion to Aristotle's Biology* (Cambridge, 2021). She has published papers on Aristotle's embryology, psychology, and ethics and on women philosophers both ancient and modern.

CAMBRIDGE CRITICAL GUIDES

Titles published in this series:

Aristotle's *Parts of Animals*
EDITED BY SOPHIA CONNELL

Hume's *Dialogues Concerning Natural Religion*
EDITED BY PAUL RUSSELL

Kierkegaard's *Works of Love*
EDITED BY JEFFREY HANSON AND WOJCIECH KAFTANSKI

Ockham's *Summa Logicae*
EDITED BY CLAUDE PANACCIO AND JENNY PELLETIER

Hume's *A Treatise of Human Nature*
EDITED BY ELIZABETH S. RADCLIFFE

Heidegger's *Being and Time*
EDITED BY AARON JAMES WENDLAND AND TOBIAS KEILING

Hume's *Essays*
EDITED BY MAX SKJÖNSBERG AND FELIX WALDMANN

Hegel's *Philosophy of Nature*
EDITED BY MARINA BYKOVA

Plato's *Gorgias*
EDITED BY J. CLERK SHAW

Boethius' *Consolation of Philosophy*
EDITED BY MICHAEL WIITALA

Wittgenstein's *Tractatus Logico-Philosophicus*
EDITED BY JOSÉ ZALABARDO

Kierkegaard's *Either/Or A Critical Guide*
EDITED BY RYAN S. KEMP AND WALTER WIETZKE

Cicero's *De Officiis*
EDITED BY RAPHAEL WOOLF

Kierkegaard's *The Sickness unto Death*
EDITED BY JEFFREY HANSON AND SHARON KRISHEK

Nietzsche's *Thus Spoke Zarathustra*
EDITED BY KEITH ANSELL-PEARSON AND PAUL S. LOEB

Aristotle's *On the Soul*
EDITED BY CALEB M. COHOE

Schopenhauer's *World as Will and Representation*
EDITED BY JUDITH NORMAN AND ALISTAIR WELCHMAN

Kant's *Prolegomena*
EDITED BY PETER THIELKE

Hegel's *Encyclopedia of the Philosophical Sciences*
EDITED BY SEBASTIAN STEIN AND JOSHUA WRETZEL

Maimonides' *Guide of the Perplexed*
EDITED BY DANIEL FRANK AND AARON SEGAL

Fichte's *System of Ethics*
EDITED BY STEFANO BACIN AND OWEN WARE

Hume's *An Enquiry Concerning the Principles of Morals*
EDITED BY ESTHER ENGELS KROEKER AND WILLEM LEMMENS

Hobbes's *On the Citizen*
EDITED BY ROBIN DOUGLASS AND JOHANN OLSTHOORN
Hegel's *Philosophy of Spirit*
EDITED BY MARINA F. BYKOVA
Kant's *Lectures on Metaphysics*
EDITED BY COURTNEY D. FUGATE
Spinoza's *Political Treatise*
EDITED BY YITZHAK Y. MELAMED AND HASANA SHARP
Aquinas's *Summa Theologiae*
EDITED BY JEFFREY HAUSE
Aristotle's *Generation of Animals*
EDITED BY ANDREA FALCON AND DAVID LEFEBVRE
Hegel's *Elements of the Philosophy of Right*
EDITED BY DAVID JAMES
Kant's *Critique of Pure Reason*
EDITED BY JAMES R. O'SHEA
Spinoza's *Ethics*
EDITED BY YITZHAK Y. MELAMED
Plato's *Symposium*
EDITED BY PIERRE DESTRÉE AND ZINA GIANNOPOULOU
Fichte's *Foundations of Natural Right*
EDITED BY GABRIEL GOTTLIEB
Aquinas's *Disputed Questions on Evil*
EDITED BY M. V. DOUGHERTY
Aristotle's *Politics*
EDITED BY THORNTON LOCKWOOD AND THANASSIS SAMARAS
Aristotle's *Physics*
EDITED BY MARISKA LEUNISSEN
Kant's *Lectures on Ethics*
EDITED BY LARA DENIS AND OLIVER SENSEN
Kierkegaard's *Fear and Trembling*
EDITED BY DANIEL CONWAY
Kant's *Lectures on Anthropology*
EDITED BY ALIX COHEN
Kant's *Religion within the Boundaries of Mere Reason*
EDITED BY GORDON MICHALSON

(*Continued after the Index*)

ARISTOTLE'S *PARTS OF ANIMALS*

A Critical Guide

EDITED BY

SOPHIA CONNELL

Birkbeck, University of London

CAMBRIDGE
UNIVERSITY PRESS

CAMBRIDGE
UNIVERSITY PRESS

Shaftesbury Road, Cambridge CB2 8EA, United Kingdom

One Liberty Plaza, 20th Floor, New York, NY 10006, USA

477 Williamstown Road, Port Melbourne, VIC 3207, Australia

314–321, 3rd Floor, Plot 3, Splendor Forum, Jasola District Centre, New Delhi – 110025, India

103 Penang Road, #05–06/07, Visioncrest Commercial, Singapore 238467

Cambridge University Press is part of Cambridge University Press & Assessment, a department of the University of Cambridge.

We share the University's mission to contribute to society through the pursuit of education, learning and research at the highest international levels of excellence.

www.cambridge.org
Information on this title: www.cambridge.org/9781316516119

DOI: 10.1017/9781009031882

When citing this work, please include a reference to the DOI 10.1017/9781009031882

First published 2025

Cover image: Unspecified - Circa 2003: On the Parts of Animals (De Partibus Animalium), by Aristotle (384-322 BC). Press, 1585. Paris, Bibliothèque Nationale De France (Library) / Photo by DeAgostini / Getty Images

A catalogue record for this publication is available from the British Library

A Cataloging-in-Publication data record for this book is available from the Library of Congress

ISBN 978-1-316-51611-9 Hardback

Contents

Contributors

SOPHIA M. CONNELL is Reader in Philosophy at Birkbeck, University of London. She is the author of *Aristotle on Female Animals: A Study of the Generation of Animals* (Cambridge, 2016) and *Aristotle on Women: Physiology, Psychology and Politics* (Cambridge, 2021) and is the editor of *The Cambridge Companion to Aristotle's Biology* (Cambridge, 2021). She has published papers on Aristotle's embryology, psychology, and ethics and on women philosophers both ancient and modern.

CHRISTOPHER FREY is the McFarlin Professor of Philosophy at the University of Tulsa. His work on Aristotle's metaphysics, natural philosophy, and psychology has appeared in *Oxford Studies in Ancient Philosophy*, *Phronesis*, and *Ancient Philosophy*. He is currently completing a book about Aristotle's concept of life, entitled *The Principle of Life: Aristotelian Souls in an Inanimate World.*

JESSICA GELBER is Associate Professor in the Philosophy department at the University of Toronto. Her current projects focus on the relation between Aristotle's conception of science and his metaphysics and on questions about the nature of causation and explanation. She has written articles on Aristotle's natural history, teleology, metaphysics, and embryology.

MARY-LOUISE GILL is David Benedict Professor of Classics and Philosophy at Brown University. She specializes in the method, natural philosophy, and metaphysics of Plato and Aristotle and is the author of *Aristotle on Substance: the Paradox of Unity* (Princeton, 1989), an introduction and co-translation of *Plato: Parmenides* (Hackett, 1996), and *Philosophos: Plato's Missing Dialogue* (Oxford, 2012). She has also coedited three collections of papers on Aristotle and ancient philosophy.

MYRTO HATZIMICHALI is Associate Professor in Classics (Philosophy) at the University of Cambridge and a Fellow of Homerton College. Her research interests and range of publications include Aristotle's biological works, the history of the Aristotelian corpus, and post-Hellenistic philosophy and the work of scholars and grammarians based in the library of Alexandria.

DEVIN HENRY is a professor in the Department of Philosophy at the University of Western Ontario. He is the author of *Matter, Form, and Moving Causes: Aristotle on Substantial Generation* (Cambridge, 2019) and coeditor of *Bridging the Gap between Aristotle's Science and Ethics* (with Karen M. Nielsen, Cambridge, 2015). He teaches courses on Plato and Aristotle and has published many articles on ancient Greek philosophy including recently "Essence and Definition in Aristotle's *PA*," "Science et méthode dans l'éthique d'Aristote," and "Aristotle on Epigenesis."

THOMAS KJELLER JOHANSEN is Professor of Philosophy in the Institute of Philosophy, Classics, History of Ideas and Art at the University of Oslo. He was previously Professor of Ancient Philosophy at the University of Oxford and Tutorial Fellow at Brasenose College. He is the author of *Aristotle on the Sense-Organs* (Cambridge, 1997), *Plato's Natural Philosophy* (Cambridge, 2004), and *The Powers of Aristotle's Soul* (Oxford, 2012).

MONTE RANSOME JOHNSON is Professor of Philosophy at the University of California, San Diego. He is the author of *Aristotle on Teleology* (Oxford, 2008) and of articles on Democritus, Aristotle, and Lucretius, including contributions to *The Cambridge Companion to Aristotle's Biology*, *The Cambridge Companion to Aristotle's* Nicomachean Ethics, *The Cambridge Companion to Ancient Greek and Roman Science*, and *The Cambridge Companion to Lucretius*. He is currently working with D. S. Hutchinson on a reconstruction of Aristotle's lost dialogue the *Protrepticus*.

GIOULI KOROBILI is an Einstein Center Chronoi Research Fellow at Humboldt University of Berlin. She has contributed to a number of collective volumes on Aristotle, ancient medicine, and Byzantine Aristotelian commentators. She has published *Aristotle.* On Youth and Old Age, Life and Death, and Respiration 1–6, *with Translation, Introduction and Interpretation* (Brill, 2022) and she is currently writing a monograph on analogies at the intersection of medicine and meteorology in Greco-Roman antiquity.

EMILY KRESS is an assistant professor in the Department of Philosophy at Brown University. Her research is centered on Aristotle's accounts of teleological causation and explanation, including their development in the biological works. She is the author of articles on Aristotle's accounts of natural teleology, luck, causation in animal generation, and spontaneous generation. She completed her PhD at Yale University in 2018.

MARISKA LEUNISSEN (PhD, Leiden University, 2007) is Professor of Philosophy at UNC Chapel Hill. She works in ancient philosophy, with special interests in Aristotelian natural philosophy and philosophy of science and its intersections with early medicine. She is the author of *Explanation and Teleology in Aristotle's Science of Nature* (Cambridge University Press, 2010) and *From Natural Character to Moral Virtue in Aristotle* (Oxford Universisty Press, 2017) and is currently working on a monograph on Aristotle's natural scientific study of women.

MICHAIL PERAMATZIS is an associate professor at the Philosophy Faculty, University of Oxford, and the Hinton Clarendon Tutorial Fellow in Philosophy at Worcester College. He works on Aristotle's and Plato's theoretical philosophies but also has interests in logic, epistemology, metaphysics, and the philosophy of science. He is the author of *Priority in Aristotle's Metaphysics* (Oxford University Press, 2011) and many articles on Aristotle and Plato.

CHRISTOF RAPP is Professor of Ancient Philosophy at Ludwig-Maximilians-Universität, Munich, since 2009. From 1993 to 2000, he was an assistant professor at the University of Tübingen. From 2001 to 2009, he held the Chair of Ancient and Contemporary Philosophy at Humboldt-Universität in Berlin. He has also held visiting positions in Berkeley (2000), Oxford (2008), and Paris (2014). Rapp has widely published on ancient philosophy, especially on Aristotle.

STASINOS STRAVRIANEAS is an assistant professor at the Hellenic Open University (School of Humanities). His research focuses on Aristotelian natural philosophy. He is the author of a modern Greek translation and commentary on Aristotle's *Parts of Animals* and coeditor of the volumes, with Andrea Falcon, *Aristotle on How Animals Move: The* de Incessu Animalium*: Text, Translation, and Interpretative Essays* (Cambridge Unversity Press, 2021) and, with Stavros Kouloumentas, *Philosophy and Medicine in Ancient Greece: Methods and Problems* (Crete University Press, 2023).

Introduction

Sophia M. Connell

I.1 Aristotle's *Parts of Animals* as Philosophy

One might wonder about the value of studying Aristotle's *Parts of Animals* (*PA*) if one's interest is chiefly philosophical. Aristotle's detailed study of the bodies of animals initially appears to be part of the history of science, an early attempt to categorize or conceptualize divergences in animal anatomy. Thought of in this manner, one might worry about contemporary irrelevance (given the relative sophistication of modern zoology) and a disconnection from the more enduring parts of Aristotle's thinking. These misgivings, although understandable, are unnecessary since this work is of the upmost importance for Aristotle's philosophical program. The *Parts of Animals* is a core part of Aristotle's project of uncovering and elucidating the structure of reality employing logical tools. Living beings, and especially animals, are the most complete substantial realities that human beings have access to in their environments. Through studying the parts of animals, which contribute to the living capacities of the whole entity, we learn about important ways in which the world works, from the inside out.

The *Parts of Animals* Book I contains one of the most powerful encomiums to the study of animal life in the Western tradition, arguing for the value of even the lowliest species due to their beauty and wonder (*PA* I.5). Philosophically-minded students of the natural world must logically order animals and their parts to gain knowledge, which applies not only to the work itself but more broadly to the entirety of Aristotle's "inquiry concerning animals and plants" (*Meteor.* I.1, 339a8). Living beings maintain themselves in existence due to their complex organizational capacities. Furthermore, the life of animals rather than of plants is to be the primary focus because their way of being alive involves sensory responses that are a kind of knowledge; given this, they also display more order and unity. Human beings are also animals and thus finding out about how the bodies of animals function provides necessary insights about our own existence; in the case of humans, we naturally seek knowledge for its own sake, for the sheer joy of understanding (*Metaph.* I.1). Inquiry leads us to understand the structure of things, including ourselves, and provides materials for the ultimate aim of philosophy, to contemplate what is true.

Almost a quarter of all Aristotle's extant works directly discuss animals; in addition to the *Parts of Animals*, the most significant of these are the *Generation of Animals* (in five books) and the *Historia Animalium* (in ten books).[1] In the last century, those working on Aristotle's philosophically motivated natural science began to recognize how crucial all these works are to understanding his thought. In terms of logic, philosophy of science, epistemology, psychology, metaphysics, and ethics, the biological works contribute significantly to our picture of Aristotle's methods and commitments.[2] This important development in scholarship on Aristotle is sometimes termed the 'biological turn.'

A reader might, however, wonder how it can make sense to view Aristotle's works on animals as 'biology,' since the term was not invented until the late eighteenth century.[3] This term need not be anachronistic if care is taken not to solely import contemporary understandings. Insofar as Aristotle undertook a systematic study of mortal living nature, and included careful empirical methodologies, there are good reasons to label his work biological. Very generally, Aristotle collects the facts (the 'that') before seeking causal explanations of these facts (the 'why') (*A. Po.* II.1, 94a20–36). He studied the basis of life and its manifestations in animals and, to a certain extent, plants.[4] On the other hand, unlike contemporary biology, Aristotle did not view his endeavors as separate from the study of philosophy. Many of his concerns are more general, for example, to find that the structure of reality aims toward ends; the ends in nature are generally to sustain the way of life of the kind across generations. There is no conceptualization of evolution but rather a recognition of the continuity of natural patterns, grounded in functional optimality.[5]

This brief introduction will explain how the *Parts of Animals* came back from obscurity to be recognized as a significant part of Aristotle's

[1] The last book is generally attested to be spurious, although Dean-Jones speculates that certain chapters count as 'embedded endoxa' (see Dean-Jones 2022).

[2] For an overview, see the essays in Gotthelf and Lennox 1987 and Connell 2021.

[3] For this objection, see Cunningham 1999. For further discussion see Connell 2021: 3–5.

[4] The systematic study of plant life was undertaken by Aristotle's student, Theophrastus. His works entitled *Inquiry Into Plants* and *Causes of Plants* maintained many of Aristotle's general structures of investigation. See Falcon 2024.

[5] See especially Henry 2013.

philosophical endeavors in the last century. It will then give a brief overview of the structure and content of the work. The final section will introduce the philosophical topics in relation to the text presented in each chapter of this book.

I.2 The Study of the Text

The *Parts of Animals* is central to Aristotle's philosophy of science. And yet for many centuries, it was sidelined by those interested in Aristotle's philosophy. While biologists and zoologists had long recognized Aristotle's prominence, the separation of science from philosophy left the *PA* behind.[6] By the nineteenth century, developments in the biological sciences saw a resurgence of interest in the text, which eventually reached the consciousness of philosophers. Two decades before the publication of Charles Darwin's *On the Origin of Species*, a famous debate raged between the French biologists George Cuvier (1760–1832) and Etienne Geoffroy Saint-Hillaire (1805–1895) over the organization of animals. For Cuvier, the emphasis had to be on the function of bodily parts and their interrelation, based on Aristotelian teleology.[7] For Geoffroy, the similarities between parts of different animals, regardless of function, was central, labeling these 'homologues.' A hundred years later, modern evolutionary science realized the necessity for and complementarity of both form and function.[8] Both aspects can be found in Aristotle's work on animal parts – the focus on the integration of parts and their functions as part of the whole, as well as a keen perception of the similarities across kinds in the shapes of body parts.

This extraordinary work comes to the consciousness of contemporary scientists and classical philosophers during this period of transition. William Ogle, a naturalist and physician, translated the text with notes in 1882 and sent it to Charles Darwin.[9] His response was as follows:

> I have rarely read anything which has interested me more; though I have not read as yet more than a quarter of the book proper. From quotations which I had seen I had a high notion of Aristotle's merits, but I had not the

6 On Aristotle's zoology in the Middle Ages and Renaissance, see Perfetti 2000.

7 Coleman 1964. Pierre Pellegrin argues that Aristotle was not a 'biologist' like Cuvier but was an important precursor to him (see Pellegrin 2025).

8 Appel 1987.

9 An earlier translation into English was Thomas Taylor's in 1810 published together with Aristotle's *Progression of Animals* (*IA*) and the pseudo-Aristotelian works, *Problemata* (*Prob.*) and *On Divisible Lines*. Taylor's main interest was in Neoplatonist uses of Aristotle. A German translation of only Book I of the work by F. N. Titze, without any notes or commentary, appeared in 1819 and another by A. Karsch in 1855.

> most remote notion what a wonderful man he was. Linnaeus and Cuvier have been my two gods, though in very different ways, but they were mere schoolboys to old Aristotle.[10]

Arthur Peck, a student of classicist and philosopher Francis Cornford in Cambridge, completed a translation with extensive introduction and notes in 1937.[11] The French edition by Pierre Louis followed in 1956.

The 1970s and early 1980s produced significant advances in understanding the contents of the *Parts of Animals* and its role in Aristotle's philosophical program.[12] There are two key aspects to this reevaluation. First and foremost, the work provides a methodology for biological investigation. This enriches and supplements the scientific and epistemological methodology detailed in works such as the *Posterior Analytics*. In *PA* I, Aristotle presents and defends various standards that circumscribe the requirements for explanatory accounts of natural phenomena. Secondly, *PA* II–IV engages directly with the application of this methodology to morphology, anatomy, and physiology in animals. We find here also the natural philosopher grappling with the application of many crucial philosophical concepts in this field of inquiry including substance (*ousia*), species/form (*eidos*), matter, analogy, nature, cause, necessity, division, and definition; each are applied with sensitivity to the subject matter and in the appropriate explanatory context.

Brief Description of the Contents of the Text

The text has two main aims, method and practice. *PA* Book I is a guide and introduction to how to study life and provides preliminary considerations about how to investigate all aspects of animal nature. *PA* Books II–IV is our most comprehensive evidence of the application of Aristotle's philosophical methodology to real world examples of substances, that is, to animals.

Book I, chapter 1 of the *Parts of Animals* sets out several key methodological principles in the study of living nature. Aristotle begins thus:

> So it is clear that for the enquiry into nature (περὶ φύσιν ἱστορία), too, there should be certain standards (ὅροι), such that by referring to them one can appraise the manner of its proofs (ἀποδέξεται). (*PA* I.1, 639a12–13)

[10] Darwin Correspondence Project, "Letter no. 13697," www.darwinproject.ac.uk/letter/?docId=letters/DCP-LETT-13697.xml. See Gotthelf 1999.

[11] The 1930s also saw a Spanish translation by F. G. Pales (1932) but without any accompanying commentary or notes.

[12] See especially Balme 1972; Kullmann 1974; Gotthelf 1976/7; Lennox 1980; Pellegrin 1982.

Whatever he means by "standards" Aristotle is clearly concerned to provide general guidelines as answers to a series of questions. The first question is one of the most important.

> I mean, for example, should one take each substantial being (οὐσίαν) singly and define it independently (καθ' αὑτήν), e.g. taking up one by one the nature of humankind, lion, ox, and any other animal as well; or should one first establish, according to something common, the attributes common to all? For many of the same attributes are present in many different kinds of animals, e.g. sleep, respiration, growth, deterioration, death, and in addition any remaining affections and dispositions such as these. (*PA* I.1, 639a15–22, trans. Lennox 2001a)

This question reflects ontological and epistemological tensions in Aristotle's philosophy. What should we look to, the particular kind, which can no longer be divided into others, or a wider kind to which several belong? In zoology, this becomes a question of what sorts of explanations are appropriate in particular instances. In some cases, such as in explaining sleep, growth, or respiration, there are no more-or-less differences between how these occur in different animals. All animals grow, deteriorate, and die, so the account can indeed cover all instances.[13] In other cases, however, one type of affection does not cover all kinds; for example, animals locomote in very different ways. Thus, locomotion although the same in *genos* is different in different animals (e.g. flying versus swimming).[14] Sometimes it is appropriate to concentrate on what is entirely distinctive of a certain kind of animal in order to explain its affections, dispositions, or parts. For example, only one species of octopus lacks a double row of suckers; Aristotle explains why thus:

> Now while the other octopuses have two rows of suckers, one kind of octopus has a single row. This is because of the length and thinness of their nature; for it is necessary that the narrow tentacle should have a single row of suckers. It is not, then, because it is best that they have this feature, but because it is necessary owing to the distinctive account of their substantial being (διὰ τὸν ἴδιον λόγον τῆς οὐσίας). (*PA* IV.9, 685b12–15)

PA I also introduces and defines other general guidelines. The philosopher of nature, like the astronomer, should seek the facts before

[13] Explanations of the causes of growth, deterioration, death, respiration, and sleep all occur in the Parva Naturalia (*Juv.*, *Long.*, *Somn.*, *Resp.*). This illustrates how *PA* I serves as an introduction to the study of living beings across the biological corpus and is not only an introduction to Books II–IV on animal parts. See Chapter 1 in this volume by Hatzimichali.

[14] On this theme, see especially Henry 2021b.

attempting to establish the reasons why (639b7–11) – but for the biologist, following this guideline is much more complicated than it is for the astronomer. In giving reasons, the biological thinker must reference final causes alongside necessity and must give priority to the final cause over material and efficient causes. In terms of necessity, they must be aware of its different manifestations in nature (639b12–640a9). Particularly important is conditional (or hypothetical) necessity, referring to what is required if natural processes are to achieve their ends; indeed, it is not the process but the end point that is the focus of this study.

> Hence it would be best to say that, since this is what it is to be a human being, on account of this it has these things; for it cannot be without these parts. (640a33–35)

The rest of *PA* I investigates other general principles in the study of animals, explaining the right and wrong ways to employ the method of division (*PA* I.2–3), how to use analogy and "the more and the less" (I.4), and why it is crucial for the philosopher to study living beings, despite feelings of disgust, since "in all natural things there is something marvellous (τι θαυμαστόν)" (I.5, 645a16–17). The end of *PA* I returns to the initial question of whether to study particular or general kinds and suggests that the study will proceed both by atomic species (ἄτομον τὸ εἶδος), the last undividable kinds, and also in broader groupings when general explanations are possible. The latter methodology avoids needless repetition (I.5, 645b13). However, the former is also required in certain cases, for example, the slimline octopus.

PA II begins with three levels of composition: elements into simple compounds, simple compounds into uniform parts of animals, and, finally, uniform parts of animals into nonuniform parts of animals. Uniform parts include, for example, flesh, bone, and blood; nonuniform parts include, for example, face, hand, and stomach. Aristotle focuses on the last two levels explaining why animals must have both types of part, why sense-organs must be uniform, and why viscera are in one way uniform and in another way nonuniform. From chapter 2, he begins the study of uniform parts, listing their texture, causes, and the differences between them. A part-by-part review begins, starting with blood, the nourishment for the whole body. The other uniform parts are arranged from more fluid to more solid. The study of nonuniform parts begins in chapter 10 with those "most necessary" to all animals, the ones that receive and expel nutriment as well as the origin of life, which lies in-between these two others (655b37). Aristotle notes an increased complexity from plants to humans and first focuses on blooded animals, treating the bloodless ones later.

Uniform parts are mostly treated from top to bottom (mapped onto the human body) – and the general pattern is to start with the most common features, those that appear in the widest class, and then to move on to differences that occur between groups of animals.[15] Animals are "grouped and regrouped" (Balme 1987b: 88) as needed to find the widest-class generalizations for each part. Having completed viscera, Aristotle turns to bloodless animals, explaining at first why they lack viscera. Grouping them into four kinds, using cephalopods as his model, he proceeds to describe their internal parts, beginning with those used for nutrition and then detailing sense-organs. *Parts of Animals* IV.5, 682a30–34 marks the end of the survey of internal parts; Aristotle announces that the external parts will be detailed starting from bloodless and then to the blooded animals. The procedure begins part-by-part, first uniform then nonuniform, proceeding from the top to the bottom (based on human uprightness), and finally moving from external to internal and back again to external. Focusing on parts, Aristotle explains why a possessor has the part it does and why one possessor has the part differently configured from another possessor.[16]

One interesting feature of *PA* II–IV is a resistance to any dogmatic attempt to draw a line between types of living being; there are cases that require the researcher to understand that nature allows for some organisms to have features usually associated with two kinds or even of two habitats. As James Lennox explains: "The entire investigation ends by looking at various blooded animals that do not fit naturally into the four extensive kinds [viviparous quadrupeds, oviparous quadrupeds (including snakes), birds and fish] recognized throughout *PA* II–IV" (2001a: 343). The use of *epamphoterizein*, meaning "tending to both sides,"[17] shows how Aristotle employs close observation and is sensitive to the actual lifestyles of animals. He even uses his knowledge of botany to help to explain how certain animals operate.

> Through the mid-line of the ascidians there is a think diaphragm, in which it is reasonable for that which controls life to preside. But those that some call 'nettles' and others 'anemones' are not hard-shelled, but fall outside the divided kinds, and tend in their nature towards both plant and animal. For

[15] Lennox 1987a: 109–114; Connell 2016: 69–72.

[16] This summary is extracted from Gotthelf 1987b: 172–175.

[17] Peck's translation of this as "dualize" gives the wrong idea. Dualizing means to make or regard as having two parts. The point about these animals is that they are of one kind but have characteristics from two seemingly opposing groups, for example, living and feeding in water but breathing air.

> by being detached and falling upon their nourishment, and by being perceptive of what they fall upon, some of them are animal-like; and further, they use the roughness of their body for self-preservation. But by being incomplete and becoming quickly attached to the rocks, and by having no apparent residue though they have a mouth, they are akin to the kind consisting of plants. (*PA* IV.5, 680b34–681b8)

Such challenges in the study of animal parts and their functions indicate to Aristotle the organizational complexity of reality. It is not a simple matter of dividing in terms of "cutting nature at its joints," as a butcher would dismember a carcass (Plato *Phaedrus* 265e). The study of animals requires a much more robust logical range, revealing the intricacy of human reasoning capabilities.

Throughout *PA* II–IV, Aristotle focuses his explanations on ends in nature and particularly the formal nature of Book I (640b28).

> Since every instrument is for the sake of something, and each of the parts of the body is for the sake of something, and what they are for the sake of is a certain action, it is apparent that the entire body too has been constituted for the sake of a certain complete action. (*PA* I.5, 645b15–18)

While scholars disagree about the exact import of this passage, it gives a clear sense of Aristotle's emphasis on the function of body parts toward the achievement of living goals.[18] In this context, the formal natures of animals have agency in shaping them in the best manner possible. This analysis can also explain how nature leads to living well over and above mere survival.[19] Furthermore, material natures contribute to the way in which animal bodies are molded so that animals can live well in their natural setting.[20] An excellent example of this type of explanation occurs in relation to the presence of horns.

> Since there is a necessary nature, we must say how the nature according to the account makes use of things present of necessity for the sake of something. First of all, what is bodily and earthen is present in greater amounts in the larger animals, while we know of no completely small horn-

[18] "Understanding living matter is anchored by means of understanding the functional goals served by an animal's parts, discovering what each of its parts is for, what it contributes to the life of the animal" (Lennox 2021b).

[19] As Aristotle explains in *De Incessu Animalium* 8, "nature does nothing in vain, but rather aims to achieve the best from among the possibilities for each animal, preserving the proper substance and essence of each of them" (708a9–11, trans. Stavrianeas). This epitomizes Aristotle's "optimality reasoning" (Henry 2013). On different interpretations of this passage, see Lennox 1997; Stravrianeas 2021.

[20] Lennox 1996 argues that "an animal's formal nature is severely constrained by its material nature" (218 n. 3). However, it is also the case that "given a certain material nature, each formal nature determines a non-random and distinctive realization of that material" (218).

> bearing animal – the smallest one known is a gazelle. And one should study nature with a view to the many; for it is what happens either in every case or for the most part that is in accordance with nature. What is bony in the bodies of animals is in origins earthen; this is also why it is most abundant in the largest animals, to speak with a view to what occurs for the most part. For the residual surplus of this sort of body, being present in the larger of the animals, is used by nature for protection and advantage, and the surplus, which flows of necessity to the upper region, in some cases it distributes to teeth and tusks, in others case to horns. (*PA* III.2, 663b22–664a35)

While there is much to connect the general philosophy of biology found in *PA* I with the rest of the work, there are few clear accounts of conditional necessity and no definitions of the essence of kinds, as one might expect in these latter books of the treatise. The philosophical norms of *PA* I are reflected in *PA* II–IV but sometimes these are under strain (Lennox 2001a: xiii). Careful treatment is required to make clear how final causes, formal agency, and definitional goals operate in Aristotle's analysis of animals' parts in *PA* II–IV.

I.3 The Philosophical Work Achieved by the Chapters in This Volume

PA I provides a program of study, focusing on formal and final causes of animals, noting that materials are for the sake of ends. Much of *PA* II–IV carries through with various more detailed attempts to explain animals' parts, both those that are common to or shared across broader groups and those that are peculiar to a certain kind of animal. Different strategies are apparent in the text; sometimes there is an emphasis on what animals have in common across kinds and on comparisons that bring out these similarities; at other points, we find particular animals' ways of life are central to explanation. Similarly, there are times when the prioritization of formal and final causes takes precedence, while at other times, the materiality of the parts is brought to the fore. The chapters in this volume take up different strands of these discussions in the text, some focusing more on broader groupings, some on particular kinds of animals; some have more to say about formal and final causes and some focus more on material ones.

The first set of chapters (1–3) concentrate on the text itself – its rationale, connections with other works in the Aristotle corpus, and its legacy. Myrto Hatzimichali (Chapter 1) considers the place of the *PA* within the biological corpus and the status and placement of *PA* I with relation to the rest of the treatise. She argues for retaining this Book within the *PA*, but as a theoretical introduction applicable to Aristotle's entire explanatory project in zoology, and also makes a case for positioning the *Historia Animalium* before *PA* I. The next chapter by Monte Ransome Johnson (Chapter 2) concerns how Aristotle sought to persuade various audiences of the philosophical and human value of studying animals and their parts. In the main, despite a few exceptions, this value is due to the way that they display order and the operation of final causes. Johnson puts the so-called protreptic section of *PA* I.5 where these ideas are expressed into the context of rhetorical methods in the Aristotle corpus. Giouli Korobili (Chapter 3) considers how the systematic elements of Aristotle's zoology began to be overlooked very soon after Theophrastus; part of the reason for this would appear to be the subsequent neglect of *PA* I. She explains how the work as a whole, and its teleological perspective, was considered carefully by Galen, even though his own rhetorical flourishes sometimes obscured its importance. In the final section, she sets out methods of analysis of the *PA* in the fascinating late Neoplatonic commentary by Michael of Ephesus, which has yet to be translated from Byzantine Greek.

The next four chapters (Chapters 4–7) are mainly concerned with *PA* I and how to understand Aristotle's preliminary principles for the study of animals. Jessica Gelber (Chapter 4) analyzes Aristotle's dissatisfaction with rival theories and argues that rather than presuppose a teleology that they may reject, his strategy is to undermine their explanations in other ways. According to Gelber, Aristotle prioritizes teleological over non-teleological explanations because the former are based on definitions. Definitions are appropriate starting points that prevent explanations from going on *ad infinitum*. Hypothetical or conditional necessity is just one kind of explanation that does provide such grounds; it is possible that there are others that might also avoid Aristotle's critiques.

For Aristotle, the predecessors who discussed life and animals did not provide correct explanatory accounts of their existence as substances. His own strategy is to emphasize organic unity where the functioning parts are for the sake of this whole. There is controversy about how this works. Aristotle does not provide definitions of species forms; how, then, might one think about animal essences? This volume presents different ideas about how form and essence work in theory and in practice in the text. Michail Peramatzis (Chapter 5) provides an account of the relationship between form and matter in animals as inextricably intertwined. This means that definitions of essences will include "subtle" matter such as being blooded or winged. This represents an innovative position on the adaptation of hylomorphism in the biological context and how important Aristotle's biology is to understandings his mature metaphysics. In Chapter 6, Devin Henry shows how in certain cases, at least, the subject matter of zoology is "atomic species" (*atoma eidos*), the end of the correct division into kinds, involving multiple differentiae. He argues that Aristotle treats atomic species as the primary objects whose essences we are attempting to grasp in zoology. Species here is meant in an Aristotelian rather than a modern sense, meaning a place where the division of kinds ends; this clarification helps to support the idea that Aristotle's explanations of animal parts depend on which atomic species they belong to. Thomas Johansen (Chapter 7) argues a related point, which is that explanations often require specificity at the level of "ultimate species" (*atoma eidos*).[21] For example, one must refer to this level of reality when a part is used in a particular manner due to the material constraints of an animal's situation. Also, the part and its function can only be understood in relation to one of the other parts of this particular species, viewing body parts as interconnected, rather than only concentrating on them in isolation. Thus, it seems that in such instances, wholes come first rather than parts, which is in keeping with the methodology recommended at the outset of the work.

Chapters 8–12 explore explanations that are comparative and focus on features shared across different kinds of animal, some of which are described as analogous. Christopher Frey (Chapter 8) wonders how analogy as discussed in Aristotle's logical works might be applied in the field of zoological research. What, if anything, unifies analogous parts? It seems that there is, in most cases, no higher genus as a point of reference; and, at times, function is not all that is indicated by the term. Stasinos Stavrianeas (Chapter 9) provides a detailed account of how animals' uniform parts are constructed. He explains that they develop from the concoction of materials and that this preserves the capacities that are required for the life of the animal. In doing so, he shows the crucial dependence of the discussion of uniform parts in *PA* II on theories espoused more fully in the chemical and physical works, especially *On Generation and Corruption* and the *Meteorologica*. Sophia Connell (Chapter 10) concentrates on one uniform part, blood, and its connection to a curious feature: animal character. Although Aristotle gives seemingly only a "material" explanation of differences in blood which underlie character dispositions, there are strong

[21] *Atoma* literally means 'uncuttable' and can be translated in a number of ways, e.g., atomic, ultimate, last, final, or indivisible (*PA* I.4, 644a28–33).

connections to formal and final causality, through the animal's material nature. Material thermodynamics, as described by Stavrianeas, are used by the animal's formal nature to determine a material basis for character (cf. *GA* II.6, 743a36–b18). However, most animals are at the mercy of the materials that sit in the blood vessels at certain times in the nutritive cycle, over which they have little control.

Mariska Leunissen's Chapter (11) marks a transition from thinking of general to more specific cases and from uniform to nonuniform parts. Aristotle sets out to explain a wide variety of different parts in all their intricacy, even odd or anomalous anatomical features. Some of these do not appear to be for the sake of anything and may even be detrimental. Thus, in accounting for the nature of animals as like craft, he comes up against certain limitations in these cases (unaware as he was of the residual forces in evolution).

The final set of Chapters (12–14) switch between a focus on a particular kind, human being, to a comparative approach whereby other animals are compared to humans. Mary Louise Gill's question (Chapter 12) is the most general one relating to human, whether intellect (*nous*) itself, as manifested in the lives of human beings, can be part of the study of nature. She argues that it is possible to read *PA* I.1, which appears to suggest that intellect is not part of the study of nature, as a dialectic exchange with certain Platonists who attempt to separate mind and nature. There is a way, then, to read the text that does not require us to think that Aristotle endorses a view at odds with other parts of his philosophy of nature, namely that embodied thinking is entirely supernatural. Christof Rapp (Chapter 13) sets out the peculiarities of human parts, morphology and orientation in space. He explains how Aristotle's account of the human body in *PA* IV.10 works and how he uses his conclusions to understand the nature of certain other animals in comparison with human morphology. Emily Kress's Chapter (14) similarly focuses on the relation of human to nonhuman animals and explains Aristotle's rationale for treating human beings first and as a model for the study of the parts of other animals. According to Kress, polymorphicity, the complexity of human parts, provides the most teleologically rich model for parts in other animals. The most articulated parts can then be mapped onto simpler ones and thus the functions in the latter discerned more clearly. Together this volume offers a rich picture of the philosophical interest of the *Parts of Animals* and invites further reflection on and argumentative engagement with this fascinating treatise.[22]

[22] I am grateful to Jim Lennox for his detailed comments on the introduction and on the volume as a whole.

CHAPTER I

Situating the Parts of Animals *in Aristotle's Biological Corpus*

Myrto Hatzimichali

Introduction

The opening paragraphs of *On the Parts of Animals* place it clearly within the scope of Aristotle's inquiry into nature (περὶ φύσιν ἱστορία), as he reflects on the norms and standards (ὅροι) that need to govern this type of investigation (*PA* I.1, 639a13–14). The fact that he envisaged a degree of unity and sequential progression in his natural philosophy as a whole is confirmed by the famous prologue to the *Meteorology* (*Meteor.* I.1, 338a20–339a9). In this sequence, a *telos* is to be reached with the examination of animals and plants (339a6–9).[1] This in turn suggests that it would be appropriate and worthwhile to pose questions about the sequence and relationships that Aristotle envisaged *within* the group of biological works that are closest to this *telos*, in particular (for present purposes) as far as the four books of *PA* are concerned.

In the space available, my main concern will be to situate *PA* vis-à-vis the other two of the 'big three' biological works, *Historia Animalium* and *Generation of Animals*. I will first address the question concerning the status of *PA* I, arguing that it belongs integrally to the treatise devoted to the parts of animals and is not a detachable general introduction of independent status. I will then turn to *PA*'s relationship with *HA*, where I will address the 'Balme hypothesis,' according to which *HA* postdates the explanatory treatises and represents a more advanced stage of inquiry. I will offer some considerations in favor of the more traditional view whereby the project undertaken in *HA* is an important and indispensable prerequisite for the explanatory procedures undertaken in *PA*. With respect to the *PA*–*GA* 'axis,' I will argue that taking seriously the claim that 'being is prior to

[1] The *Meteorology* prologue and the precise nature of the proposed order it advocates has been the subject of extensive scholarly discussion. See Burnyeat 2004; Falcon 2005: esp. 9–16; Falcon 2015. On the question of scientific norms across the different domains mentioned in the *Meteor.* prologue, see especially Lennox 2010a, 2021: esp. chapter 4.

coming-to-be' (and therefore *PA* precedes *GA*) has important implications for our interpretation of the explanations offered by Aristotle in *PA* II–IV. In particular, it speaks against interpretations that appear to attribute a significant explanatory role to processes. Finally, I will conclude with some brief remarks about the 'biological corpus' as a whole.

1.1 The Status of *PA* I

From the outset, it must be made clear that there is no attempt here to deny that *PA* I is a theoretical introduction whose results are relevant throughout the biological works, offering Aristotle's "philosophy of biology." The question is whether this theoretical introduction belongs integrally with the study of animal parts or is a self-standing set of investigations, as Balme suggested:

> The book in its present form, which must go back at least to Andronicus' edition in the first century BC, is a general introduction to zoology, complete in itself and independent of the remaining three books of *PA*, which are factual studies. The abrupt differences of style and lack of overall continuity suggest that it is a collection of five separate papers. (Balme 1992: 69)

Lennox more recently made a case for the internal unity and coherence of *PA* I (Lennox 2010b), but the consensus has been that it is "generally recognised to be independent of the rest of *PA*."[2] This is reflected in two important detailed discussions of the sequence of the biological corpus by Kullmann and Rashed, which separate *PA* I from *PA* II–IV, placing the former before *HA* and any other works that are thought to precede *PA* II–IV. Their proposed orders are as follows:

Kullmann (2007: 146–147): *Parts of Animals* I, *Historia Animalium*, *De Anima*, *Parva Naturalia*, *Movement of Animals*, *Progression of Animals*, *Parts of Animals* II–IV, *Generation of Animals*.

Rashed (2004: 200–202): *Parts of Animals* I, *Historia Animalium*, *Parts of Animals* II–IV, *Progression of Animals*, *De Anima*, *Parva Naturalia* 1 (i.e. up to *On divination in sleep*), *Movement of Animals*, *Generation of Animals*, *On Plants*, *Parva Naturalia* 2.

The three main arguments in favor of this practice are (i) the view represented by the quotation from Balme cited above, that is, that the

[2] Lennox 2001a: 119.

contents of *PA* I represent a general introduction to the biological project in the context of natural philosophy, posing methodological questions that apply well beyond the causal investigations of II–IV;[3] (ii) that *PA* I diverges philosophically from II–IV, especially in terms of a more prominent role of body-soul hylomorphism; and (iii) the textual point that the opening of *PA* II appears to follow directly from *HA*:

> From which parts and from how many parts each of the animals is constituted has been exhibited more clearly in the enquiries about them; it is the causes owing to which each animal has this character that must now be examined, on their own and apart from what was said in those enquiries.[4] (II.1, 646a8–12, trans. Lennox)[5]

The latter point is not a great obstacle to a unified reading of *PA* I + II–IV, because the perfect-tense reference to *HA* can be taken as more remote than it perhaps appears at first glance (taking it to mean "we have discussed elsewhere . . ." rather than "we have just completed the discussion . . ."). Point (ii) is more complicated, with a lot of weight falling on *PA* I.1., 641a17–32, where the natural philosopher is expected to be talking primarily about soul (opening the question of whether this is meant to include *all* soul), and I.5, 645b19–20, where the body is said to be for the sake of the soul and each of its parts for the sake of some function (ἔργον). Even this idea, however, is consistent with the (teleological) explanations of *PA* II–IV where, for the most part, the ultimate explanans (the final cause) is some soul function, even if indirectly: for example, at II.8, 653b30–36 we learn that bones, skin, sinews, blood-vessels, hair, nails, etc. are for the sake of flesh (τούτου χάριν), which in turn is the primary organ of perception; at IV.10, 686a27–32, the upright posture of human beings is explained through their "divine nature and being," which amounts to their capacity to think (νοεῖν καὶ φρονεῖν).[6]

We are therefore left with the major question whether *PA* I's function as a theoretical introduction precludes it from being meaningfully attached to II–IV. One aspect of this is the perception that *PA* II–IV does not quite live up to the scientific goals of *PA* I and does not fulfill certain investigative promises made there. I will not have much to say on this aspect of

[3] See Kullmann 2007: 154, who also adds a chronological argument.

[4] ἐκ τίνων μὲν οὖν μορίων καὶ πόσων συνέστηκεν ἕκαστον τῶν ζῴων, ἐν ταῖς ἱστορίαις ταῖς περὶ αὐτῶν δεδήλωται σαφέστερον· δι' ἃς δ' αἰτίας ἕκαστον τοῦτον ἔχει τὸν τρόπον, ἐπισκεπτέον νῦν, χωρίσαντας καθ' αὑτὰ τῶν ἐν ταῖς ἱστορίαις εἰρημένων.

[5] All translations of *PA* passages are taken from Lennox 2001a, unless otherwise indicated.

[6] Moreover, it would not be easy to argue that II.1, 646a15–b27 is less 'hylomorphic' than anything we find in *PA* I. See also the next paragraphs on Henry's account of *PA* II–IV.

the problem because it has been very recently taken up by Henry (2021a) specifically on the topic of definition. Henry addresses a worry arising from the fact that *PA* I advertises what sounds like a search for definitions as one of the goals of the natural scientist: one of the possible ways of conducting the inquiry is "taking each substantial being singly and defining it independently" (λαμβάνοντας μίαν ἑκάστην οὐσίαν περὶ ταύτης διορίζειν καθ' αὑτήν, 639a15). There is also a suggestion that "grasping indivisible kinds (λαβεῖν τὰ ἄτομα εἴδη)" is the goal of a successful division, a goal that cannot be reached by Platonic-style dichotomous division (643a17, 644a11). And yet, there are no definitions forthcoming in *PA* II–IV, raising questions about consistency across the four books. Henry makes a new suggestion for what the definiendum is meant to be, moving away both from *Metaphysics*-style 'pure forms' and from 'matter-involving impure forms.' According to Henry, *PA* I announces a project whose aim is to "define the matter of the composite," based on *PA* I.1, 640b15–29.[7]

'Defining the matter' is not as controversial as it might sound, because there are degrees of matter and the 'matter' of the composite animal is in fact its parts (καὶ ἡ ὕλη τοῖς ζῴοις τὰ μέρη, most explicitly at *GA* I.1, 715a9–12). Animal parts are 'informed' enough to be defined, always with reference to teleological principles arising from the formal natures of the composite animals involved.[8] Henry is able to point to a number of such definitions across *PA* II–IV, for example:

> On account of what (διὰ τί), then, those animals with marrow have it has been stated. And from these considerations it is also evident what marrow is (τί ἔστι) – the enclosed, concocted residue of the sanguineous nourishment apportioned to bones and fish-spine. (*PA* II.6, 652a19–23)[9]

It is more complicated to explain how this procedure can also account for the desideratum of "grasping the indivisible (composite) kinds" – Henry leaves it quite open, envisaging a scenario where we will be able to discern which of these parts belong in the essence of particular animal kinds – but Aristotle offers us no help here. We have therefore seen one set of grounds advocating for the unity of *PA* I with the rest of *PA*, namely that the accounts of animal parts given in *PA* II–IV speak to the goal of "defining the matter of composite" that is introduced in *PA* I.1.

7 Henry 2021a: 765–773.
8 Cf. Henry 2021a: 774–775 for the ways in which form is always in the picture.
9 See Henry 2021a: 776 for more examples, including *PA* II.5, 651b17–18; III.5, 667b13–15; IV.10, 685b29–33.

However, when arguing for a meaningful attachment of *PA* I to the rest, we must also step back and ask whether it also serves as an introduction to *HA*, especially given that it is placed immediately before *HA* by both Kullmann and Rashed (above p. 14). I will try to address this in parallel with further thematic links between *PA* I and II–IV, such that require an uninterrupted transition. First of all, let us recall that *PA* I begins by giving examples of some issues that, if one is trained (πεπαιδευμένος) in thinking through them, he is able to assess whether a science (in this case, natural philosophy, περὶ φύσιν ἱστορία, 639a13) is conducted properly, even without being familiar with its facts (i.e. without having the relevant *epistēmē*). The first such question is whether one should proceed by defining substances individually (639a15, cited above), or by establishing the attributes common to all the relevant substances. This question as posed here is not pertinent for *HA*, because the focus of the latter is on the ways in which animals differ, not their common characteristics or the establishment of individual *ousiai*. Moreover, one could argue that *HA* in fact provides the groundwork for establishing which features are common (and on what levels of generality) and which are individual. This groundwork consists in the major insight, which I take to originate in *HA* (cf. *HA* I.1, 486a14–487a10), that some animals differ by excess and defect or 'the more and the less' while possessing basically the same feature, whereas in other cases the similarity is only by analogy.[10] This underlies the eventual endorsement in *PA* I.4 of a resolution to this first methodological question:

> Perhaps, then, the right course is this. In some cases – whenever kinds are spoken of by people in a clearly defined manner and have both a single common nature and forms in them not too distant – we should speak in common according to kinds, like bird and fish and any other there may be that, though it is unnamed, embraces, like a kind, the forms within it. But whenever they are not such as this, we should speak one by one, e.g. about mankind and any other such kind. (*PA* I.4, 644b1–7)[11]

The second theoretical question raised in *PA* I.1 is whether the natural scientist should follow the mathematical astronomers in their practice of first ascertaining the phenomena and then proceeding to determine their causes (639b8–11). This is a surprising question to anyone familiar with Aristotle's practice (cf. *APo.* II.1–2), especially as a few lines later he takes

[10] On these differences, see Lennox [1987b] 2001b: 160–181.

[11] Cf. 644a16–18: "For those animals that differ by degree and the more and the less have been brought together under one kind, while those that are analogous have been kept apart."

an affirmative answer for granted (640a13–15).[12] We must remember, however, that we are dealing with broad methodological questions that fall within the remit of 'general training,' not science, and that Aristotle's point is that the discovery of biological facts is a necessary prerequisite for the investigation of causes, with more emphasis on its indispensability than on its chronological priority. Furthermore, as Kullmann suggests, this could be a query about the separation of the two steps into separate written works, a practice Aristotle adopts only in the case of biology (presumably because of the sheer volume and complexity of the amassed facts, which was not present in e.g. *Meteorology* or *De Caelo*). This would add more nuance to the reference to astronomers, because Eudoxus had apparently separated the treatment of phenomena from that of causes.[13]

Thus far we have seen that the first two methodological questions of *PA* I.1 do not seem to apply as an introduction to *HA*. Once we consider the third major methodological question posed in *PA* I.1, which comes to dominate the rest of that chapter, it becomes clear that *PA* I is an introduction to a *causal* investigation of animals, indeed a causal investigation of animal *parts*, not to an investigation of animals in general. This third question concerns the relative priority of the four causes, arguing in favor of the final cause (639b12–15), and throws new light on the two earlier methodological questions, which, in the light of the overall explanatory project, can now be read as follows: "at what level of generality should we explain animal attributes?" and "does the investigation of causes depend on a separate establishment of the phenomena?"[14] All three questions and their respective answers do not serve to introduce the research project of *HA* as such but rather explore what is the best and most appropriate use of its results in all the subsequent causal investigations.

Thus far I have given some reasons for why *PA* I is not an introduction to *HA*. It remains to consider the same issue from an alternative angle, asking whether it can be reasonably detached from II–IV. In my view, Aristotle presents all the major philosophical insights of *PA* I (including

[12] It is also treated as uncontroversial at *HA* I.6, 491a9–13.

[13] Kullmann 2007: 144 (cf. Hipparch. *In Arat.* I.2, 1). Lennox 2021: 160–162 offers a different account of the comparison with the way in which mathematicians explain astronomical phenomena. He draws attention to the 'subordinate science' situation whereby mathematical proofs are used for natural (astronomical) phenomena and to the *PA* I.5 distinction between the study of eternal vs. perishable beings that come to be and change in highly complex ways. Both considerations make it far from obvious that the biologist can proceed in a similar way to the mathematician/astronomer.

[14] See also Lennox 2021: 162–173 for how the third methodological question has an important impact on the final answers to the other questions.

the point about definition highlighted by Henry) in a way that links them very specifically to the explanation of animal *parts* and leads the reader to expect this explanation to follow directly. First of all, the claim that substantial being (οὐσία) has explanatory priority over the process of generation (γένεσις) is put forward based on the priority of the final cause and is argued for in a way that culminates in the correct manner of accounting for the presence and sequential coming-to-be of parts (640a10–b4, cf. especially the references to "the parts" (τῶν μορίων) at 640a35 and b2). Secondly, in the major theoretical standoff with his predecessors, where they are found wanting in their overemphasis on matter and neglect of form, Aristotle makes constant reference to parts. It does not suffice to speak of the four elements as what animals are 'made of'; one has to refer to both uniform and nonuniform parts (as we saw above, they are the matter of living things). This is a step that points in the direction of form (saying "in virtue of what each of them is such as it is, and in respect of what sort of potential," 640b21–22). On several occasions where Aristotle pushes for an account that goes beyond matter or 'mere' shape, this requirement is explicitly said to apply both to the animal as a whole and to its parts (I.1, 640b30; 641a17–18; 641a20; I.3, 643a25; I.5, 645a30–36).

These connections are of course not accidental: the explanation of animal parts is an arena where the emphasis on function and adaptation enables Aristotle to showcase the superiority of the formal/final explanations promoted in *PA* I (by which I mean explanations that *take proper account of* formal/final causes and of the relevant senses of necessity) over the limited material/efficient ones of his predecessors. In addition, a proper understanding of animal parts is the key step toward the (ultimately unfulfilled) desideratum of "grasping the indivisible kinds" (I.3, 643a17), as indicated in the chapters on division, where most of the *differentiae* mentioned concern parts. Other types of differentiae, such as, the "common functions of the body and soul" – for example, dividing into walkers and flyers – are deemed unhelpful. The material in I 5 that follows after the famous but oddly placed protreptic has a new focus on "activities" (πράξεις), not seen earlier in the book – but these too are said to map onto the differences of parts. For these reasons, I believe that, while *PA* I introduces a number of concepts and priorities that are relevant across the biological corpus and thus can helpfully be read as a general introduction to all *causal* investigations in biology, its main role is to introduce a method for the study of animal parts, and should therefore not be treated as separate or separable from *PA* II–IV.

1.2 Facts and Explanations: *PA* and *HA*

One consequence of the discussion above is that it leaves *HA* "dangling" at the beginning of the biological corpus, without any theoretical preliminaries that would preestablish its position and relationship with the other works. As scholars have long noticed, however, there is sufficient information in the opening chapters of *HA* itself, outlining its place within Aristotle's biological enterprise:[15]

> Our objective is to determine first of all the differences that exist and the attributes that belong to all of them. Having done this, we must attempt to discover the causes. For this is the natural way to pursue the investigation, once we have done the research on each one. For this will give us a clear indication both about which things (περὶ ὧν) the demonstration must be and from which things (ἐξ ὧν) it must proceed. (*HA* I.6, 491a6–14)

This makes quite clear that the project of *HA* is to sort out which features are different and which ones are common to all animals, in a stage that precedes the discovery of causes and enables the scientist to distinguish explananda from explanantia (περὶ ὧν, ἐξ ὧν, cf. *APo.* I.10, 76b20–22). This two-step approach is advertised again explicitly at the beginning of *PA* II (cited above) and at *IA* 1, 704b10 and is implicit in a number of cross-references to *HA* across the biological corpus. Nevertheless, this picture has been challenged by Balme, who argued that *HA* is not presupposed by, but rather presupposes, the contents of the other works.

This 'Balme hypothesis' has been criticized by other interpreters of *PA*[16] but was taken very seriously and was analyzed in a balanced way by Lennox (1996a).[17]

Balme's argument is a chronological one, applicable to the date of composition of the written treatises and not to a putative reading or lecturing order that could have been imposed at a different stage. As a result, his arguments are primarily textual, arising from a careful comparison of instances where the same information appears (or does not appear) in both *HA* and the explanatory treatises (for present purposes, I will concentrate only on *PA* II–IV).[18] His main observation is that *PA* contains in itself all the facts that are necessary for its causal investigation and in fact presents them in a fuller form than *HA*'s "less informative, more

[15] See Lennox 2001a: 53–60. [16] See Kullmann 2007: 152–156; Feola 2017: 45–47.

[17] Falcon 2021a: 5 n. 4 also finds some grounds in favor of the Balme hypothesis.

[18] Balme's argument takes other works into account, with examples from *IA*, *GA*, *Mem.*, *Resp.*, cf. Balme 1991: 22.

condensed, less intelligible" versions, which, on Balme's view, are excerpted from *PA*.[19] While this is claimed for facts that appear both in *PA* and *HA*, Balme also remarks on the greater richness of *HA* in other respects, this time attributing it to discoveries that were not available at the time of writing *PA*.[20] The most important subcategory in the second group comprises data that conflict with *PA* and "should have modified the conclusions stated,"[21] for instance, the existence of a four-legged insect (*HA* I.5, 490a32–b3; V.19, 552b17–23) and the fact that the chameleon's brain is connected to its eyes (*HA* II.11, 503b18). In both cases, *PA* has sweeping statements to the effect that "all insects have multiple legs" (IV.6, 682a35–b6; "multiple" means more than four, cf. *IA* 16, 713a26–28), and that "the brain has no continuity with the sensory parts" (II.7, 652b3). I am not convinced, however, that these two exceptional animals should have modified Aristotle's general claims, especially given his expressed focus on what happens "for the most part" (cf. *PA* III.2, 663b28–29).

Another case where there is a contradiction between *HA* and *PA* concerns the liminal species of the sponge, which is mentioned at *PA* IV.5, 681a9–17 as an illustration of nature's gradation from lifeless things to animals, which creates "fuzzy" borderlines – in the case of the sponge between plant and animal. Sponges "have in every respect the function of a plant" especially because they live only by being attached to something. But we should not take this to mean that Aristotle thinks they *are* plants (otherwise their appearance in this context would be pointless). Therefore, there is not as much contradiction as Balme thought with *HA*'s hearsay reports that they have sense-perception, in fact this is what confirms their liminal status (I.1, 487b9–11; V.16, 548b10–12; 549a4–9).[22]

Finally, while *PA* insists that only blooded animals have a brain (II.7, 652b7–32), there are reports in *HA* that cephalopods (μαλάκια) also have one (I.16, 494b26–29), contained in a small piece of cartilage between

[19] Balme 1991: 21; 1987a: 13.

[20] Balme 1987a: 15–16. Lennox 1996a: 242–243 comments that such a combination of arguments renders the theory "immune to disconfirmation" as it can account in a potentially ad hoc manner for cases where *HA* is either more or less expansive. According to Lennox, this can be addressed by distinguishing cases where the two works involve "precisely the same data," from cases where *HA* provides "important and relevant data not discussed at all in the other treatises." However, it could prove far from straightforward to assess what type of data counts as "important and relevant."

[21] Balme 1987a: 15; 1991: 23.

[22] These *HA* reports make it clear that there was dispute among Aristotle's sources ("some say," "others dispute") over whether sponges really do perceive (the alleged evidence being that they are hard to detach unless one approaches them suddenly). *HA* VII (VIII).1, 588b19, which is also preoccupied with "fuzzy natures," is in agreement with *PA* on sponges' resemblance to plants. On all this material, see Lloyd 1996: 67–82, esp. 75–76.

their eyes (IV.1, 524b3–4). As Lennox points out, there are important theoretical implications here, which could potentially affect Aristotle's view of the brain as a heat-moderator: if cold, bloodless animals turn out to have one, this explanation is in jeopardy.[23] Once again, however, I do not believe that these discrepancies necessarily point to a belated discovery that would have caused Aristotle to revise his view: we must remember that observations that *we* know to be more accurate need not have appeared so to Aristotle. It is equally plausible that he was aware of an organ in cephalopods that looked very much like a brain, but when he formulated his explanatory account of the brain as a cooling part he decided that this cannot possibly be the case:[24] in fact, at *PA* II.7, 652b24–25, he acknowledges the presence of such a part, which is described as a brain-analogue in the octopus: "All the blooded animals have a brain, while virtually none of the other animals does, except by analogy, such as the octopus; for they all have little heat on account of being bloodless."

Thus far we have seen that the evidence used by Balme to show that *HA* postdates *PA* because it contains material that would have altered the explanatory course of *PA* can be interpreted in ways that do not necessitate this conclusion. Similar points can be made about the textual arguments invoking 'condensation' of data in a process that has *HA* drawing on *PA*. The relevant instances picked out by Balme include[25] the information on external parts of birds at *HA* II.12, 503b29–504b9, which is supplemented at *PA* II.13, 657a28–b22; 14, 658a12; 17, 660a29; IV.12; the reverse relationship between abundance of flesh and size of blood vessels at *HA* III.16, 520a1–6, which is also covered at *PA* III.4, 667a23–32; and the 'tongue' of insects and testacea at *HA* IV.7, 532a6–10, discussed also at *PA* II.17, 661a15–25. In most of these cases, where *PA* contains more detail than *HA*, the discrepancy can be equally well explained by the different contexts and different purposes of each work: the 'extra' observations are required for the explanatory account but would not be pertinent in the *HA* context of collecting and comparing differences between animals. As an example, I propose to look more closely at the two corresponding accounts of the kidneys, which are also cited by Balme as a case where *HA* draws on *PA* and condenses its information:

> Passages lead to the kidneys both from the great blood-vessel and from the aorta, though they do not lead into the cavity. For kidneys have a cavity in

[23] Lennox 1996a: 240.

[24] It would not be the first time that Aristotle's theoretical commitments get the better of his observations, cf. Lloyd 1987.

[25] See Balme 1991: 22.

the middle; in some animals it is larger, in others smaller; but there is none in the seal's kidneys. These are similar to those of oxen, and are more solid than those of all other animals. The passages which lead to the kidneys peter out in the body of the kidneys. That they do not continue through the kidneys *is shown by the fact* (σημεῖον δέ) that they contain no blood, nor does it congeal in them. [They contain, however, as has been said, a small cavity.] From the cavity of the kidneys two vigorous channels lead into the bladder, and other strong continuous ones come from the aorta. (*HA* I.17, 497a4–15)

The channel which runs from the blood-vessel to the kidneys does not terminate in the cavity of the kidneys, but peters out in their body; *this is why* (διόπερ) no blood goes into the cavities, and at death none congeals there. From the cavity of the kidneys two vigorous, bloodless channels lead into the bladder, one from each. Other strong, continuous channels run from the aorta (to the kidneys). This arrangement is *in order to* (ὅπως) enable the residue from the moisture to travel from the blood-vessel into the kidneys, *and so that* when the fluid is filtered through the body of the kidneys the excretion that results may collect into the middle of the kidneys, where the cavity is in most cases. *This is why* (διό) the kidney is the worst-smelling of all the viscera. From the middle of the kidney the fluid is passed off through these channels into the bladder, by which time it is more like a residue. (*PA* III.9, 671b13–23)

The two passages demonstrate very nicely the different priorities of the two works: *HA* is preoccupied with the differences between animals, citing different sizes in the kidneys' cavities and singling out the seal as an exceptional case. The section quoted is followed by some more anatomical detail on blood-vessels that run to the bladder, which is transitional to a discussion of the latter. By contrast, the *PA* passage is characterized by the vocabulary of explanation; note also how the relationship between absence of blood and dispersal of blood vessels in the body of the kidneys is presented as an inference from a sign in *HA* but as a causal connection in *PA*. The 'extra' material in *PA* (from "This arrangement . . ." to the end of the passage) is not additional factual information but the teleological explanation for the particular structure of the kidneys, with the porous body and the cavity in the middle.

The fact that there are alternative explanations for the phenomena noticed by Balme across the two texts is not sufficient to dismiss his hypothesis, but it makes it less compelling. A more pressing question still remains, concerning the two-step approach of (i) establishing facts and (ii) investigating causes, which is, as we saw, advertised explicitly in both *HA* and *PA* and is also promoted in *APo.* This is not a matter of relative dates of composition of written works but concerns the conduct of the scientific investigation itself. Balme's hypothesis does not dispute the two-step approach, but locates both (i) and (ii) within *PA*,[26] implying that all the pre-explanatory groundwork required to produce explanations of animal parts was in place independently of the writing of *HA*. But such a scenario sits very uncomfortably with the most widely accepted (and for good reason) interpretation of *HA*, which Balme himself was instrumental in formulating: on this interpretation, *HA* is not a mere collection of facts but a sophisticated and well-organized survey of the ways in which animals differ. Lennox has written persuasively about how *HA* represents the 'pre-demonstrative' stage of the science, as outlined in *APo.* (and called *historia* explicitly in *A. Pr.* 46a22–27).[27] The systematic study of differentiae in *HA* makes possible a "rendering of the true propositions currently known about animals for the purpose of causal investigation" and helps specify the right level of generality at which different parts must be explained.[28] It emerges, therefore, that the explanatory project of *PA* II–IV is based on and presupposes a preexisting, highly complex establishment of facts and connections, with groupings according to *differentiae*. It is not based on a mere collection of data: in fact, the project presupposes precisely the kind of sophisticated groundwork that is conducted in *HA*. I find it very uneconomical to suggest that all this required work was available in some other form prior to the writing of *HA* (a written form would certainly be required given the amount and complexity of the material). Finally, there is the fact that *HA* announces a causal investigation as a subsequent step (see above p. 20): Balme is mindful of this and even gives examples of how causal accounts appeal to the interrelationships of differentiae revealed in *HA*. But instead of identifying the announced investigation with our existing treatises, he posits a missing discussion: "and so the discussion of causes that *HA* is to precede was never reached."[29] I find this just as uneconomical as a putative pre-demonstrative work that is separate from *HA*, and for these reasons I believe that we should treat the explanations of *PA* II–IV as drawing on *HA* rather than vice versa.

[26] Cf. Balme 1987a: 13, "the other treatises, which are concerned with the causal explanations, themselves contain all the facts they need."

[27] Lennox [1991] 2001b: 40–43.

[28] Lennox [1991] 2001b: 53–58. He also shows how *PA* II–IV succeeds in providing explanations at the level of primitive universality (recommended in *APo.* II.14–18), once this has been identified through *HA*'s procedure of picking out coextensive differentiae. Cf. 58–64, with examples.

[29] Balme 1987b: 86–89. His main reasons for doubting that *PA* and *GA* are at least parts of this discussion are the internal chronological evidence already discussed and the fact that "it is the causes of *all differences*, not merely of parts or generation, that will be discussed according to 491a19."

1.3 Being and Coming-to-Be: *PA* and *GA*

It is now time to address what is in many ways *PA*'s closest and most 'organic' relationship within the biological corpus, that with *GA*. The connection is established explicitly at the very end of *PA*, which reads as follows: "Having determined these matters, the step that immediately follows is to examine the ways in which animals are generated" (697b29–30). There are also numerous future-tensed cross-references within the text.[30] *GA* itself begins with an account of reproductive parts, which functions as a transitional 'joint' to *PA*, thus reinforcing the continuity.[31] In what follows, I shall focus on Aristotle's programmatic description of the two inquiries and the implications of this for our interpretation of his explanations in *PA* II–IV. I have attempted above to show that *PA* I belongs immediately before these books, mainly because it sets up an inquiry specifically into *parts*. It must be stressed, however, that from the outset of *PA* I, a prime concern of the natural scientist seems to be *generation*: at 639b12–13, the priority of final over efficient cause is asserted with respect to "natural generation" (περὶ τὴν γένεσιν τὴν φυσικήν); at 639b25, the key concept of conditional necessity is introduced with respect to "things generated" (τὰ ἐν γενέσει); at 641b27, the teleological regularity of nature is shown by the non-accidental replication through seed. This focus on generation at the outset of an explanatory project on biology is not surprising: living beings, Aristotle's substances "most of all" (*Metaph.* Z.8, 1034a4), "undergo an especially complex form of unqualified coming-to-be and passing away," which requires more specific principles to be established in a new methodological introduction here at the last stage of Aristotle's inquiry into nature.[32]

Perhaps the ultimate explanandum is indeed animal generation, the coming-to-be of individual substances that are the basic items of Aristotle's ontology; this would confirm *GA*'s status as the 'culmination' of the biological project.[33] It is certainly the case that generation presents Aristotle with the more challenging problems, as well as with the greater

[30] Notably at *PA* IV.10, 689a9–15 Aristotle assumes as a hypothesis (ὑποκείσθω) that semen and menstrual blood are residues and undertakes to "prove it later" (ὕστερον δειχθήσεται). See also IV.4, 678a19; II.3, 650b10).

[31] Henry 2019: 107–108. Lefebvre 2018: 45–55 explores the problem of placing the study of generative parts (which belong to mature organisms) in *GA*, and shows how the parts that contribute to generation are an ongoing (and unifying) concern of *GA*, going beyond the discussion of reproductive organs in *GA* I.3–16. This serves to integrate the study of generative parts into the program of *GA* ("*GA* I is not a *PA*-like inquiry," 54).

[32] Lennox 2010a; 2010b: 59; 2021, esp. chapter 6. [33] Connell 2016: 65.

range of *endoxa* from predecessors to contend with, compared to the teleological study of animal parts. This is also borne out by the fact that in all of *PA* II–IV, there is only one occurrence of *aporia* or related vocabulary (III.7, 669b15 and 26, on the question whether the liver and spleen are "single-natured" or "double-natured"), compared to twenty-four in *GA*.[34]

Returning to the *PA* I.1 context that put *genesis* front and center, we must recall Aristotle's emphatic stipulation about the explanatory priority of the end product (previously identified with the final cause, cf. 639b15–19) over the process:

> Another matter which must not be passed over without consideration is, whether the proper subject of our exposition is that with which the earlier writers concerned themselves, namely the way each thing is naturally generated (πῶς ἕκαστον γίνεσθαι πέφυκε), or rather the way it is (πῶς ἔστιν). For there is no small difference between these two views. The best course appears to be that we should follow the method already mentioned (begin with the phenomena presented by each group of animals, and proceed afterwards to state their causes) in the case of generation too. For in house building too, these things come about because the form of the house is such and such, rather than its being the case that the house is such and such because it comes about thus. For the generation is for the sake of being (ἕνεκα τῆς οὐσίας), but being is not for the sake of the generation. (640a10–19)[35]

Empedocles famously failed to realize this, and thought that the nature of the spine can be explained by events (twisting and breaking) taking place during incubation, not realising that the full form (the animal's 'being') pre-exists in the seed (640a20–25). To avoid Empedocles' mistake, we must get clear on what the animal is like ('the way it is') before we can proceed to account for its process of generation: "it is because the parent is such and such that the generation of the child is thus and so" (640a26–27).

Accordingly, scholars have applied this distinction and the attendant relationship of priority to the two major works, explaining that *PA* constitutes the discussion of being and, as such, precedes and prepares *GA*.[36] This is confirmed by the remarks in the opening chapter of *GA*, where Aristotle declares that he has already dealt with the final and formal

[34] On *aporiai* in *GA*, see Gelber 2017.

[35] The same ideas are expressed, with added emphasis on their implications for matter/form relationships, at *PA* II.1, 646a23–b10.

[36] Lennox 2005a: 57; Connell 2016: 66; Henry 2019: 107. Lennox 2021: 167: "In order to understand biological generation, one actually needs to have a systematic understanding of the facts about the fully developed animals one is studying and the lives they live."

causes of animals, which are more or less (σχεδόν) one and the same, as well as with the material cause, inasmuch as nonuniform parts are the matter for the entire animal, uniform ones for the nonuniform, and the so-called elements are the matter for the uniform parts. It remains to consider the moving cause, which is in a way (τρόπον τινά) the same as examining their generation (I.1, 715a3–15).[37] This distribution of causes corresponds very well with a passage of *PA* I.1 where Aristotle lists the types of 'being-focused' explanations that one should establish *before* proceeding to examine the process of generation (640a33–b1):[38] these explanations concern parts, and should state either (1) that the part is specified in the animal's (e.g. the human being's) definition (ἐπειδὴ τοῦτ' ἦν τὸ ἀνθρώπῳ εἶναι), or (2a) more generally that it is impossible for the part to be absent or to be different (sc. for reasons other than the definition), or (2b) that it is better for it to be present. Remarks about the process of generation should follow after these considerations (ταῦτα δ' ἕπεται). All three types of explanation will involve formal, material, and final causes (or combinations thereof), but not efficient ones, in accordance with the *GA* opening. Furthermore, the recommendations make clear that the 'how it is' account of *PA* is not restricted to explaining essential features only, as is abundantly clear in *PA* II–IV – although these books leave us largely on our own to work out which features are essential and which are not.[39]

These programmatic considerations on the relationship between *PA* and *GA* create certain expectations about *PA* explanations, arising from the focus on the outcome as that which determines the process rather than vice versa: we expect these explanations to be static rather than dynamic and less preoccupied with the source of the motion/process (i.e. the efficient cause).[40] All this is tied up with the fact that *PA* explanations operate on the level of animal groupings, determined each time at the appropriate level of generality and not on the level of individuals. Each of these groupings (e.g. lung possessors, blooded but not live-bearing, viviparous

[37] For a persuasive interpretation of this seeming distribution of causal labor, see Lefebvre 2018: 36–44.

[38] For a very good discussion of this passage, stressing (unlike many commentators) its role in the argument that explanation of an animal's being should precede the account of generation, see Code 1997: 139–141.

[39] See above p. 16. Gotthelf 1985 offers a study of those few instances where essence (*ousia*) is invoked in explanations as suggested in (1) above.

[40] Code 1997: 135: "A scientific account of the efficient causes of living things requires a prior account of the nature towards which such causes are directed." See also Lefebvre 2018: 42–44, and Lennox 2021: 168: "Understanding the facts about animal generation requires that we already have an understanding of what the parts that are coming to be are coming to be for."

quadrupeds, blooded oviparous land animals) comprises kinds that, according to Aristotle, remain unchanged thanks to the unbroken continuity of animal reproduction and form replication.

Thus, it would be absurd to ask, why, for example, goats as a kind *came to possess* horns (as opposed to why they *have* horns), because the former question would imply that there was a stage when goats were hornless. As Depew puts it: "Successful transmission makes questions about the historical emergence of traits themselves into non-starters."[41] Change, on the other hand, (which remains of course the primary concern of the natural scientist) takes place on the level of the individual substance, and this is the level on which *GA*'s key insights operate.[42] Aristotle expresses this point briefly and in passing in *Metaph.* A.1, 981a17: "All actions and comings-to-be are concerned with the particular" (αἱ δὲ πράξεις καὶ αἱ γενέσεις πᾶσαι περὶ τὸ καθ' ἕκαστόν εἰσιν).

Within *PA* I.1 itself, however, Aristotle invites us to modify these expectations and to allow a place for coming-to-be even in the explanation of features that belong to in common to groups of animals:

> This is how the demonstration must proceed: for example, that respiration is for the sake of X (ἔστι μὲν τουδὶ χάριν), and that this comes to be on account of Y by necessity (γίνεται διὰ τάδε ἐξ ἀνάγκης). By 'necessity' we sometimes mean that these things must be there, if the final end is to be reached; and sometimes that things are like this by nature. (642a32–35)[43]

The example here picks out an activity (respiration) rather than a part, but throughout II–IV, there are examples where parts or features of parts are explicitly said to *be* (ἔστιν, ὑπάρχει) for the sake of something and *come to be* (γίνεσθαι, γέγονε) by necessity, either conditional or material: eyelashes (II.15, 658b22–25); the location of viscera (III.10, 673a33–b3); the mesentery (IV.4, 678a4–12);[44] the distribution of talons, spurs, and webbed feet in different kinds of birds (IV.12, 694a23–b8). I believe that this rationale applies by and large to all the 'double-barrelled' explanations of *PA* II–IV, even in the majority that contain no reference to coming-to-be:

41 Cf. Depew 1997: 217.

42 Again, Depew puts it nicely (1997: 215): "In *GA* Aristotle inquires extensively into this ontogenetic process, introducing into this treatise a diachronic (and individual-centered) dimension that is not fully present in *HA* or *PA*."

43 This procedure is in fact an application of the recommendations of 640a10–19 (cited above) to a particular example.

44 The situation appears to be more complicated with the omentum, because we do not have there a statement that it is for the sake of something, only that nature makes use of it. See Henry 2019: 186–187.

the primary cause for the existence of a part is formal/final, but inquiry must not stop there, it must also uncover the necessary factors that make the realization of this end possible and also explain the mechanics, the *how* as opposed to the *why*.[45] We may take *PA* II.1, 646b20–35 as an indicative example of how material explanations supplement teleological ones, because it applies generally to the relationship between uniform and nonuniform parts in all animals. The passage states that nonuniform parts are composed of uniform parts and not vice versa because animals perform complex actions that require instruments with combined *dunameis* of, for example, both hardness and softness in a paw or hand. This is the main, final cause explanation (ἕνεκά τινος), but "since we are also inquiring into how it is necessary for things to be this way" (ζητεῖται καὶ πῶς ἀναγκαῖον), it is pointed out that it is mechanically possible for nonuniform bodies to be composed from uniform ones, but not vice versa, as this would mean that the uniform part would (absurdly) consist of many nonuniform parts.

It remains to ask how this interpretation that privileges end product over process can account for a different type of "dynamic" explanation in *PA*, namely the frequent references to nature in anthropomorphic terms, as a power that "devises" (μηχανᾶται, II.7, 652a30); "makes use of"/"finds uses" (καταχρῆται, παραχρῆται, παρακαταχρῆται, III.9, 671b1, IV.4, 679a30, IV.10, 688a24, 690a2); "distributes" (διανέμει, 687a11); "is unable" (ἀδυνατεῖ, 655a27). These verbs are treated as metaphors when it comes to denying Aristotle's belief in a cosmic, "demiurgic" Nature.[46] But they seem to be taken more seriously insofar as they are thought to describe the "actions" of an animal's formal nature, operating as a "goal-directed efficient cause."[47] Given the remarks above, I believe we must be cautious about such dynamic explanations: undoubtedly there is a place for efficient causes, both external and internal,[48] in Aristotelian biology, but these operate on the level of the individual animal. They account for the process by which, for example, an individual goat comes to possess horns, but do not explain why goats in general have horns – this has to be explained with reference to the mature goat and its functions.[49]

45 Thus also Johnson 2005: 195–196. See, e.g., *PA* II.14, 658b2–10 (humans have the hairiest head); III.9, 672a2–22 (kidneys are fatty); IV.11, 692a2–8 (snakes can turn their head backwards).

46 Lennox [1996b] 2001b: 184. See also Preus 1975: 225–235.

47 Lennox [1996b] 2001b: 183, 202 n. 9; Leunissen 2010: 19, 36 and *passim*.

48 On this issue see Connell 2016: 141–170; Henry 2019: 143–155.

49 As is in fact done at *PA* III.2, 662b23–663b21.

Moreover, when discussing the distinction between *being present* for a purpose/function and *coming to be* as a result of material necessity, it should be made clear that Aristotle considers the two complementary for a full explanatory account, but does not consider them co-ordinate, because coming to be is for the sake of being and not vice versa. In the light of the programmatic statements quoted above, I am not sure whether Aristotle would endorse qualified teleological explanations based on a temporal sequence of events whereby the animal's formal nature opportunistically assigns a purpose to parts or materials *after* they have come to be by material necessity.[50] The objection is not to the causal and explanatory value of material necessity itself but to the implication that a *process* can be explanatory of the end result – this would amount to *genesis* being explanatory of *ousia*.[51]

1.4 Conclusions: The Biological "Corpus"

The preceding discussion has situated *PA* at the head of Aristotle's explanatory project on animals. This entire project follows *HA* and the *Dissections*, in the sense that it presupposes their results, but also runs parallel to them as the reader is frequently referred there for more detail.[52] Cross-references in *PA* to *HA* and the *Dissections* indicate existing materials (cf. past-tense references at II.1, 646a9, II.16, 660a8, II.17, 660b1), but for the most part they refer the reader not to an earlier stage of the investigation but to a parallel, synchronous source of more detailed factual information (cf. III.2, 663b3, III.5, 668b30, III.14, 674b16–17, IV.8, 684b2, IV.10, 689a17, IV.13, 696b15). I also argued that there are strong reasons for keeping *PA* I organically attached to II–IV. Another connection that emerged very strongly was that between *PA* and *GA*, especially given the programmatic comments about generation in *PA* I and the 'division of labor' between the works found in *GA* I.1 (see above 000). In this connection, I argued that

[50] This impression is given by Leunissen's account (2010a, esp. 95–96) and Leunissen 2012. Leunissen's approach appears to bring the ontogenetical perspective of *GA* into the explanations of *PA* (cf. the prominent role played by *GA* II.6, 744b11–27 in her analysis of teleology and the status of various parts in Leunissen 2010a: 81–99).

[51] In the popular example of horns (*PA* III.2, 663b21–664a3), it is the fact that some animals are such that they have horns for self-defense that determines the process whereby the earthy residue is directed there. They do not have horns because at some stage nature 'discovered' a surplus and assigned a purpose for it. Material necessity explains only in what way large and earthy animals come to grow hard means of defense: the portion of earthy surplus is carried upwards and provides the matter for either teeth/tusks or horns but is not enough for both.

[52] Thus also Lefebvre 2018: 41; Feola 2017: 52 uses the image of *HA* as a tree trunk with the other treatises as sprouting branches.

references to changes or processes in *PA* are either presented as the necessary 'mechanics' underpinning the functions of purposeful parts or should be taken metaphorically whenever they portray nature's 'actions' as determining outcomes.

The central "axis" of *HA* – *PA* – *GA* is also present in an influential ancient catalog that had a great impact on the manuscript tradition and subsequent editions,[53] which, immediately following *Meteor.* and before *Metaph.*, reads as follows:

> *On the Soul,* 3 books
> *Sense and Sensibilia,* 1 book
> *On Memory and Sleep,* 1 book
> *Movement of Animals,* <1 book>
> *Dissections,* 7 books
> *On the Characters of Animals / Research into Animals,* 10 books
> *Parts of Animals,* 4 books
> *Generation of Animals,* 5 books
> *Progression of Animals,* 1 book
> *On Length and Shortness of Life,* 1 book
> *On Life and Death,* 1 book
> *On Plants,* 2 books

Aristotle's texts, however, contain traces of alternative sequences, which point in different directions, and thus force us to speculate about phases when the biological corpus might have taken different forms. First of all, the opening of *IA* identifies it as a discussion of parts, and its conclusion makes an even stronger claim to the place immediately following *PA*:[54]

> So much for our discussion of the parts of animals and particularly those which have to do with progression and all change from place to place. Now that these points have been settled, our next task is to consider soul. (*IA* 19.714b20–22)

It is a reasonable hypothesis (but no more than that), that Aristotle sought to insert *IA* after the original axis *PA* – *GA* was already in place (whether because he wrote it later or because he conceived of a more expansive and

[53] This catalog, preserved in Arabic sources only, is often thought to reflect the corpus-organizing activities of Andronicus of Rhodes in the first century BCE, see, e.g., Hatzimichali 2016. This has now come under scrutiny in Rashed 2021.

[54] Some commentators disregard such transitional passages as spurious, but here (as in the one from *MA* below) I am following the experienced textual critics who take them seriously, Rashed 2004; Primavesi 2020.

integrated biological corpus at a later stage).[55] The order suggested in the conclusion of *IA* is not adopted in the ancient catalog cited above, but it is confirmed by the equivalent conclusion of *MA*:

> So then, concerning the parts of each animal, and concerning soul, and again also concerning perception and sleep and memory and motion in general, we have stated the causes. (*MA* 11, 704b1–3, trans. Morison)

The significance of this passage for present purposes is that it "inserts" yet more works between *PA* and *GA*: these are *DA* again, with the treatises on the senses (*Sens.*), memory (*Mem.*) and sleep (*Somn.*, probably taken together with *Insomn.* and *Div. Somn.*), which all now form part of the *PN*, as well as *MA* (*IA* may be included under the reference to parts, since locomotive parts are one of its main considerations). This complex picture of different layers/stages is in evidence also if we consider the internal cross-references from *PA* to other works, where we find past-tense references to the results of 'psychological' research and observation (i.e. soul functions and their bodily location, including perception and motion, e.g. II.7, 653a19; IV.13, 696b1) but also similar past-tense references from these works back to *PA* (*Somn.* 2, 455b34; *Resp.* 7, 473a25–28).[56]

These varied pieces of evidence point toward the tentative conclusions that (i) *PA* was composed as part of a "core" biological corpus consisting of *HA*, *PA*, *GA*; (ii) as Rashed suggests,[57] there was at some stage an effort to integrate further works with this core, works whose investigations were part of the same broad inquiry into life and living beings but were not written in a way or a chronological sequence that facilitated the compilation of a corpus free from anomalies. The 'full' corpus, with all its problems and inconsistencies would look like this:[58] *HA*, *PA*, *IA*, *DA*, *Sens.*, *Mem.*, *Somn.*, *Insomn.*, *Div.Somn.*, *MA*, *GA*, *Long.*, *Juv.*, *Resp.* One way in which this sequence makes sense is that all the works before *GA* can be deemed also to treat the 'being' of animals; conversely, the *PN* works concerning the preservation and eventual loss of life (*Long.*, *Juv.* + *Resp.*)

[55] For the problems with placing *IA*, see Kullmann 2007: 146; and Falcon 2021b on the complex picture of *IA*'s status relative to the other works as it emerges from early printed editions. See further Falcon 2021c: 109 on Aristotle's tactics for integrating *IA* into a broader explanatory project.

[56] See Kullmann 2007: 148; Feola 2017: 55.

[57] Rashed 2004 offers a very helpful discussion of the constitution of *PN*. He makes the crucial distinction between the process of writing the treatises and that of integrating them into a sequence. He concludes that the "full" version of *PN*, as presented in our modern editions and sanctioned by the prologue of *Sens.* is the earlier version and associates the split of *PN* with a subsequent effort to integrate the more "psychological" writings into a single biological project.

[58] Thus Rashed 2004, except on the issue of *PA* I (see above p. 14). He also considers the status of *On Plants*, which I have not included here, cf. Falcon 2015.

are reasonably placed after *GA*. In fact, the closing words of *Long.* read as follows: "This is our account of the reasons for the duration of life and for short life in animals. It remains for us to discuss youth and age, and life and death. To come to a definite understanding about these matters would complete our course of study on animals" (τέλος ἂν ἡ περὶ τῶν ζῴων ἔχοι μέθοδος; 6, 467b5–9, trans. Ross). We shall never have definitive proof as to whether the integration of all these works into an extended biological corpus was done by Aristotle himself or someone else. All we can say is that taken together they are a fitting and magnificent completion (*telos*) of Aristotle's inquiry into nature.

CHAPTER 2

Rank Ordering of Skills and Sciences in Aristotle's Protreptic to Biology

Monte Ransome Johnson

Introduction

In the first part of *De Partibus Animalium* I.5 (644b22–645a36), Aristotle encourages his audience to engage in a novel kind of philosophy:[1] the scientific inquiry into animals and plants (hereafter: biology): "Since we have completed stating the way things appear to us about divine things, it remains to speak about animal nature, omitting nothing in our power" (645a4–6); "one should approach research about each of the animals without disgust, since in every one there is something natural and beautiful" (645a21–23). The kind of discourse here, to use Aristotle's own terminology in the *Rhetoric*, is protreptic ("exhortation," in Latin), that is, speech encouraging the audience to do something. What Aristotle is exhorting his readers to do, biology, is newly and originally conceived, but the literary technique employed – protreptic speech or argumentation – is one of the oldest and most traditional kinds of discourse, a genre for which Aristotle had already written a general account (in the *Rhetoric*) and a popular work, the lost dialogue *Protrepticus*, which was an exhortation to philosophy itself.

The protreptic section in *De Partibus Animalium* I.5 can be divided into three separate arguments.[2] In the first (644b22–645a7), Aristotle compares the value of two theoretical sciences, astronomy and biology, concluding that they are comparable in worth and so should both be valued: "Each study has its attractions ... perishable things, however, take the prize in respect of understanding ... they provide a certain compensation compared with the philosophy concerned with divine things"

[1] The second part of I.5 (645a36–646a4) then encapsulates the method of how, once one has been convinced to do biology, the inquiry is to be conducted.

[2] The third is a self-refutation argument leveled against an anonymous or hypothetical objector discouraging students from the study of biology (645a26–36). Unfortunately, I will not have the space to discuss it here, although we have elsewhere discussed the parallel self-refutation argument in the *Protrepticus* (in Hutchinson and Johnson 2018: 27–143), and I discuss the *PA* version in Johnson, 2025: 109–114.

(644b31–645a4). The second protreptic argument (645a7–26) involves comparing the value of the objects of biology with products of technical skills such as sculpture or painting: "It would be unreasonable, even absurd, for us to enjoy studying likenesses of animals on the ground that we are at the same time studying the skill that made them, such as painting or sculpture, while not prizing even more the study of things constituted by nature" (645a10–14).

The argumentative technique employed in both of these arguments, comparing and ranking technical skills or arts and sciences, is discussed in general in both the *Rhetoric* and *Topics* and then applied in several well-known openings of works in the Aristotle Corpus, including *De Anima* I.1, *Metaphysics* I.1, and *Nicomachean Ethics* I.1. The same technique was also developed and deployed heavily in the early exoteric dialogue entitled *Protrepticus*.[3] This chapter will largely be concerned with these other texts in order to bring out what is distinctive and original about Aristotle's exhortation to biology.

Examining Aristotle's explicit methodological remarks and application of the argumentative technique to parallel texts affords insight into his method of writing and composition: Aristotle applied what he considered established and generic rhetorical and dialectical techniques, now collected in *Topics* and *Rhetoric*, in writing the protreptic openings to several of his own works. In fact, a stronger claim can be supported: not only did Aristotle apply such established techniques in those works but, having already employed them with great success early on in his *Protrepticus*, he sometimes later reused or adapted arguments already written in the *Protrepticus* in the protreptic sections (usually openings) of the more specialized works of the Aristotle Corpus.[4]

[3] I here assume the results of previously published research that argues that large, continuous, and well-ordered fragments of Aristotle's lost *Protrepticus* survive in Iamblichus' *Protrepticus epi philosophian* (Hutchinson and Johnson 2005). The same is true of another textbook of Iamblichus' entitled *De communi mathematica scientia* (*DCMS*); two chapters have been identified and confirmed as containing source material that originated in Aristotle's *Protrepticus*: chapter XXIII (Merlan 1953; Festugière 1956), and chapter XXVI (for the history and rightness of this attribution, see Hutchinson and Johnson 2018: 144–154). *DCMS* XXVII has not hitherto been ascribed to the lost work, but I make the attribution (in Johnson, 2025), on the basis of the fact that the first thirty lines of *PA* I.1 is a text reuse and adaptation of the source cited by Iamblichus in *DCMS* XXVII, and that source text can be shown to be Aristotle's *Protrepticus*. In our reconstruction (in Hutchinson and Johnson, unpublished manuscript), we present the whole sequence of chapters XXII–XXVII as a set of excerpts from the lost dialogue.

[4] The most widely discussed case is *Metaphysics* I: "To teach us to understand it (sc. knowledge) in this profound sense was what Aristotle aimed at in the *Protrepticus*, and the famous introduction to the *Metaphysics* is in essence nothing but an abbreviated version of his classical exposition of the matter there" (Jaeger, trans. Robinson 1941: 69). On *Nicomachean Ethics*, see Hutchinson and Johnson

In the present chapter, I will show that Aristotle applied a general technique of rank-ordering skills and sciences in expounding the protreptic arguments of *De Partibus Animalium* I.5: a technique explicitly discussed, and strongly paralleled, in both the earlier *Protrepticus* and some of the most familiar works of the Aristotle Corpus. I will begin by introducing Aristotle's discussion of comparing sciences in the *Rhetoric* and *Topics*. Next, I will discuss parallel applications of this argumentative technique in the *Nicomachean Ethics* I, *Metaphysics* I, *Protrepticus*, and *De Anima* I. In Section 2.3, I will show how Aristotle deployed the technique in the first two protreptic arguments of *De Partibus Animalium* I.5.

2.1 Rhetorical and Dialectical Method

Deliberative rhetoric concerns "either *protreptic* (exhortation or encouragement) or *apotreptic* (dissuasion or discouragement)" (1358b8–9). Three features define protreptic speech. First, protreptic aims to persuade people to choose to undertake some activity; second, and as a result, protreptic is oriented toward future action, "for both protreptic and apotreptic deliberate about things that will happen" (1358b14–15). Third, the basis for encouraging or discouraging future action is "the advantage or disadvantage of a certain course of action" (1358b22).

Thus, Aristotle next gives an outline of the concept of advantage.[5] He begins by asserting that the universal aim to which all people look when choosing or avoiding some activity is *eudaimonia*. This is so "both publicly and privately" he says, indicating that protreptic rhetoric is applicable in both political and personal contexts. Thus, an analysis of *eudaimonia*, which Aristotle also offers in the *Ethics* and *Politics*, appears in the *Rhetoric*, because "all forms of protreptic and apotreptic are concerned with this and the things that contribute or are opposed to it" (1360b9–11). In the *Rhetoric*, Aristotle does not, as he does in the *Ethics*, point out that philosophers differ from the majority about what constitutes *eudaimonia*.[6]

2014. The protreptic opening of the *De anima* will be discussed briefly below. The first thirty lines of *De Partibus Animalium* I.1 is a text reuse and adaptation from the *Protrepticus* (see previous note).

[5] Aristotle initially explores the advantageous with respect to the various political and geopolitical subjects that are the usual subjects of rhetorical speeches. He then states that he will return to the "the sources from which arguments of protreptic or apotreptic about these *and other matters* should be supplied" (I.4, 1360b1–2).

[6] Aristotle initially defines *eudaimonia* as follows: "Let success (*eudaimonia*) be defined as: (a) good action combined with virtue; or (b) self-sufficiency in life; or (c) the most pleasant life accompanied by security or abundance of possessions and bodies (e.g., animals and slaves), with the ability to keep secure and use these things; for all people agree that success is mostly one or more of these things"

This is because his purpose in the *Rhetoric* is not to determine a normatively or even descriptively adequate definition of *eudaimonia* but rather to provide a rhetorical analysis of *eudaimonia* that will be useful in designing speeches that aim to persuade people to choose to undertake a certain activity.[7]

Although *eudaimonia* is the ultimate end, the proximate goal of all choices and avoidances is the advantageous or the good, and so, since deliberation is (generally) about means rather than ends, the focus of protreptic speeches is (usually) the means to *eudaimonia*, that is, the goods that either constitute or bring about *eudaimonia*. Accordingly, Aristotle provides both a general definition of the good[8] and an enumeration of many specific goods in *Rhetoric* I.6, where he goes into detail about how to argue whether certain things are good or not. But because "in many cases both sides in a debate agree that two things are advantageous but disagree about which is the more so" (1363b5–6), Aristotle goes on to discusses "the greater good and the more advantageous" in *Rhetoric* I.7. In this context, Aristotle reiterates his definition of good and expands on it in order to indicate how there might be a comparison of goods.[9]

The basis of this comparison is: "That which the intelligent men (οἱ φρόνιμοι) would judge or has judged a greater good, either all or most of them, must be so; either absolutely, or insofar as they judge in accordance

(*Rhet.* I.5, 1360b14–18). For alternative definitions of *eudaimonia*, see *EN* I.4 (1095a12–30), *EN* I.8, 1098b22–25, VII.13, 1153b9–21, and X.6-8, 1176a30–1179b23.

[7] Aristotle goes on to give a laundry list of the conventional "parts" of success: good birth (i.e. nobility), numerous and worthy friendships, wealth, good and numerous children, good old age, health and other bodily virtues, reputation, honor, and even good luck. Aristotle works all of these into the arguments of the *Protrepticus* (Hutchinson and Johnson 2018: 121).

[8] "Let good be defined as: (1) what is chosen for itself, and for the sake of which we choose other things; and (2) what everything aims at, or everything having sensation or intellect, or that could get intellect; and what intellect would give to each, and what intellect does give to each in individual cases (καὶ οὗ ἐφίεται πάντα, ἢ πάντα τὰ αἴσθησιν ἔχοντα ἢ νοῦν ἢ εἰ λάβοι νοῦν, καὶ ὅσα ὁ νοῦς ἂν ἑκάστῳ ἀποδοίη, καὶ ὅσα ὁ περὶ ἕκαστον νοῦς ἀποδίδωσιν ἑκάστῳ), for this is the good for each thing; and (3) what by its presence causes one to be well off and self-sufficient; and self-sufficiency itself; and what is productive or preservative of such things; and what such things follow upon; and what is preventive of their opposites and destroys them" (*Rhet.* I.6, 1362a21–29).

[9] "We call good: (1) what is chosen for itself and not for the sake of something else; and (2) what everything aims at, and would choose if they have intellect and intelligence (καὶ οὗ πάντ' ἐφίεται, καὶ ὃ νοῦν ἂν καὶ φρόνησιν λαβόντα ἕλοιτο); and (3) what is productive and preservative of such things, or is always accompanied by them. That for the sake of which things are done is the end (an end being that for the sake of which all else is done) and for each individual that thing is a good which fulfills these conditions in regard to himself. It is necessary, then, that a greater amount is greater than one or than any of the smaller numbers (ἀνάγκη τά γε πλείω τοῦ ἑνὸς καὶ τῶν ἐλαττόνων), if that one or smaller number is included in the count; for then the larger number surpasses the smaller, and the smaller quantity is surpassed as being contained in the larger" (*Rhet.* I.7, 1363b12–21).

with their intelligence (κατὰ τὴν φρόνησιν)" (1364b11–14). Thus, the criterion of the comparative "better" is epistemic: what the person with an epistemically superior state of mind would choose is ipso facto the more choiceworthy and the better. Of course, the question then becomes what criterion the intelligent person uses to choose among goods, and this leads Aristotle to discuss the comparative good and the relatively advantageous. Aristotle mentions dozens of ways to show how one good thing may be argued to produce a greater amount of good than other, similar good things. Two of these indicate how to argue that one science may be considered comparatively better than another science.

> **<a>** And: of the sciences that are more beautiful or more respectable (αἱ ἐπιστῆμαι καλλίους ἢ σπουδαιότεραι), their objects too are more beautiful and more respectable. For as the science has this condition, so too does the reality; but each commands its own. **<b>** And: of the things more respectable and more beautiful, their sciences are analogous because of this. (**T1:** *Rhetoric* I.7, 1364b7–11)

Let us represent symbolically the analogy contained in this dense passage. First of all, we are to assume a relationship between a given science (S_1 or S_2) and its object (O_1 or O_2), and then draw an analogy or proportion between these objects and sciences, as follows:

$$S_1 : O_1 :: S_2 : O_2$$

Unfortunately, in the *Rhetoric*, Aristotle gives no example, but the following one will serve our overall purpose:

e.g. astronomy: stars :: biology: animals

The single colon (:) indicates the relation between an object and its science, and the double colon (::) indicates a second order relationship of similarity (or difference) with respect to some predicate X between the two relations, where the predicate X is some term that admits of degrees (e.g. beautiful, respectable, or choiceworthy). The relation could be "more X than" or "less X than"; these will be indicated by the comparative sign, as in the expression $S_1 > S_2$, meaning "Science 1 is more X than Science 2."[10] This corresponds to Aristotle's idea that one kind of science (or object) may be "more beautiful and more respectable"

[10] I follow Gambra Gutiérrez 2012: 430 in using (in addition to conventional logical symbols) the comparative signs ">" and "<" for "more" and "less" of some predicate X, e.g. "A > B" means "A is more X (e.g. beautiful, respectable, or choiceworthy) than B."

than another. Now, we can represent Aristotle's argument in **T1a** as follows:[11]

$$(S_1 > S_2) \rightarrow (O_1 > O_2)$$

e.g. If astronomy is more beautiful and respectable than biology, then stars are more beautiful and respectable than animals.

I will call this formula "rank-ordering objects by the epistemic criterion," because it rank-orders objects based on the comparative beauty or respectability of the science. But Aristotle also seems to countenance the following formula, in **T1b**:

$$(O_1 > O_2) \rightarrow (S_1 > S_2)$$

e.g. If stars are more beautiful and respectable than animals, then astronomy is more beautiful and respectable than biology.

I call this formula "rank-ordering sciences by the objective criterion" because it rank-orders the value of certain sciences based on the comparative beauty or respectability of their objects.

So far, Aristotle seems to assume that there will be a perfect correspondence between rank-orderings according to epistemic and objective criteria. Although Aristotle does not explicitly say so, another formula seems to follow as a corollary of the previous two:

$$(S_1 = S_2) \leftrightarrow (O_1 = O_2)$$

e.g. If astronomy and biology are equally beautiful and respectable, then stars and animals and plants are equally beautiful and respectable. And if stars and animals and plants are equally beautiful and respectable, then astronomy and biology are equally beautiful and respectable.

I call this "rank-leveling" (by either the epistemic or objective criterion), because if two sciences are considered equally valuable, then it seems that their objects must be as well, and if their objects, then the sciences accordingly. Thus, the bidirectional arrow in the rank-leveling formula.

Aristotle's technique of the comparison of sciences in the *Rhetoric* is extremely brief, but it appears again in *Topics* III when he treats the topic of the "more choiceworthy" (*hairetōteron*). That book opens with the following observation.

> And the question which is more choiceworthy or better of two or more things should be examined upon the following lines. But first it must be

[11] This operation has been given the name *alternando*; cf. *APo*. II.17, 99a12–14.

> clearly laid down that the inquiry we are making concerns not things that are widely divergent and that exhibit great differences from one another (for nobody raises any doubt whether happiness or wealth is more desirable), but things that are nearly related and about which we dispute for which of the two we ought rather to vote, because we do not see any advantage on either side as compared with the other. Clearly, then, in such cases if we can show a single advantage, or more than one, our judgement will record our assent that this is more choiceworthy, whichever of these sides turns out to have the advantage. (**T2**: *Topics* III.1, 116a3–12, trans. Pickard-Cambridge, adapted)

In any dialectical contest, one needs to set up a choice between very few things and preferably only two. Although more are possible, Aristotle seems to envision reducing these cases to successive pairwise comparisons. Each pairwise comparison should involve things that are already as close together as possible. This will enable one to establish that something is more choiceworthy even if it can only be shown that it has a single, small advantage. The rest of the chapter, in parallel to *Rhetoric* I.7, articulates several modes of arguments about comparative advantage. The first one that Aristotle mentions is the following.

> First, then, that which is more long-lasting or more secure is more choiceworthy than that which is less so. And that which the intelligent person or the good man would choose more, or the right law, or which the people most respectable concerning each thing would choose, insofar as they are these kind of people or those having scientific knowledge in each genus, either most or all of them; for example: in medicine or carpentry, that which most or all of the medical doctors would choose, or, generally, whatever things most or all people or all things would choose, e.g. the good. For all things aim at the good. (**T3**: *Topics* III.1, 116a13–20)

The first mode is clear enough: what is more long-lasting or more secure is more choiceworthy.[12] The second mode returns to the idea, which we saw was also mentioned in *Rhetoric* I.7, that the more choiceworthy depends on what the more intelligent person would choose.[13] Aristotle says that what the most intelligent person in a given domain would choose is better than what the generally intelligent or generally good person would choose; what the generally intelligent or good person would choose is better than

[12] One may argue, for example, that wisdom is superior to other goods by being more long-lasting and secure than other goods such as health or wealth, and Aristotle employs such an argument in the *Protrepticus*, apud Iamblichus, *Protr.* VIII, 47.16–48.13 Pistelli.

[13] See above note 8 regarding an even more expansive version of this, given in *Rhet.* I.6, including: what all things with intellect would choose, all things with sensation would choose, or just "everything" would choose.

what most or all people would choose; and what most or all would choose is better than what few or no one would choose. Notice that this technique invokes an epistemic criterion: those with a certain kind of scientific knowledge set the standard for what is better to choose.

The examples given in the *Topics* depend on domain specificity: if the question is about building, then the standard is what the best carpenters would choose; and if the question is about medicine, the standard is what the best doctors would choose. So far, this is clear and uncontroversial. But what Aristotle goes on to say is not so clear, and he provides no examples:

> And one must lead the discourse towards what is needed. And absolutely better and more choiceworthy is the one in accordance with the better science; but for a particular thing it is the one in accordance with its appropriate science. (**T4:** *Topics* III.1, 116a20–22)

The point about choosing "the one in accordance with its appropriate science" (τὸ κατὰ τὴν οἰκείαν) is presumably the same point that has just been made about domain specificity and instanced by carpentry and medicine. But Aristotle also mentions the "absolutely better and more choiceworthy" (ἁπλῶς μὲν βέλτιον καὶ αἱρετώτερον), and he says that this is "the one in accordance with the better science" (τὸ κατὰ τὴν βελτίω ἐπιστήμην). One small problem with this is how, by means of the *comparative* notion of a "better science" we are able to choose the "absolutely better" and so (presumably) superlative or best. I think this problem can be resolved by adhering to the dialectical setup that Aristotle stipulated at the beginning of *Topics* III.1 (**T2**): if we restrict the comparison to only two things, then whichever is the better of the two will be the best superlatively and absolutely.

The deeper problem is how one might compare sciences outside of the context of domain specificity. How would one choose what is better between, for example, carpentry and medicine? Or between medicine and physics? Or, getting to the overall point of the present inquiry, between astronomy and biology? As we saw, the *Rhetoric* suggests an approach to this kind of question: apply the objective criterion: those sciences are better which have better objects. So, if the good of medicine (i.e. health) is better than the good of carpentry (e.g. convenience or comfort), then medicine has a better object than carpentry and is thus a better kind of knowledge. Or if humans are better than animals, then the medical knowledge of how to treat humans would be superior to veterinary medicine.

Aristotle does not mention the objective criterion in the *Topics*, only the epistemic one. According to **T4**, the better science determines what is the better object, and Aristotle does not mention that this also goes in the other direction (as we have suggested is a corollary of **T1**). But as we will see, this is exactly the dialectical technique that Aristotle pursues in the first protreptic argument of *De Partibus Animalium* I.5, where he argues that the *epistemic* superiority of biology indicates the respectability of its *objects*, the animals and plants.

Before turning to that argument, let us examine Aristotle's application of this argumentative technique of comparing sciences in some other texts. What he says about comparing sciences in *Topics* and *Rhetoric* have only hinted at how one chooses the better when a certain domain is specified, not how one is to choose among, and thus rank-order, sciences in general. In fact, so far, we have little more than the terminology "the better science" to go on. But what exactly is the "better science" is bound to be controversial, and a fortiori what is "the absolutely better" science. Does Aristotle nevertheless offer a way to rank-order all these sciences?

2.2 Applications in the Aristotle Corpus and the *Protrepticus*

It is well-known that Aristotle does envision a way to rank-order the technical skills and practical sciences: on the basis of their ends. In fact, this is the very point with which Aristotle opens the *Nicomachean Ethics*.

> Every technical skill and every investigation, and similarly every action and also choice, seems to aim at some good; for this reason the good has been rightly said to be that at which all things aim. But a certain difference appears among ends: for some are the activities, but others are functions beside the activities. In those cases in which there are ends besides the activities, the functions are by nature better than the activities. But, as there are many actions, technical skills, and sciences, their ends too are many; the end of the technical skill of medicine is health, that of ship-building a vessel, that of strategy victory, that of economics wealth.
>
> But where such technical skills fall under a single capability – as bridle making and the other technical skills concerned with the equipment of horses fall under the technical skill of riding, and this and every military action under strategy, in the same way other technical skills fall under yet others – in all of these the ends of the architectonic technical skills are to be preferred to all the subordinate ends; for it is for the sake of the former that the latter are pursued. It makes no difference whether the activities themselves are the ends of the actions, or something else apart from the activities, as in the case of the sciences just mentioned. (**T5:** *Nicomachean Ethics* I.1, 1094a1–18)

Thus, in those cases where "technical skills fall under a single capacity," it is possible to arrange technical skills in an "architectonic" structure: we can discern the superior science by ordering the sciences according to ends, and we can do this fairly easily. We treat the ends of these practical sciences as their objects, and then we can, following Aristotle's example, establish an analogy between these sciences and their corresponding objects. This will be the first step in a program that will determine a rank-ordering of these sciences, which I specify in the following steps.

1. Draw an analogy (with 3 terms this time, instead of 2), treating the ends of technical skills that fall under a single capability as the objects of those technical skills.

$$O_1 : S_1 :: O_2 : S_2 :: O_3 : S_3$$

e.g. victory : strategy :: victory of horseback fighters : tactics on horseback :: equipping horseback fighters : bridle-making.

2. Rank-order the objects (ends).

$$O_1 > O_2 > O_3$$

e.g. victory > victory of horseback fighters > equipping horseback fighters.

3. Apply the objective rank-ordering criterion.

$$(O_1 > O_2 > O_3) \rightarrow (S_1 > S_2 > S_3)$$

e.g. if victory > victory of horseback fighters > equipping horseback fighters, then strategy > tactics on horseback > bridle-making.

4. Derive the rank-ordering of technical skills or sciences.

$$S_1 > S_2 > S_3$$

e.g. strategy > tactics on horseback > bridle-making.

Rank-ordering in the technical skills or practical sciences is relatively easy, because the objects of practical sciences are just their ends, and we are used to thinking about how ends are related to and subordinated to one another in our own everyday decision-making. For example: I want to keep my job, so I need to show up for work, so I need to get out of bed: the getting out of bed and going to work, no doubt, is for the sake of keeping my job. We are constantly and even subconsciously applying this kind of reasoning. Aristotle suggests that it can, at least in theory, be represented syllogistically; this would be the so-called practical syllogism.

But since the theoretical sciences, which are our concern here, have no end other than knowledge itself, this kind of easy intuitive solution is unavailable to us. It remains unclear how theoretical sciences can be rank-ordered either relative to the practical-productive sciences (such as politics and architecture), or among themselves. Nevertheless, Aristotle does believe that it is possible not only to rank order the theoretical sciences relative to the practical ones but also to rank-order the theoretical ones themselves. This is made clear in *Metaphysics* I.2.

> We have said in the *Ethics* what the difference is between technical skill and science and the other similar things; but the point of present discussion is this, that all men suppose what is called wisdom to deal with the first causes and principles of things. This is why, as has been said before, **<a>** the man of experience is thought to be more wise than the possessors of any perception whatever, **<b>** the technician more wise than the men of experience, **<c>** the architect more wise than the manual laborer, and **<d>** the theoretical sciences more wise than the productive sciences. Clearly then wisdom is knowledge about certain causes and principles. (**T6:** *Metaphysics* I.2, 981b25–982a3, trans. Ross, adapted)

Aristotle introduces a number of rank-orderings among epistemic representatives here, including **<a>** the person with experience is more wise than the person with perception; **<b>** the technician is more wise than the person with experience; **<c>** the architect is more wise than the manual laborer; and **<d>** the theoretical sciences are more wise than the productive sciences.[14] The third evaluation follows from the account of overlapping technical skills with which Aristotle opens the *Nicomachean Ethics* (**T5**) discussed above. The basis for the fourth evaluation is not yet clear: one may think, and Aristotle himself raises the possibility early in the *Nicomachean Ethics*, that politics might be the most architectonic and superior science, since it issues orders to all the other technical skills.[15] But in the *Metaphysics*, Aristotle explicitly rejects this view and asserts the superiority of the theoretical sciences to the practical ones. The first and second evaluations, since they share a common term, may be combined into a single one: the skilled technician is more wise than the person with mere experience, and the person with experience is more wise than the person with mere perception. The reason for all of these evaluations becomes clear in the following passage.

> Since we are seeking this knowledge, we must inquire of what kind are the causes and principles, the knowledge of which is wisdom. If we were to take

[14] These relationships could be unified as follows: the theoretical scientist > the skilled technician > the person with experience > the person with perception. But, in accordance with his methodological suggestion in **T2**, Aristotle sticks to pairwise comparisons.

[15] *EN* I.2, 1094a25–b7 (see further Johnson 2015: 167–170).

> the notions we have about the wise man, this might perhaps make the answer more evident.
>
> We suppose, **<a>** first, that the wise man knows all things so far as possible, though not having knowledge of each thing individually; **<b>** second this wisdom is the capability to understand the hard things and not things easy for a human to understand (for to perceive is common to all, for which reason it is easy and not wise); **<c>** again, the one who is more clear and more able to teach the causes is more wise concerning each science; **<d>** and of the sciences, also, the one which is chosen for its own sake and thanks to the knowing is more wise than the for the sake of their results; **<e>** and the more architectonic sciences are more wise than the subordinate. For the wise man should not receive orders, but give orders, and he must not obey someone else, rather the less wise must obey him. (**T7:** *Metaphysics* I.2, 982a4–19, trans. Ross, adapted)

In this passage, Aristotle utilizes both objective and epistemic criteria for the sake of determining the absolutely best science, the acquisition of which corresponds to wisdom. The objective rank orderings are that **<a>** the one with scientific knowledge of all things in general is more wise than the one with scientific knowledge of particular things; and **<b>** the one with scientific knowledge of hard things is more wise than the one with scientific knowledge of easy things. These are interrelated: perception of particular things is relatively easy, and so this is part of the reason why the person with experience is better than the person with mere perception, and the person with technical knowledge is even better: because the objects of experience and technical skill are harder to grasp than those of perception.

I must digress here to clarify because, in a sense, perception, experience, and technical skill all have the same objects, but because they are apprehended differently (some more clearly and in a way that makes their causes more teachable), Aristotle sometimes speaks as if their objects were different, as we commonly do. For example, the objects of music and painting, melodies and portraits, are harder to grasp than those of hearing and seeing, sounds and colors, even though the objects of music and painting involve and are in a sense reducible to sounds and colors. This is because melodies and portraits are more complex objects that cannot be grasped simply by hearing sounds or seeing colors, and so one needs education in playing musical instruments or drawing in order to appreciate them in a way that other animals and the uneducated cannot.[16] In this way the objects of music and painting are themselves harder to grasp, and according to Aristotle better than the objects of mere hearing and seeing.

[16] Education in drawing: *Pol.* VIII.3, 1338b1–2; in playing music: VIII.6, 1341a14–15.

At this point we can move on to discuss Aristotle's other epistemic rank-orderings. The third evaluation in **<7c>** is key to understanding Aristotle's entire rank-ordering scheme. As Aristotle puts it here, three comparative expressions are used: "The one who is *more clear* and *more able to teach of the causes* is *more wise* concerning each science" (ἔτι τὸν ἀκριβέστερον καὶτὸν διδασκαλικώτερον τῶν αἰτιῶν σοφώτερον εἶναι περὶ πᾶσαν ἐπιστήμην). The overarching comparison is between two people, A and B, where A is said to be more wise than B, because A's epistemic state is comparatively better in two ways: A is both "more clear" and is "more able to teach the causes" than B.

Aristotle does not explain either of these superior epistemic states here, and Jonathan Barnes may be correct that "in general being *akribēs* seems to amount (vaguely enough) to being of good epistemic quality" (Barnes 1993: 189). But an examination of Aristotle's use of the term *akribesteron* can be of a little more help. Aristotle himself glosses the root term *akribēs* with "clear" (σᾰφής, *Topics* II.4, 111a9), and so I will use that translation in preference to the more usual "exact," "accurate," or "precise" (although certainly allowing it may mean any or all of those things as well). A crucial text for an interpretation of the comparative notion is *Analytica Posteriora* I.24, in which Aristotle says: "If a demonstration which proceeds from a principle is more clear (ἀκριβεστέρα) than one which does not proceed from a principle, then a demonstration which proceeds more from a principle (ἡ μᾶλλον ἐξ ἀρχῆς . . . ἀπόδειξις) is more clear than one that is less so" (86a16–17, trans. Barnes, adapted; cf. I.27). This passage allows us to connect the two epistemic states mentioned in the passage of the *Metaphysics* we are examining: being "more clear" and "more able to teach of the causes." The science that is more clear is the one whose demonstrations "proceed more from a principle," and this is presumably the reason why those who possess such sciences are more able to teach the causes, because they can discern how these causes are related to a principle.

Memory is produced from perception, and experience from memory, "for many memories of the same thing produce finally the capacity for experience" (980b29–1a1). A person with experience must be "more wise" than a person with mere perception because the person with experience has "more clear" knowledge and is "more able to teach of the causes." The person who has merely perceived some object, but not repeatedly, and not in connection with several memories of it, has less clear knowledge and is less able to teach its causes.

That seems a very reasonable and consistent position, but one then wonders why the person with experience is less wise than the person with technical skill or even knowledge, since the comparison with perception shows that even the person with mere experience must have some kind of clear knowledge and some ability to teach of the causes. Aristotle's position is that the person with technical skill not only has experience, experience that makes the object more clear and enables him to teach more of its causes than mere perception would enable him to do, but this technically skilled person also can offer demonstrations that "proceed more from a principle," whereas the person with mere experience, even if they have some grasp of causes, do not apprehend the principles, cannot offer a demonstration from its first principles, and so their ability to teach of its causes is weak, based only on perceptions and memories of the object. Again, the person with experience grasps the causes more than the person with mere perception but less than the person who has technical knowledge, whose demonstrations "proceed more from a principle." Thus, the person with technical or scientific knowledge grasps the objects more clearly and is more able to teach their causes than the person with mere experience. The same analysis can be used to show why the person who has theoretical science is said to be more wise than one who has practical science: the one who has theoretical knowledge grasps his objects more clearly, offers demonstrations "more based on a principle," and so is more able to teach of the causes than the one who has mere practical knowledge, experience, or perception.

As I have interpreted this passage, then, the third criterion in **<7c>** is really the key to the overall rank-ordering: those kinds of scientific knowledge are better whose objects are grasped more clearly and whose causes are more able to be taught. The more architectonic practical sciences involve both grasping objects more clearly and more in terms of their causes than those that are more subordinate, and certainly more than mere experience or perception affords; but in general, the theoretical sciences involve grasping objects that are both more clear and more teachable in terms of their principles than those of the practical sciences. There is every reason to think that, among the theoretical sciences as well, some must afford grasping their objects more clearly, and more in terms of their principles, than others.

In support of this interpretation, let me turn to the neglected passages of Aristotle's *Protrepticus* identified by Merlan and confirmed by Festugière. In the first, Aristotle discusses two criteria by means of which "we judge one science to be more valuable than another":

> Now then, it is agreed that there are certain sciences that are chosen for themselves and not only for what results from them; and this is possible either only or especially for sciences that are theoretical, because their end is

> nothing other than the theory. But these are the same thing: those things by means of which we put down one science as being more choiceworthy than another, and those things by means of which we choose each thing. And we choose one science over another either because of its clarity or because what it theorizes about is better and more honorable. (**T8:** *Protrepticus*, apud Iamblichus, *DCMS* XXIII, 72.2–10 Festa)

Aristotle here is focused on theoretical sciences, which he defines as those sciences chosen for their own sake, not for something else beyond the knowledge itself. He says that we can use what makes each of them choiceworthy as a basis for determining which among them is more choiceworthy, and this is done on the basis of one of two criteria: either "because of their clarity" (διὰ τὴν αὐτῆς ἀκρίβειαν) or "because they are better and more honorable" (διὰ τὸ βελτιόνων καὶ τιμιωτέρων). Recall that in *Rhetoric* I.7 (**T1**) Aristotle mentioned sciences that are "more beautiful or more respectable" and in *Topics* III.1 (**T4**) sciences that are "absolutely better and more choiceworthy." Although there is a slight variation in terminology, Aristotle appeals to the very same epistemic and objective criteria in this text that I argued are key to his overall rank-ordering scheme in *Metaphysics* I.2 and that underlie those rhetorical and dialectical argumentative techniques recommended in *Rhetoric* I.6 and *Topics* III.1. But in the *Protrepticus*, unlike in any of those texts, Aristotle offered an example to illustrate this: astronomy, which he says is both especially clear and has the most honorable and most divine objects: the stars.

> Of these sciences, everyone would agree with us that clarity belongs especially to the sciences that are mathematical, but there are several who assign the aforementioned place of honor to the principles that are first, but they hypothesize that the nature of the principle is proper to numbers and lines and their qualities, because of the simplicity of its substance. Again, the objects that are observed in the sky have the most honorable and most divine arrangement of the things perceptible to us and are naturally cognized by the science of astronomy, which happens to be one of the mathematical sciences. (**T9:** *Protrepticus*, apud Iamblichus, *DCMS* XXIII, 72.11–20 Festa)

By the use of these superlatives "the most honorable and most divine," Aristotle argues, protreptically, that astronomy is the best kind of science, since it has both the greatest clarity and the best objects. The argument is concluded by bringing the method and criteria of comparison we have been discussing to bear on the relative value of the practical and theoretical sciences.

> But it would seem to be absurd and entirely unacceptable for those who assert that the philosopher is related to the truth to think it necessary for him to seek some fruit other than from these kinds of observable objects, which share in the highest truth, and for those who are in love with spectacular sights to think it right to acquire such sciences as these for something else, sciences which are about the most common things in nature as well as about the most divine of the things perceptible to us, sciences which, being full of the most numerous as well as the most amazing observations (θαυμασιωτάτων θεαμάτων), have a clarity not molded out of empty arguments, but one of their own, one that is solid and secure from their underlying nature. (**T10:** *Protrepticus*, apud Iamblichus, *DCMS* XXIII, 72.20–73.3 Festa)

This passage from the *Protrepticus* helps illuminate the cryptic methodological remarks of *Rhetoric* I.7 **(T1)**, *Topics* III.1 **(T4)**, and *Metaphysics* I.2 **(T8)**, by providing something utterly absent from all those texts: a concrete example. It is convenient for our present purposes that the example happens to be astronomy.

But not all philosophers in the classical period were convinced that students should be encouraged to seriously pursue the science of astronomy. Aristotle portrays this brilliantly by devising an "Isocratean" apotreptic to the mathematical and theoretical occupations of Academic philosophy. Isocrates had argued in the *Antidosis*, and then was portrayed by Aristotle as arguing in the dialogue of the *Protrepticus*, that only those sciences that are useful are valuable and should be chosen.[17]

> So it should not be overlooked by someone who is going to scrutinize these subjects that everything that is good and beneficial for the life of humans consists in being used and put into action, and not in the mere knowledge. For we are not healthy by being acquainted with what produces health, but rather by applying it to our bodies, nor are we wealthy by knowing about wealth, but by possessing a very substantial amount nor, most important of all, do we live well by knowing certain sorts of beings, but by acting well, for this is truly what it is to be successful. Hence it is appropriate for philosophy as well, if indeed it is beneficial, to be either a practice of good things or else useful for those sorts of practices. (**T11:** *Protrepticus*, apud Iamblichus, *De communi mathematica scientia* XXVI, 79.15–80.1 Festa)

In the *Protreptricus*, Aristotle portrayed "Isocrates" arguing this point with reference to three pairs of examples, crucially including astronomy.

[17] Isocrates, *Antidosis* 261–269; Aristotle, *Protrepticus*, apud *DCMS* XXVI (on the attribution and interpretation, see Hutchinson and Johnson 2018: 144–154).

> We have the greatest example of this **<a>** in the sciences that are similar to it and the opinions that underlie them, for we see the geometers being able to do none of those things that they observe by demonstration; and yet to divide an estate, and all the other properties of quantities as well as locations, is something that the land-surveyors can do on the basis of experience, whereas those who know about the mathematical subjects and the arguments about them know how they should act, but are not able to act.
>
> **<b>** The case is similar with music and the other sciences in which the cognitive aspect is divided off from the empirical. For those who determine the proofs and the arguments about harmony and other things like that are accustomed to enquiring, but take part in none of their practical functions, just like those who do philosophy. In fact, even if they happen to be capable of handling something in them, when they learn the proofs, they automatically do it worse, as if on purpose, whereas those who have no knowledge of the arguments, if they are trained and have correct opinions, are altogether superior for all practical purposes.
>
> **<c>** So too with the subject matter of astronomy such as sun and moon and the other stars; those whose training has been in the causes and the arguments have no knowledge of what is useful for humans, whereas those who have what are called navigational sciences about them are capable of predicting for us storms and winds and many of these events. Hence, for practical activities, sciences like this will be entirely useless; and if of the activities they miss out on the correct ones, the love of learning misses out on the greatest goods. (**T12:** *Protrepticus*, apud Iamblichus, *De communi mathematica scientia* XXVI, 80.5–81.4 Festa)

The Isocratean position involves the following evaluations: musical performance is superior to harmonics, and navigation to astronomy, and land-surveying to geometry. In each case, the practical science (or technical skill) is preferred because of its usefulness and productivity of goods, compared to the uselessness of their theoretical counterparts.

Aristotle responds to this argument in two paradigmatic ways. First, he argues that the theoretical sciences do in fact turn out to be useful for the sake of the practical sciences and their ends. The general point had been made earlier in the dialogue,[18] but later Aristotle applies the argument directly to the field in which Isocrates considered himself the unrivaled master: political science.[19] The second response to the Isocratean position, and the more important one for present purposes, is the argument that the

[18] The usefulness of theoretical sciences is discussed in the *Protrepticus*, apud Iamblichus, *DCMS* XXIII, 71.16–24 and 71.26–72.2 Festa.

[19] Aristotle, *Protrepticus*, apud Iamblichus, *Protr.* X, 55.7–56.2 Pistelli. See also Johnson 2015: 179–183.

theoretical sciences are intrinsically valuable and as such are better than the corresponding practical sciences, which are only instrumentally valuable. This argument had already been discussed earlier in the dialogue (in texts **T8** and **T10**), but Aristotle later amplifies his praise of theoretical sciences to a rhetorical crescendo, which outdoes Isocrates at his own game.[20] Part of the argument is rhetorically supported by appealing to the wisdom of Anaxagoras, who is described in the *Protrepticus* as holding that astronomy is in fact *the* greatest science and that nothing else is even worthwhile.

> And they claim that when somebody asked Anaxagoras for what reason anyone might choose to be born and to live, he replied to the question by saying that it was to be a spectator of the sky and the stars around it, as well as moon and sun, since everything else at any rate is worth nothing. (**T13:** *Protrepticus*, apud Iamblichus, *Protr.* IX, 51,11–15 Pistelli)

Such an extreme position makes a reappearance in Aristotle's protreptic to biology. We shall return to it in due course, although it should be noted immediately that this is probably less likely to be a view attributable to the historical Anaxagoras than a rhetorical provocation on the part of one of the characters in Aristotle's *Protrepticus.*[21] But before turning to *De Partibus Animalium* I.5, let us consider one last application of the method of comparing sciences, one that does not yield the same "Anaxagorean" emphasis on the absolute supremacy of astronomy.

I have in mind the opening of the *De Anima*, for here Aristotle argues for the superiority of the inquiry concerning the soul, but seems to apply the very same two criteria that in the *Protrepticus* were used to argue for the superiority of astronomy.

> It is assumed that cognition is among the beautiful and honorable things, but since one cognition may be greater than another in accordance with either **<a>** its clarity, or **<b>** its being about better and more amazing things, on account of both of these we would be reasonable in placing the inquiry concerning the soul in the first rank. (**T14:** *De Anima* I.1, 402a1–4).

According to this protreptic argument, the study of the soul (hereafter: psychology) is ranked first among theoretical sciences for two reasons: first, in accordance with its clarity (κατ' ἀκρίβειαν); and second, because its objects are better and more amazing (τῷ βελτιόνων τε καὶ θαυμασιωτέρων

[20] Aristotle, *Protreptcius*, apud Iamblichus, *Protr.* IX, especially 52.16-54.5 Pistelli. See also Hutchinson and Johnson 2018: 148–151.

[21] Perhaps this is the reason why this piece of evidence (59A30 Diels) is absent from recent editions of Anaxagoras, including Curd 2007 (cf. 89–90, 102).

εἶναι). Notice that this comparative term "more amazing" (θαυμασιωτέρων) is directly parallel to the superlative term "most amazing" (θαυμασιωτάτων) used by Aristotle with reference to astronomy in the *Protrepticus* (**T10**).

This extremely tight parallel raises the following issue: if the study of the soul is of the first rank, how can it be that astronomy has the "most amazing," "most divine," and, in a word, the best objects, as Aristotle so vigorously argued in the *Protrepticus*? The answer requires us to take seriously Aristotle's position that the objects of astronomy, the stars, are alive and have souls, an issue about which Aristotle says in *De Caelo*:

> We think of the stars as mere bodies, and as units with a serial order indeed but entirely inanimate; but we should rather conceive them as enjoying life and action. On this view the facts cease to appear surprising. (**T15:** *De Caelo* II.12, 292a18–22, trans. Stocks)

There would indeed be a conflict between the supremacy of astronomy and psychology if we were to assume that the study of the soul is restricted to the plants and animals down here, and the stars up there were lifeless objects. In that case, either astronomy or psychology would have to be ranked one above the other, and Aristotle's views would have to be interpreted as either contradictory or as having developed between the writing on the *Protrepticus* and the *De Anima*. But neither interpretation is necessary, because in Aristotle's view the stars have souls and are alive (they are, in fact, extraterrestrial life and intelligence), and the study of the stars depends, at least partly, on psychology. So, in a sense, Aristotle does, no doubt, place psychology, broadly conceived, as in first rank among the theoretical sciences. This remains the case even if we take into consideration the theology of the *Metaphysics*. The object of theology, like biology, is ultimately soul and living things: the unmoved mover or god is a soul and a living thing. In fact, theology, astronomy, psychology, and biology overlap in ways that cannot be pulled apart, and psychology is the most general of all these sciences.[22]

But in the *Metaphysics*, theology is explicitly called "first philosophy," and this clearly implies an overall rank-ordering of all theoretical sciences if anything does. Indeed, Aristotle says that natural philosophy is in second place, although it would be in first place if there turned out to be no immovable and invisible objects.[23] This indicates that Aristotle's criterion

[22] On the implication of this biological doctrine for Aristotle's astronomy and theology, see further Johnson 2021.

[23] *Metaph.* 6.1, 1026a6–33; cf. *Metaph.* 11.7, 1064b1–6 (see further Johnson 2015: 173–174).

for rank-ordering in the *Metaphysics* is the objective one: the objects of theology are better than the objects of natural science. But, natural science, insofar as it is a theoretical science, is still superior to the productive and practical sciences, by virtue not only of its better objects but also its greater clarity.

An open question, however, is to what extent it is possible to consistently rank-order the natural sciences themselves. The passages from the *Protrepticus* about astronomy **(T8–10, T13)** and the passage from *De Anima* about psychology **(T14)** suggest that Aristotle believes that it is possible, but it would be wrong to suggest that Aristotle is perfectly clear from all these texts what exactly the ultimate rank-order would be. That brings us, then, to the discussion of rank-ordering of sciences in *De Partibus Animalium* I.5.

2.3 Application in *De partibus animalium* I.5

The first protreptic argument of *De Partibus Animalium* I.5 involves the comparison of two different sciences of living things, astronomy and biology.

> **<a>** Among the substantial beings constituted by nature, some are ungenerated and imperishable throughout all eternity, while others partake of generation and perishing.
>
> **<b>** Yet it has turned out that our studies of the former, though they are honorable and divine, are fewer; for, as regards both those things on the basis of which one would examine them and those things about them which we long to know, the perceptual phenomena are altogether few.
>
> **<c>** We are, however, much better provided in relation to understanding about the perishable plants and animals, because we live among them. For anyone willing to labor sufficiently can grasp many things about each kind.
>
> **<d>** Each of these is delightful. Although our contact with eternal beings is slight, nonetheless because of its surpassing value this knowledge is more pleasurable than all other things around us, just as to briefly catch a glimpse of those we love is more pleasurable than seeing other even great things with clarity.
>
> **<e>** Perishable things, however, excel in respect of science because we have knowledge of more of them, and we know them more fully. Further, because they are nearer to us and more of our own nature, they afford a certain compensation compared with the philosophy concerned with the divine things.
>
> **<f>** Since we have explained how things appear to us about divine things, it remains to discuss the animal nature, omitting nothing in our power, whether of lesser or greater value. For even in the study of animals

> disagreeable to perception, the nature that crafted them likewise provides extraordinary pleasures to those who are able to know their causes and are by nature philosophers. (**T16:** *PA* I.5, 644b23–645a10)

In **T16a**, Aristotle introduces the two things that are to be compared: astronomy and biology. Aristotle describes them in such a way that their essential similarity is made clear: both sciences have as objects substantial (living) beings constituted by nature. Notice that this close comparison between two objects follows the dialectical procedure stipulated in *Topics* III.1 (**T2**), according to which the things being compared should be brought as close together as possible so that any comparative advantage whatsoever may be identified. The key difference between their objects is that those of astronomy are indestructible and immortal, while those of biology partake of birth and death.

In part of **T16b** and **d**, Aristotle acknowledges that the imperishable things (the heavenly bodies – stars, sun, and moon) are "honorable and divine" and have "surpassing value" (τιμιότητα). His neat comparison of doing astronomy with catching "a glimpse of those we love" suggests that the "surpassing value" of the stars is due to their beauty, and so the value of science relates to its objects being "more beautiful," just as suggested in the *Rhetoric* (**T1**). And it is very interesting that this kind of knowledge is said not only to be "delightful" but in fact "more pleasurable than all the things around us." For the objective and epistemic criteria that we have been discussing are also employed by Aristotle in his theory of pleasure, as when he describes the best pleasure as resulting from the cognitive organ being in the best condition and in relation to the best kind of object.

> In the case of each sense the best activity is that of the best-conditioned organ in relation to the finest of its objects. And this activity will be most complete and pleasant. For, while there is pleasure in respect of any sense, and in respect of thought and contemplation no less, the most complete is pleasantest, and that of a well-conditioned organ in relation to the most respectable of its objects (πρὸς τὸ σπουδαιότατον τῶν ὑπ᾽ αὐτήν) is the most complete; and the pleasure completes the activity. (**T17:** *Nicomachean Ethics* X.4, 1174b18–23, trans. Ross, adapted)

According to this analysis, astronomy could be considered the most pleasurable of all activities, as it involves the best thought organ in relation to the most beautiful and respectable of all objects: the stars. Such a superlative conception of astronomy would be strongly resonant with the position attributed to Anaxagoras in the *Protrepticus* (**T13**), according to which the objects of astronomy are of such a surpassing value that nothing else is even worth considering or makes living worthwhile. And it is just this position that Aristotle seems to refute when he goes on in **T16b** and **c** to explain a key comparative advantage of the study of mundane living things: the study of the heavenly bodies has fewer and more remote objects. Thus, there can be less perception, memory, and experience of them. The basis for astronomical sciences is thus relatively impoverished, and necessarily and inalterably so, because "our contact with eternal beings is slight"; by contrast, "we are much better provided in relation to knowledge about the perishable plants and animals."[24] Aristotle goes so far as to say that *anyone* willing to labor can grasp many things about each kind of animal and plant,[25] no one could ever say the same about the stars. We not only know more of the animals and plants, but we know them much more fully. Despite the great "clarity" acknowledged in mathematical sciences like astronomy,[26] here Aristotle says that the objects of astronomy are "more pleasurable than seeing other even great things with clarity," which suggests that the objects of biology (the "other, even great things") can actually be seen with greater clarity than those of astronomy. This may be difficult to accept if we are thinking of the clarity (or accuracy, precision, or exactness) of modern mathematical astronomy, but it is easy to see if we instead think of the relative inexactness of Aristotle's own concept of astronomy. More importantly, in accordance with what has been said in the *Metaphysics*, more clear knowledge belongs to the one who is more able to teach the causes, and it turns out to be more possible to teach the causes of animals and plants than it is of the stars. In fact, in *De Caelo* II.12, Aristotle attempts to explain the cause of the variable motions of the earth, moon, sun, planets, and stars by relating all of them to

[24] Aristotle assumes that detailed investigation of the heavenly bodies, including even the moon, is impossible (*MA* 4, 699b17–21); he fails to predict telescopic vision. In *GA* III.11, Aristotle states that since there are living things in earth, water, and air, there ought to be some in fire as well, most likely in the region of the moon; this speculation is apparently unconfirmable because of the impossibility of examining the region of the moon adequately (761b12–22).

[25] Aristotle's reference to willingness to labor (διαπονεῖν) resonates with the argument in the *Protrepticus* according to which the fact that philosophers spend their leisure doing philosophy is a sure sign that it comes with pleasure, since without such compensation "no one is willing to labor for a long time (πονεῖν γὰρ οὐδεὶς ἐθέλει πολὺν χρόνον)" (*apud* Iamblichus *Protr.* VI, 40.23). Feasibility is defined in the *Rhetoric* in terms of what is possible, easy, and beneficial, and the author of *PA* I.5 is arguing that the study is both possible and easy, in the sense of something voluntarily done that comes with pleasure. The same author proceeds to show that knowledge of living things is beneficial because of the insight it produces and because it is accompanied by great intellectual pleasures.

[26] In *Metaphysics* II.3, 995a15–20, Aristotle associates clarity with mathematical sciences as opposed to natural sciences. But **T14** and **T16d** both show that Aristotle is prepared to describe natural sciences as enjoying great and even superlative clarity. The position of *Metaphysics* II.3 is very close to that of the *Protrepticus* (in **T8** and **T9**).

different kinds of animals and plants: the explanatory priority employed there is perfectly consistent with Aristotle's rank-ordering in *De Partibus Animalium*. Another epistemic advantage of biology is that, because the animals and plants are closer to us in kind, this science provides us with a valuable kind of self-knowledge, the kind of knowledge mentioned in the protreptic opening of the *De Anima* (**T14**), knowledge of the kind of things we ourselves are (terrestrial things, animals, things that grow, use nutrition, reproduce, have sensation, move around by means of organs, mature, decay, die, etc.).

What we have, then, is a conflict between two different criteria, the objective and epistemic ones, with the following outcome.

Criteria for rank-ordering sciences	Astronomy	Biology
Objective criterion (objects more beautiful)	better	worse
Epistemic criterion (knowledge more clear)	worse	better

Now even if Aristotle had to acknowledge that stars are better (more beautiful, etc.) than animals, he could still maintain that animals are better than stars *as objects of human scientific knowledge*: this follows from their relative abundance, proximity, and similarity to ourselves. Within the horizon of human knowledge, animals and plants are arguably better as objects of knowledge than the stars, and in this sense, biology is arguably a more choiceworthy science than astronomy.

According to the objective criterion, then, astronomy is better than biology, but according to the epistemic criterion, biology is better. In the network of evidence that we have examined, Aristotle has nowhere given us a means of prioritizing either the objective or the epistemic criteria: these have always been presented together, as if in every case the more clear science always has better objects than the less clear one. But here those two criteria are taken apart, and applying them to various theoretical sciences produces different rank-orderings.

Now Aristotle could have left it at that, but I think that what he does instead is argue that, because each science is better than the other with respect to one of the criteria for rank-ordering, we should emphasize not the difference between them but instead their similarity: in terms of their overall value, both sciences along with their objects are essentially comparable. This is why Aristotle goes so far as to invoke the obscure remark that "there are gods here too" in a passage that we will examine momentarily (**T17**): in a sense the objects of biology are, like the stars, living and divine, and they are also beautiful, and honorable, and even "amazing," as Aristotle says, again using the word *thaumaston*, a term that we saw he used with reference to astronomy in the *Protrepticus* (**T10**) and in the protreptic to psychology in *De Anima* (**T14**). Thus, on the basis of the superiority of its knowledge, and the nearly equal excellence of its objects, biology is comparable in value to astronomy.

Before the protreptic conclusion can be reached, however, another premise is needed. This argument can only succeed if an interlocutor (or reader) accepts that astronomy is in fact valuable. We have seen that Anaxagoras is someone who would readily agree to that, since he says that everything else (apparently including all animals and plants, even human beings) is worthless. The "Anaxagoras" of the *Protrepticus* thus stands refuted by Aristotle in *De Partibus Animalium* I.5: if astronomy is worth doing, as Anaxagoras thinks, then so too is biology. But recall that the "Isocrates" of the *Protrepticus* does not accept that the theoretical and mathematical science of astronomy is worth choosing, because he holds that only the practical sciences or technical skills with their beautiful or useful effects, like navigational astronomy, have value (**T12c**). It is against this kind of view that the second protreptic argument of *De Partibus Animalium* I.5 is directed.

> Surely it would be unreasonable, even absurd, for us to enjoy studying likenesses of animals – on the ground that we are at the same time studying the technical skill, such as painting or sculpture, that made them – while not prizing even more the study of things constituted by nature, at least when we can behold their causes.
>
> For this reason, we must not be childishly disgusted at the examination of the less honorable animals. For in all natural things there is something amazing (θαυμαστόν). Even as Heraclitus is said to have spoken to those strangers who wished to meet him but stopped as they saw him warming himself by the oven – he bade them to enter without fear, "for there are gods here too" – so too one should approach research about each of the animals without disgust, since in every single one there is something natural and beautiful. For what is not haphazard but rather for the sake of something is in fact present most of all in the works of nature; the end for the sake of which each animal has been constituted or born assumes the place of the beautiful. (**T17:** *PA* I.5, 645a10–26)

The key to the assumed superiority of the objects of astronomy is their obvious and undeniable beauty, and so Aristotle offers an additional argument to show that the objects of biology are also beautiful, even if less obviously so than the stars. And even if not beautiful, an appreciation of their teleological organization "assumes the place of the beautiful." But the argument also works against an interlocutor who would admit that the objects of fine art or technical skill (such as painting and sculpture) are beautiful and therefore good, but would not admit that the theoretical sciences are. This, as we saw, is the position of Isocrates, who accepts the value of a practical technical skill like musical performance (**T12b**) but rejects theoretical sciences like harmonics, astronomy, and geometry.

Aristotle now argues that such a person must concede that the objects of biology are superior to those of painting and sculpture. This is because the objects of painting and sculpture are imitations of living things, and it is presupposed that imitations are inferior to the natural objects of which they are imitations. This follows from the principle, applied extensively in the *Protrepticus* and key texts in the Aristotle Corpus, according to which every technical skill exists to imitate nature or fill in where nature falls short.[27] Further, imitations of what is not beautiful cannot be beautiful,[28] and so if the objects of such imitative art or technical skill – pictures of plants and animals – are beautiful, then a fortiori the objects of biology must be beautiful. Furthermore, the works of imitative arts or technical skills are enjoyed, according to Aristotle, just because they enable us to learn more about the actual objects of which they are the imitations.[29] Assuming that the interlocutor concedes that the fine arts or technical skills are good, then, the protreptic conclusion follows: if the objects of painting and sculpture are enjoyable, then a fortiori the objects of biology are.

The protreptic to biology of *De Partibus Animalium* I.5 has a strongly dialogical feel, which seems to be why commentators have frequently pointed out its stylistic differences from the rest of the biological works.[30] Aristotle seems to refute several anonymous objections to biology but by appealing to different criteria for rank-ordering the choiceworthiness of sciences. The first refutation is of someone who, like Anaxagoras, would accept the value of astronomy but reject the value of biology. Aristotle invokes the epistemic criterion to show a way in which biology is superior to astronomy, so that anyone who recognizes the value of astronomy should after all engage in biological research.

[27] See *Protrepticus*, apud Iamblichus, *Protr.* IX 49.28–50.1, 50.12; X 54.22–23; and in the Corpus: *Phys.* II 2, 194a21–22; II.8,199a15–17; *Meteor.* IV 3, 381b6; *PA* I.5, 645a10–15; *Metaph.* VII 9, 1034a33–34; *Pol.* VII 17, 1337a1; cf. [Ar.] *Mund.* 396b1–12.

[28] This questionable principle is stated in the *Protrepticus*: "An imitation of what is not beautiful cannot be beautiful, nor can an imitation of what is not divine and stable in nature be immortal and stable" (apud Iamblichus, *Protr.* X, 55.22–23 Pistelli).

[29] This questionable principle is stated in the *Poetics*: "It is natural for all to delight in works of imitation .... The explanation is to be found in a further fact: to be learning something is the greatest is the greatest of pleasures not only to the philosopher but also to everyone else, however small their capacity for it; the reason of delight in seeing the picture is that one is at the same time learning – gathering the meaning of things" (*Poetics* 4, 1448b8–19, trans. Bywater, adapted).

[30] Balme 1972: 122–123; Lennox 2001a: 172; see also Lennox 2010a: 60–62. Their sentiments are endorsed in the recent study of Thein 2014: 209.

But since that dialogical context has been removed, we can interpret this argument as a declarative one, using the formalization introduced at the beginning of this chapter.

1. $S_1 : O_1 :: S_2 : O_2$ (analogy)

Astronomy is to stars as biology is to animals and plants.

2. $(S_1 > S_2) \longrightarrow (O_1 > O_2)$ (**T1a**; *alternando*). Epistemic rank-ordering

If astronomy is better than biology, then stars are better than animals and plants.

3. $(S_1 = S_2) \longleftrightarrow (O_1 = O_2)$ (Corollary of **T1**). Epistemic and Objective rank-leveling

If astronomy and biology are equally valuable, then stars and animals and plants are equally valuable. And if stars and animals and plants are equally valuable, the astronomy and biology are equally valuable.

4. $\sim (S_1 > S_2)$ (**T16b, d**)

It is not the case that astronomy is better than biology.
Warrant: the epistemic criterion; greater clarity is possible in biology because animals and plants are closer to us and more abundant than the stars, so we can have more perception, memory, and experience of them, and are able to teach their causes more.

5. $\sim (S_1 > S_2) \longrightarrow (S_2 > S_1)$ or $(S_2 = S_1)$ (assumption)

If astronomy is not better than biology, then either biology must be better than astronomy or biology and astronomy must be equally valuable.

6. $\sim (S_2 > S_1)$ (assumption; cf. **T16d**)

It is not the case that biology is better than astronomy.

7. $S_2 = S_1$ (by 4, 5, 6; cf. **T14e**)

Therefore, biology and astronomy are equally valuable.

8. $O_1 = O_2$ (by 3, 7; cf. **T14e**)

Therefore, animals and plants and stars are equally valuable.

9. Astronomy and the stars are valuable.

This is a shared assumption of an unnamed interlocutor, like "Anaxagoras" in **T13**, but not Isocrates in **T12c**.

10. Therefore, biology and animals and plants are valuable (by 7, 8 & 9).

This argument gives us the first protreptic conclusions of *De Partibus Animalium* I.5: that the animals themselves and the theoretical science of them are valuable and worthy of attention, contrary to the view of philosophers like Anaxagoras, and perhaps other mathematically oriented philosophers, including certain Academics, who might be described as holding the position that only "mathematical" sciences like astronomy are worth pursuing.

In the second protreptic argument, Aristotle also argues against someone who, like "Isocrates," accepts the value of the fine arts or technical skills like painting and sculpture (and their imitations of animals) but would deny the value of the activity of studying actual living things in detail. We can represent that as a declarative argument as follows:

1. $O_1 : S_1 :: O_2 : S_2$ (analogy)

Animals and plants are to biology as imitations of animals and plants are to painting and sculpture.

2. $(O_1 > O_2) \longrightarrow (S_1 > S_2)$ (**T1b**; *alternando*). Objective rank-ordering

If animals and plants are better than imitations of animals and plants, then biology is better than painting and sculpture.

3. $O_1 > O_2$ (**T16**)

Animals and plants are better than imitations of animals and plants.
Warrant: technical art imitates nature, not vice versa; an imitation of that which is not beautiful cannot be beautiful; and the value in observing imitations of animals is in learning about actual, living animals.

4. $S_1 > S_2$ (by 2, 3).

Therefore, biology is more valuable than painting and sculpture.

5. Painting and sculpture is valuable.

(Shared assumption of unnamed interlocutor, like "Isocrates" in **T12b**).

6. <u>Therefore, biology is valuable</u> (by 4, 5).

And here we reach the second protreptic conclusion of *De Partibus Animalium* I.5. It is important to notice that the validity of both arguments depends on the premise that either astronomy or painting are worthwhile activities. Although these seem like easy points to agree to, the arguments themselves do not work without those assumptions and cannot stand on their own to convince someone unwilling to admit either one of those claims that they should do biology and value animal and plant life for itself. It seems that the third protreptic argument, the one involving a self-refutation argument, might be aimed at such an intransigent interlocutor. But that argument will have to be examined elsewhere.

2.4 Conclusion

I think that the two protreptic conclusions from *De Partibus Animalium* I.5 I have discussed here are sound on the basis of Aristotle's views about the technical skills and fine arts, theology, astronomy, psychology, and biology. But I think they are interesting even in relation to our own, very different modern views. Suppose that (contrary to Aristotle's views, but true as far as we know), all stars and planets beyond earth are lifeless, and either only on earth does anything live, or we will never discover anything living beyond earth. In that case, it would be reasonable to hold that the study of animals and plants on Earth is just as, or even more, valuable than astronomy, and on both objective and epistemic grounds. And even if we did, like Aristotle, believe in the existence of extraterrestrial life or intelligence, we might still, like Aristotle, consider terrestrial biology and psychology to be more valuable, and its objects more beautiful, than astrobiology, exobiology, and their objects. For one thing, psychology and biology provide us with a kind of self-knowledge that may be more valuable than any knowledge we might gain of alien lifeforms. So, there is something right and profound about Aristotle's rank-ordering the study of the soul and biology as in first place, a ranking stressed by Aristotle in the opening of *De Anima* I.1 and expanded upon in *De Partibus Animalium* I.5.

Fortunately, it is no longer necessary to defend Aristotle's protreptic to biology – biology and psychology are very much being pursued, and it is not these fields against which the modern apotreptics against philosophy are directed. Probably biology and psychology no longer need any protreptic coming from philosophy. But this was not always the case, and the esteemed status that these sciences now enjoy is due in no small part to the very protreptic arguments of Aristotle that we have been examining.[31]

[31] I would like to thank the following people for written and oral feedback at various stages in the composition of this chapter: D. S. Hutchinson (my collaborator on all *Protrepticus* research), the audience at the conference in Cambridge in August 2019, Klaus Corcilius, David Sedley, James Warren, Pierre Destrée, Katerina Ierodiakonou, Robert Bolton, Philip Mitsis, Ronja Hildebrandt, Pierre-Marie Morel, James G. Lennox, and most especially Sophia Connell (also as organizer of the conference and editor of this volume).

CHAPTER 3

On the Parts of Animals: *From Aristotle to Michael of Ephesus*

*Giouli Korobili**

Introduction

The *Parts of Animals* (*PA*) deals mainly with the anatomy and physiology of "the living animal in its actualized, functional form."[1] *PA* I is a general introduction to the study of zoology. *PA* II opens with the three modes of composition of living things, that is, out of the elements, the uniform parts and the nonuniform parts, focusing subsequently on the uniform parts. From *PA* II.10 until the end of book IV, Aristotle examines the nonuniform parts of blooded and bloodless animals. His main concern in this work is to identify the causes responsible for the composition of animal parts as things that already *are*. In this respect, he seeks to substantially revise his predecessors' methodological approaches to zoological matters and their respective accounts, on the grounds that they have been mainly concerned with the material out of which a thing comes-to-be (material cause). The *PA* is just one part – though admittedly a brilliant one – of his zoological research program and at the same time an integral part of his inquiry into nature. As such, one would expect its content and methodology to have been widely studied and discussed by the following generations. This is not the case, however.

Aristotle's successors in the Peripatos are, broadly speaking, interested in raising questions regarding various aspects of animal activity, yet they do not undertake to write treatises devoted to the explanation of animal physiology and anatomy. What is more, they do not seem to have understood the overarching goal of Aristotle's zoological program and

* This chapter has been written thanks to the generous support of Berlin-Brandenburgische Akademie der Wissenschaften during the CAGB project "Theodorus Metochites's Commentary on Aristotle's De Partibus Animalium I". I would like to sincerely thank Sophia Connell, Christian Brockmann, Jim Lennox, and Petros Bouras-Vallianatos for their constructive feedback and suggestions on an earlier version of this chapter, as well as Myrto Hatzimichali, Andrea Falcon, and Teun Tieleman for sharing with me their unpublished works.

[1] Lennox 2001a: xi.

the central axes of his research project.[2] In Hellenistic times and Late Antiquity, at least as far as the evidence that has reached us is concerned, the zoological research program introduced by Aristotle and structured by the *Post. An.* framework disappears. This view has long been established by James Lennox, who describes the situation as a "mystery."[3] After Lennox's seminal paper, many scholars verify in one way or another his views. Yet what becomes clear when one assesses the nature of the data provided by these contributions is that

1. Aristotle's *History of Animals* (*HA*) is cited, referred to by name,[4] and discussed by later thinkers, who nevertheless use it for their own purposes (discussed below). The same holds for his *Generation of Animals* (*GA*) to some degree.[5]
2. Aristotle's *PA* is absent from any zoological and medical discussion, at least until the time of Galen.[6]

The first commentary *stricto sensu* on the *PA* was written by Michael of Ephesus, a Byzantine intellectual of the twelfth century. The *PA* was translated into Arabic in the ninth century[7] and from Arabic into Latin in the eleventh century by Michael Scot (*Liber de animalibus*).[8]

In what follows, I will discuss the lack of sufficient evidence by which to trace the intellectual transmission of the *PA* in the period up to Galen and try to identify the reasons for its marginalization compared to the other major zoological treatises of Aristotle. In Section 3.2, I will consider passages from Galen's *On the Usefulness of the Parts* (*UP*), which provide evidence about the status and readership of the *PA* until his time. In Section 3.3, I will focus on Michael's commentary on the *PA*. I shall first present the main features of this commentary and then look into the

[2] For Peripatetic zoology, see Hellmann 2006: 329–332. See also references in n. 5 below.

[3] Lennox 1994.

[4] Although not exactly the same name as that we currently use – Aristotle too is not consistent in using a single title when referring to his *HA* in other works; see Louis 1964: xviii–xix; Lennox 1994: 15. To be sure, some of Aristotle's treatises were acquisitions of the libraries of Alexandria and Pergamon and also of personal libraries (Tutrone 2013; Dorandi 2017: 80).

[5] See now Falcon 2021; Hatzimichali 2021.

[6] Galen quotes *PA* I and II directly although he, like his predecessors, seems not to have understood Aristotle's zoological project; see Lennox 1994: 18–22.

[7] The *PA* was translated along with the *HA* and the *GA* as a single work entitled *On Animals* (*Kitāb al-Ḥayawān*). Books XI–XIV of the *Kitāb* contained the *PA*, for which a critical edition was published in 1979 by R. Kruk. The identification of the Arabic translator of the *PA* remains a matter of controversy. Some scholars have proposed Ibn al-Biṭrīq, but this suggestion is not without problems; see Kruk 1979: 18ff.

[8] See van Oppenraaij 1998.

case of marrow and the brain, which is particularly illuminating for two reasons: first it helps, in the absence of any relevant treatment in Aristotle's account, to contextualize these terms, especially with reference to their conceptual history; and second, it testifies to the way in which Michael's contemporaries understood and used them.

3.1 From Aristotle to Galen

Observation (2) above seems to prevent any further discussion of the reception of the *PA* until Galen and his *UP*. At the same time, observation (1) suggests not only that the interest in zoological inquiry was still alive during this period but also that Aristotle's *HA* and *GA* were standard reference points for those embarking upon such inquiry. Already in the third century BCE, and while the imprint of Theophrastus' research was still fresh,[9] Aristotle's zoological doctrines were circulated in the form of compendia, which were very popular, at least as compared with his original treatises.[10] Only one of them has survived, the *Epitomē* (*Epitomē peri zōōn*) of Aristophanes of Byzantium, whose references to Aristotelian zoological material might well have been a source for the later treatments of Apollonius (*Amazing Stories*),[11] Aelian (*On the Characteristics of Animals*),[12] and Artemidorus (first century CE; *Geography*).[13] Books I and II of the *Epitomē* are preserved as parts of a collection (*syllogē*) of zoological writings, gathered together at the command of Constantinos Porphyrogennetos, emperor of Byzantium during the first half of the tenth century. This *syllogē* incorporates zoological information drawn from various sources, including Aristotle's *PA*. According to the opening of book II of the *syllogē*, Aristophanes summarized such topics as those found in the *PA* in his *Epitomē* II–IV. Regrettably, books III and IV are now lost. In the

9 Theophrastus actually worked together with Aristotle, so I have decided to leave him out of the current discussion. Diogenes Laertius attributes to him several titles of animal books: see Hatzimichali 2021: 230.

10 Balme 1991: 4–5; Hellmann 2006: 345 and 353–354.

11 Scholars are still puzzled about the identity and chronology of the compiler of the *Ἱστορίαι θαυμασίαι*. See in detail Spittler 2016.

12 Ca. 170–222/230 CE. His 17-book compilation Περὶ ζῴων ἰδιότητος does not aim at a scientific zoological account. In combining amusement with moral instructions, Aelian attempts to teach thoughtless and selfish humankind a lesson driven from animal behavior. See further Fögen 2007b: 57–61.

13 Lennox 1994: 14; Hellmann 2015: 1251. On Aristophanes (fl. 255–180 BCE), see Hellmann 2006.

surviving books, I–II, one can identify material drawn not only from the *PA* but also from the *HA* and the *GA*.[14]

However, the very nature of this work (being a synopsis of various zoological ideas), as well as its structure (organized by treating single animal species)[15] and aim[16] had in all probability disorientated ancient readers from the true content and purpose of Aristotle's *PA*.[17] Consider the beginning of *Epitomē* II where Aristophanes justifies the reason why he wrote a compendium:

> I attempted this task so that you would not have to browse through Aristotle's discussion of animals, which is spread into many parts, but you would have the information (*historia*) on each individual animal gathered in one place.[18]

Aristophanes presents his *Epitomē* as a *historia*, the accumulation of a vast amount of knowledge yielding all sort of information about animal life, various characteristics, habits, and activities of animals into a single source data that is easily accessed and handled.[19] Yet the undertaking of authoring a *historia* of animals as virtually a research project not only blurs the aims of Aristotle's own project but could also easily lead to the erroneous assumption that the book arising from such a project would cover a broader research domain in comparison to a narrower investigation into the parts of animals. In other words, for Aristophanes, any project oriented toward a *historia* of animals seems to incorporate all kinds of information about animals, including detailed reports about their parts, which is why it might be thought of as an all-inclusive book. Thus, consulting Aristotle's

[14] See Berger 2012. According to Cerami and Falcon 2014: 40, it is highly probable that Aristophanes "engaged directly with Aristotle's text," while at the same time incorporating material found in intermediate sources.

[15] This is a practice to which Aristotle clearly states his opposition on the grounds that it leads unavoidably to repetition (*PA* I.1, 639a15–b6). In Hellmann's words, this "rearrangement" of the material "from an order by 'parts', as in Aristotle, to an order by species, as Aristophanes prefers" is an "important innovation," which, however, had "major consequences for the field of biology" (2006: 335).

[16] Rather than being concerned with animal identification after observation, this work is intended, according to Hellmann, to be used by a learned circle "as a reference book or lexicon" and, hence, to be read only partially from those who are looking for information "quickly" (2006: 354).

[17] Lennox 1994: 14–15: Kullmann 1998: 129; Berger 2012: 3 shows that the structure of the second book of the *Epitomē* is roughly based on Aristotle's zoological classification of *megista genē*. We nonetheless have to acknowledge that such a feature could not have been easily identified by ancient readership, if they had never read the *PA* (or been aware of Aristotle's purpose there) and had had only the *Epitomē* at their disposal.

[18] *Epit.* II.1, 36.3–5, Lambros 1885; trans. Hatzimichali 2021: 235–236.

[19] "This epitome reads like a randomly arranged discussion of various interesting animals. It bears no evidence of the organization of the work it allegedly epitomized" (Lennox 1994: 15).

HA, or a collection transmitting its views, like the *Epitomē*, might have appeared sufficient when ancient readers merely wished to have access to Aristotle's zoological observations about animal parts.[20] Hence, Aristophanes' own characterization of his work might have been one of the reasons for the shift of interest toward Aristotle's *HA*, to which Aristophanes gives preference[21] throughout the part of the collection that has come down to us. In this sense, Aristophanes' *Epitomē* not only becomes a crucial factor determining the later reception of Aristotle's *HA* but is also to a degree responsible for creating expectations about its content and scope.[22]

Closely related to this point, the obvious question arises as to why the *PA* should not have had at least the fate of the *HA*. Is it its topic that later thinkers considered unworthy of attention or its structure and line of argument? This question is hard to answer as there is no direct evidence as to the ancients' evaluation of the quality of the *PA* as a natural philosophical discussion of the causes of animal parts. We only find limited comments on Aristotle's zoological enterprise in general and interspersed references to his books as the outcomes of his zoological inquiry. Given that *PA* I had the same fate as the rest of the work, it is not only the zoological treatment of *PA* II–IV that can be easily misinterpreted, if not neglected, but Aristotle's methodological commitments about how to pursue zoology properly. Besides Aristophanes, evidence of committed involvement in the examination of animal life can be found in the works of numerous other ancient Greek and Roman authors:[23] Callimachus of Cyrene,[24] Antigonus of Carystus,[25] Lucretius,[26] Nicolaus of Damascus,[27]

[20] As Balme (1991: 21) has shown, Books I–IV and IX (VII), which take up 40 percent of the *HA*, consist of information reported also in other biological works of Aristotle.

[21] Hellmann 2006: 342.

[22] Cf. Hellmann 2006: 355–356.

[23] See Cerami and Falcon 2014; Sassi, Coda and Feola 2017; Hatzimichali 2021.

[24] The famous poet, active in Ptolemaic Alexandria between ca. 300 and 240 BCE. He is said to have written two works on animals now lost, *On Birds* (*Peri orneōn*) and *On the Changes in the Names of Fish* (*Περὶ μετονομασίας ἰχθύων*) (*Suid. s.v.* Καλλίμαχος, vol. 3, p. 19 Adler). While Pfeiffer places *On Birds* in the section dedicated to *Fragmenta Grammatica* (1949: 340–344), Hellmann has shown that this work was not meant to be just a catalog of names of birds but an organized collection whose material was useful for the philological and poetical preoccupations in Hellenistic Alexandria (2015: 1245–1247). Cf. Dorandi 2017: 72–73.

[25] Antigonus (third century BCE) wrote a *synagogē* about the life of animals (περὶ τοὺς βίους τῶν ζῴων) and a treatise *Περὶ ζῴων* (Dorandi 2017: 61–73). For the surviving fragments, see Dorandi 1999: 39–41.

[26] Ca. 99–55 BCE. For a number of passages from his *On the Nature of Things* reminiscent of Aristotle's *GA*, see Hatzimichali 2021: 239–240. See also n. 28 below.

[27] In a philosophical compendium, fragments of which are preserved only in a Syriac manuscript, a certain Nicolaus dealt with Aristotle's natural philosophy and devoted his book IX to the *PA*. The fragments of book IX are still unpublished (for a description of the single MS and other sources

Seneca,[28] Pliny the Elder,[29] Plutarch,[30] Aelian, Athenaeus,[31] to mention just a few frequently quoted examples. These thinkers are aware of Aristotle's long-term zoological enterprise and most of them have read at least one compilation encapsulating Aristotle's zoological doctrines. However, as is the case with Aristophanes, the works of these thinkers transmit ideas that stem principally from the *HA* and not the *PA*, prompting us to assume that Aristotle's mastery in zoology was acknowledged mainly due to the ancients' preoccupation with the *HA*. It is plain that they are not interested in the sort of investigation of animals that the *PA* develops.[32] Their observations are often used in ethical contexts to serve didactic purposes. In paying particular attention to animal habits and behaviors as well as to their activities and socialization as parts of the animal kingdom, some of these authors attempted to show that humankind can learn lessons from the examination of animals. This becomes all the more evident, if we look at the way they present their observations, that is, embedded in an anecdotal, comical, or miraculous framework, in which it is still possible for them to identify particular animal features. Such a practice, however, does not emerge from an organized effort on their part to carry out zoological inquiries modeled on Aristotle; and we

preserving fragments, see Drossaart-Lulofs 1957 and 1965, especially 6–7, 11–15, and 24–25). This author has been traditionally identified as Nicolaus of Damascus, who lived in the second half of the first century BCE. However, this identification was recently contested by Fazzo 2008, who changed the date of the author of the compendium from the first to the fourth century CE. See Cerami and Falcon 2014: 42–43 and Hellmann 2015: 1257–1258.

28 First century CE. On Lucretius' and Seneca's conception of animal, see Tutrone 2012. Cf. Cerami and Falcon 2014: 41 n. 3.

29 Writing for a broader group of laymen rather than for specialists in zoology, Pliny (24–79 CE) incorporated detailed zoological material in his *Natural History* VIII–XI. In Fögen's words, "Pliny often combines scientific exposition with anecdotal digressions, in particular with stories from Roman history, and provides his narrative with a moral dimension" (2007a: 192).

30 Plutarch (46–119 CE) wrote two works (*Whether Land or Sea Animals are Wiser*; *Whether Beasts are Rational*) his primary aim being ethical, not biological. While his understanding of and access to the Aristotelian corpus remain a matter of controversy, it has been argued that Plutarch knew the *HA* very well. See Sandbach 1982; Michalewski 2016: 223. Around 50 CE, Philo of Alexandria wrote a dialogue on animals in the pattern of Plato's dialogues (extant only in Armenian) arguing for and against rationality as an animal trait. See Jazdzewska 2015.

31 In his *Deipnosophistae* (*Deipn.*), the compiler Athenaeus of Naucratis (end of second- beginning of the third century CE) took recourse to intermediate sources combining zoological information from both Aristotle's and Theophrastus' works. See Moraux 1985: 340 and 344 n. 34; Balme 1991: 4–5.

32 Cf. Hellmann 2015: 1246–1247, 1250–1251, and 1259–1260; Dorandi 2017: 80. As Andrea Falcon (2021) explains, until the end of antiquity engagement with Aristotle's natural philosophy followed the practice of doing "physics without biology." That is to say, thinkers were concerned with studying nature and the natural world, while at the same time neglecting or marginalizing the study of life and living beings. Indeed, as well as showing no signs of readiness to embed such discussions within their works, they appear rather selective when dealing with Aristotle's natural-philosophical writings.

can see this clearly exemplified in Aristophanes: "I will omit the genus of animals which produce grubs, as they are small and weak and not worth observation."[33] This statement stands in direct contrast with Aristotle's famous protreptic in his *PA* I.5 to the study of all kinds of living beings.[34]

If we assume that Aristophanes' views at the opening of *Epitomē* II (1, 36.3–5) reflect the general consensus of his time, ancient thinkers seem to have found it a burdensome task to classify and arrange the material of Aristotle's zoological works in such a way as to pursue their own goals. To put it another way, thinkers were not that eager to spend their time searching within different texts to answer their questions. Consider also the following statement by Pliny, who seems to suggest that a critical engagement with Aristotle's zoology was thought a demanding task calling for well-established evaluative and analytical skills:

> We leave to each reader how to judge these facts; our aim is the description (*indicare*) of the apparent phenomena, not the pursuit of their obscure causes (*causas . . . dubias*).[35]

In this passage, Pliny admits that insects do not possess the same sort of blood as blooded animals do, while also announcing his refusal to undertake the difficult task of investigating the causes of this phenomenon. This seems to be a clear hint both at certain views held by Aristotle and at the chief feature distinguishing his project from other contemporary or earlier zoological treatments.[36] Thus, while Pliny seems to recognize, even indirectly, Aristotle's authority in the field of zoology/biology, he does not embrace one of the core aims of Aristotle's zoological project, the search for causes.

Plutarch's *Whether Land or Sea Animals are Wiser*, written in a dialogic form, launches into an examination of whether nonhuman animals are endowed with intellect. Before the discussion turns to a comparison between land animals and sea-dwellers regarding their intellectual capacities, the interlocutor Autobulus welcomes Optatus in the following way:

> Sit with us, so that if we need a witness, we will not trouble the works (*bibliois*) of Aristotle but will come away with a true verdict by following your expertise (*empeirian*).[37]

[33] *Epit.* 2.3, 36.13–14 Lambros; trans. Hellmann in Hellmann 2006: 337.

[34] See Chapter 2 in this volume.

[35] *NH* XI 8.6 Mayhoff cited by Moraux 1985: 340 and 344 n. 33. Cf. Hellmann 2015: 1241.

[36] In this chapter, Pliny focuses on insects and targets Aristotle's theory of respiration by putting in the cross-hairs of his criticism those who believe that animals without lungs cannot breathe. In all probability, he alludes to Aristotle's treatment in *De Juventute et Senectute, Vita et Morte, de Respiratione* 15. In this latter work, Aristotle mentions both the *PA* and the *HA*; see 3.468b32 and 22.478a27–28.

[37] 965D7–10; trans. Newmyer 2020. Newmyer remarks that here Autobulus has the *HA* in mind (2020: 71 n. 93).

Optatus has been picked up to act as a referee in the discussion not because he is an expert in zoological matters but because he is objective and unbiased.[38] In this passage, Autobulus does not question Aristotle's views on animal reasoning.[39] What he seems to be tacitly proposing to his companions is the examination of the relevant topic in a more palatable manner in comparison to what is pursued in Aristotle's books.[40] Thus, according to this text, while the current discussion on animal reasoning could very well take its cue from Aristotle's works, this is nonetheless not the option to be preferred, especially seeing that the aim of the interlocutors is to reach an agreement as fast and as smoothly as possible.

Evidence of familiarity with some of the ideas developed exclusively in the *PA* can be traced in Hellenistic medicine. Yet this is not surprising given that the *PA* addresses a topic kindred to medicine, that is, animal (and human) physiology and anatomy. Before turning to Galen, I will briefly dwell on an example from the *De Spiritu* (*DS*), a treatise of Peripatetic origin, which combines natural-philosophical and medical ideas on *pneuma*, written probably in the early third century BCE.[41] O. Lewis and P. Gregoric have singled out three cases in which the author might have reformulated or expanded upon ideas found in the *PA*. Two of them, in my view, do not provide a compelling reason to argue that the author of the *DS* transmits certain views found exclusively in the *PA*, and it is with these that I will start my discussion. However, the third one might point us in this direction.

Lewis and Gregoric find it likely that the theory of *arteria*, *phleps*, and *neuron* as the three constituents of the skin, and of *arteria* and *phleps* as the two constituents of the flesh, which the author of the *DS* proposes, could be an expansion of previous ideas concerning the composition of tissues.[42] They particularly cite *PA* III.5, 668a30–32. The idea proposed in this context is that flesh consists of blood vessels that are not always visible due to their small size, yet it is possible to assume their existence since flesh bleeds when it is cut. While we cannot rule out that the author has this vivid passage in mind, we should notice that the same idea is stated at *HA* III.19, 520b17–18 and can be inferred or extracted from several other passages in the same treatise.

[38] Cf. "true verdict." According to Jones 1972: 264, Optatus was a Delphic official; he is depicted in the dialogue as not forcing any particular position but as a neutral interlocutor.

[39] In this dialogue, Plutarch endorses Aristotle's view that nonhuman animals possess intellectual capacities that differ only quantitatively, in respect to the more and the less (*HA* VII.1, 588a16–31).

[40] As Mossman 2005: 154 puts it: "This seems to be less a rejection of Aristotle's views on animal reason (though Autobulus would indeed wish to reject them) and more a marker of the change in direction towards a more practical and anecdotal set of arguments."

[41] According to Lewis and Gregoric 2015: 147.

[42] Lewis and Gregoric 2015: 135 n. 31.

The second example relates to Aristotle's belief that *phlebes* are capable of effusing their content (i.e. blood) not only from their mouths but also from their sides (*GA* II.6, 743a8–10). Aristotle's explanation of this characteristic feature of *phlebes* rests on his belief that blood is potentially flesh in the sense that the thinner the *phlebes* become, the more their *poroi* decrease in size, the result being the formation of flesh (*PA* III.5, 668a33–b1). As with the first example, both of Aristotle's present views can be found in the *HA* as well – in fact Lewis and Gregoric mention *HA* III.5, 515b1–3.[43]

I now turn to the example that might be taken to show more certain signs of influence from the *PA*. The author of the *DS* believes that in respiration *pneuma* is delivered to the stomach, an idea that is not held by Aristotle. The way in which the author frames his argument guides us to think that he aims to displace earlier views according to which *pneuma* is transmitted to the stomach through the esophagus:[44] he actually claims that this is impossible (*adunaton*, 483a21), and he is confident that his own view (i.e. that *pneuma* is delivered to the stomach "through a passage along the loin") is verified by empirical observation (*tē aisthēsei phaneron*, 483a23–24).

At *PA* III.3, 664b6–8, Aristotle proceeds to reject the view of those who believe that the lung receives liquids via the windpipe. He finds it ridiculous in many respects (664b9–10), justifying his opinion by appealing to empirical evidence (*phanerōs*: 664b8, *phainetai*: 664b9–10). His argument begins as follows: "For there is no tube into the stomach from the lung, as we see with the oesophagus coming from the mouth" (664b10–12).[45] Lewis and Gregoric propose that this statement is likely to have "motivated the author of *De spiritu* to note that the passage distributing *pneuma* to the stomach splits off from the windpipe before the bronchi, that is, *before* the lung" (emphasis in the original).[46] In further support of this proposal, Aristotle's insistence on the information provided by perceptual data, which is utterly endorsed by the author of the *DS*, should not go unnoticed. The passage abounds with various expressions of disapproval against those who, due to their negligence of what is obvious, believe that food or drink can actually reach the lung.

The reason behind the decision to include the case of the *DS* within the scope of this study is that even if we accept that some piece of knowledge

[43] Lewis and Gregoric 2015: 138 n. 46.

[44] "These lines . . . attest to the existence of a debate concerning the entry route of *pneuma* to the stomach" (Lewis-Gregoric 2015: 143).

[45] Translations of the *PA* are taken from Lennox 2001a.

[46] Lewis and Gregoric 2015: 143 n. 62.

originating from the *PA* was circulating in Hellenistic philosophical or medical circles, which is probable, this was done in the manner exemplified by the *DS*, that is, through individual ideas that later thinkers found attractive for one reason or another, singled out from their context, and incorporated into their own treatments without even mentioning Aristotle's name. If the interpretation offered by Lewis and Gregoric is correct, which is not implausible, then, for the author of the *DS* the *PA* turns out to be a text that goes beyond the status of a mere conveyor of ideas. Its material can be particularly useful in the hands of a thinker who critically engages with the tradition and aims to show himself as an eager contributor to the transformation of existing knowledge. Still this assessment is somewhat limited, while the whole subject certainly requires further study in order to gain a better picture of the *PA*'s inherent capacity to trigger further reflection.

3.2 Galen and His *On the Usefulness of the Parts*

Galen (129–ca. 216 CE) devoted a seventeen-book work to human anatomy and physiology entitled *Peri chreias moriōn*. Although the immediate topic addressed in this work seems to overlap with that of the *PA* only partially, that is, only so far as explanation of human bodily structure is concerned, there are to be found in it certain significant passages where Galen discusses the various functions of animal species. It was among Galen's intentions to write a separate treatise devoted to animal parts and the providence of nature about them.[47]

In the *UP*, many outstanding figures of the past, such as Hippocrates and Plato, are acknowledged as influential in various degrees. Aristotle belongs to that group.[48] In fact, Galen himself presents the *UP* as a work following the tradition of Aristotle's *PA*:

> *Function of the Parts* was by now quite widely read, on account of the enthusiasm of most doctors with a traditional medical training, as well that

[47] See, e.g., *UP* IV 17 (I, 241 H): "But Aristotle has an excellent discussion of these matters, saying among other things that Nature, diverging gradually from her custom in creating plants, makes one animal after another, each more perfect than the preceding, until she comes to the most perfect of all, the one that is the theme of my present discourse [. . .] If life were not all too short for the investigation of the noblest subjects, perhaps some day I should supply what remains [to complete his observations] in this field." Cf. *UP* XIV 3 (II, 286 H): "Some day you may hear from me about the other animals, when I add what Aristotle has omitted from his treatise." Translations of the *UP* are taken from May 1968.

[48] Galen is eager to move from strong criticism to expressions of gratitude toward his predecessor. As Hankinson 2016: 240 puts it, Aristotle was an "ambivalent figure" for Galen.

> of philosophers of the Aristotelian school; for Aristotle himself had written a treatise (*pragmateia*) of a similar kind.[49]

Here Galen informs us that a great number of physicians and the philosophers from the Peripatos studied the *UP* due to their zeal for their respective disciplines. Among them, the Peripatetics were much interested in acquiring the work because they recognized in it similarities to a particular treatise of Aristotle, in all probability the *PA*.[50] This means that they must have been familiar not only with the title of the *PA* or the subject of its inquiry but also with its content – so much so as to be able to proceed in a comparative examination of the two works. Presented as showing an unrestrained eagerness to get to know the content of the *UP*, these thinkers were trained to work with 'scientific' or 'technical' texts as the term *pragmateia* suggests.[51] Leaving aside the obvious overstatement on Galen's part regarding the 'popularity' of his work already at the time of its first circulation, what matters most is his outlining of the level and quality of the readership of the *UP*, an indirect comment also for the *PA*'s readership, in the tradition of which the *UP* seeks to claim its place.

Physicians' and Peripatetic natural philosophers' strong desire to consult the *UP* is further reported in another passage where Galen admits that the topic of the usefulness of the parts was one among those lively discussed and debated in antiquity. This 'ancient quarrel' was actually one of the main reasons that triggered Galen to compose the *UP*:

> I have mentioned Aristotle and Plato not because I wished to confute (*elegksai*) what they have said wrongly (*ta kakōs autois eirēmena*), but in order to point out why I have felt impelled to begin a discussion of these matters. For since there was much difference of opinion (*diaphōnias*) among the physicians and philosophers of old concerning the usefulness of the parts (the former believed that our bodies are not formed for the sake of anything or with any skill at all, and the latter, that they are formed for some purpose and skillfully, one philosopher claiming one use for each of the parts, and another, another), I sought first to discover a standard (*ezētēsa kritērion … heurein*) for judging this difference of opinion and then to devise some one universal method (*methodon*) which will enable us to find the usefulness of each part and its attributes. (*UP* I 8 [I, 12 H])

49 *Lib. Prop.* XIX 20–21 K; trans. Singer, cited by Hankinson 2016: 248.

50 A direct reference to the similarities between the two works is to be found in Galen's *On Seed* ([Kühn, IV] 4.611.1–4).

51 See van der Eijk 2013.

While acknowledging the existing controversy around a topic of investigation is a common 'excuse' for launching a new work, in this particular case, Galen seems to be honest. Besides Aristotle's work, the *UP* is said to have taken into account the doctrines of Plato (probably those put forth in the *Timaeus*) and of course Hippocrates, whose name is mentioned a bit further down in the text. Other authorities, who are acknowledged as active participants in this debate include Herophilus and Erasistratus. These thinkers' views comprise the *archaeology* of the *UP* and, when applicable, the *Nachleben* of Aristotle's *PA*. Their main shortcoming is qualified as methodological, seeing that they fail to employ a universal criterion according to which the usefulness of a part can be identified properly. However, a methodological failure has serious consequences in terms of organization of the material and the appropriate data to be included in its exposition. Hence, Galen criticizes earlier physicians for failing to investigate for what purpose a particular part functions, as well as earlier natural philosophers for not being consistent in their accounts of the function of single parts.

A bit further down the text continues as follows:

> Therefore, since neither Aristotle nor any of the ancients mention all (*hapasas*) the actions of the instruments, we cannot approve their writings (*suggrammata*) on the usefulness of the parts [. . .] In fact, the best philosophers are plainly ignorant of their usefulness, nor, as I was saying, do they understand the writings of Hippocrates [. . .] I have mentioned [these questions] here, however, only to show why I have undertaken this treatise on the usefulness of the parts when Aristotle has written so fully and so well on this subject, and no small number of other physicians and philosophers, among whom, certainly, is Herophilus of Chalcedon, have also expressed themselves well, though perhaps more briefly than Aristotle. (*UP* I 8 [I, 14–15 H])

In this passage, Galen alludes to the *PA* as if it were the chief reference work in antiquity regarding the usefulness of the parts of the body. Interestingly, he refers to the 'writings' of the ancients on the topic, so we must assume that he has specific passages in mind if not larger pieces of text. However, the accounts contained in them are said to lack completeness, a virtue that Galen attributes to Aristotle's treatment of the subject not only in the current passage but in several places in his *UP*. After Aristotle, many thinkers, both philosophers and physicians, engaged with the topic at hand, though only briefly, among whom Galen mentions Herophilus (ca. 335–280 BCE), one of the greatest anatomists of Alexandria. This point has its significance since it implies, on the one hand, that the doctrines of the *PA* were coterminous with subjects that medical practitioners and authors with medical inquisitiveness were examining and, on the other hand, that inquiry into animal physiology was still at that time taken into serious account by those who investigated human physiology.

Galen's statement that his *UP* follows the tradition of Aristotle's *PA* does not seem to change radically the fate of the latter's reception during the Middle Ages. While Galen and Aristotle are sometimes mentioned in the same context by authors, this is not based on doctrines that they discuss in these two treatises. After Galen, the picture we draw of the reception status of the *PA* resembles the one drawn of the period before him. Neoplatonic commentators do not engage with the *PA* or any other biological treatise because they conceive of them as "intermediate" writings (*metaxy*) between the particular writings (*merika*) and the general ones (*katholou*), both of which should be prioritized in the order of study of Aristotle's works.[52] In the late fourth century, Nemesius of Emesa authored *On the Nature of Man*, but the passages identified so far as stemming from Aristotle's *PA* are in all probability drawn from intermediate sources.[53] Similarly, information reflecting Aristotle's *HA* is incorporated in the work of Meletius the monk (date uncertain, perhaps ninth century), yet through medium sources, among other thinkers Aristophanes of Byzantium, Galen, and Nemesius.[54] In one of his shorter treatments on natural phenomena, Michael Psellus (eleventh century) devotes considerable space to answering, among others, a series of biological questions that, however, pertain more to the topic of Aristotle's *GA* and *HA* than to that of his *PA*.[55] One of his students, Michael of Ephesus, goes one step further writing the first Greek commentary on the *PA*, so far as we know.

[52] Hellmann 2015: 1259–1260.

[53] Morani 1987: 138 has identified three passages in his *Index Locorum* that seem to stem from the *PA*: For 10.10 Morani (perhaps recalling *PA* III.10, 673a8, 28), which depicts a commonplace after Aristotle, see Pollux *Onomasticon* 6.200.2–3 (and *passim*), Sextus Empiricus *Pyrrh. Hypot.* 2.211.6 and Porphyry *Isagōgē* 20.11–12, 19–22; for 66.2 Morani (perhaps reminiscent of *PA* II.10, 656b28), see Pseud-Arist. *Problemata* XI 58. On Nemesius' use of Aristotle's biology, with a particular focus on passages from the *GA* and *HA*, see now Tieleman 2021.

[54] See Gielen 2018. For a collection of Aristotelian biological passages found in Nemesius' and Meletius' works (which, however, do not stem from the *PA*), see Kalogeridou 2010: 66. Cf. the previous note.

[55] The relevant section bears the name "Other solutions to physical puzzles addressed to my students and others who have queries" and is included in Duffy 1992: 47–56. As far as I am aware, there exists no translation of this work into any modern language.

3.3 Michael of Ephesus, the First Commentator of the *PA*

Michael of Ephesus is one of the most characteristic examples of what we may call "a Byzantine philosopher," an Aristotelian commentator who was a "philosopher and political man" (*philosophos kai hama politikos*), a professional working in the circle of Anna Comnena.[56] Living in Constantinople during the first half of the twelfth century, along with Eustratius – Metropolitan of Nicaea – he played an active role in a new tendency that connected the Neoplatonic commentary tradition with the Byzantine practice of interpretation for the purpose of commenting on those Aristotelian works that were less known at that time. While little is known about Michael's life[57] and his activity as professor of philosophy in Constantinople, it is by now generally agreed that he was a unique case of a commentator working nearly with the whole Aristotelian *corpus*.[58] He seems to be the first who comments on Aristotle's zoological works, his *Rhetoric* (now lost) and *Pol.* (of which we have only passages). Besides the commentary on the *PA* (*In PA*), he produced commentaries on the *GA* (*In GA*, wrongly attributed to Philoponus), the *Movement of Animals* (*In MA*), the *Progression of Animals* (*In IA*), the *PN*, the *Nicomachean Ethics* (*In EN*, books V, IX and X), *Metaphysics* Z–N, and the *Sophistical Refutations* (both wrongly attributed to Alexander of Aphrodisias). He also commented on the *Physics* (*Phys.*), *On the Heaven*, *Topics* (*Top.*), *Prior* and *Posterior Analytics* (all now lost), and the pseudo-Aristotelian *On Colours* (*Col.*).[59]

[56] Trizio 2007: 291–292.

[57] Little albeit significant information about Michael's life, work, and his companions is included in his commentaries. For four passages from the commentary on *Politics* (*Pol.*) referring to the commentator's own times, see Barker 1957: 140–141. For two passages in his commentary on the *GA*, see Kalogeridou 2010: 81–82. For a collection of many autobiographical passages (with translation and notes) from Michael's commentaries, see Coughlin at www.academia.edu/38778680/Michael_of_Ephesus_autobiographical_passages_with_brief_notes_and_English_translation_revised_Nov_2021_.

[58] Cf. Koch 2015: 2–4. In Koch's view, Michael was working on several of his commentaries at the same time and he was probably revising them regularly based on his lectures. According to Barker, Michael "would seem to have lectured on the *Politics* (and also the *Ethics*) of Aristotle in the school of philosophy founded by Constantine IX Monomachus in 1045. He had also lectured, at an earlier date, on the physical and metaphysical writings of Aristotle" (1957: 136).

[59] Cf. Michael's reference that he has completed commentaries on Aristotle's biological works and on *Metaphysics* and intended to work on *Col.* (*In PN* 149,8–16). Michael's inclination toward Aristotle's biological writings may not be due solely to his personal interest on this field of studies. Anna Comnena encouraged and employed commentators to undertake the exegesis of Aristotelian works on which no commentary existed from antiquity. In her funeral oration, George Tornikes famously mentions "the wise man from Ephesus," whom Browning has identified as Michael: "I have myself heard the wise man from Ephesus attribute the cause of his blindness to her,

As a commentator of the *PA*, Michael follows Aristotle's text closely, adding philological observations and remarks on language (punctuation, use of synonyms, syntax, grammar, elliptical sentences, proposal for obelism etc.).[60] As a rule, Michael cites passages from the *PA* without mistakes or dysorthografia, while in some cases he comments on different readings or errors he detects in the manuscript(s) before him.[61] Whether his commentaries were meant to guide students or to circulate privately within Comnena's circle is still an open question; yet his general aim in choosing to buttress Aristotle's text seems to be to make it more comprehensible for the reader. The examples used are frequently drawn from the sector of animal parts in compliance with the main topic of Aristotle's *PA*. Michael sometimes incorporates observations taken from the fields of physiology and psychology that are not found in the original text. Based on these observations, it has been argued that Michael could have been a physician, yet the surviving sources are not sufficient to establish this without doubts. It seems more probable that Michael had medical interests in the broader sense, showing an attitude that is not unprecedented for intellectuals of his time.[62]

For the most part Michael avoids the complex style of writing while his comments are clear and intelligible, having a direct purpose: to decode Aristotle's meaning in a twofold manner, namely by interpreting his train of thought (*dianoia*) and by spelling out the vocabulary used (*lexis*). This practice of commenting is reminiscent of the one followed by the Neoplatonic commentators.[63] However, Michael's exegesis is not exhaustive: he paraphrases Aristotle's meaning by picking up shorter parts of text and by elucidating vocabulary that might seem abstruse to his readers. Hence, for a complete comprehension of the *PA*, the parallel use of Aristotle's text is mandatory. Rather than merely rewording Aristotle's text, Michael's ultimate aim is to cultivate in the Byzantine reader a deep

because he spent sleepless nights over commentaries on Aristotle at her command, whence came the damage done to his eyes by candles through desiccation" (trans. Browning 1990: 406). Furthermore, Michael is located in the sphere of influence of Michael Psellus (who might have been his master: Barker 1957: 136; cf. Arabatzis 2009: 170) and John Italus (Koch 2015: 1), both of whom had already expressed an interest in Aristotle.

60 In his comment on *In Resp.* 142,5–15, Koch emphasizes that commenting denotes above all a 'translation' of the text into a syntactic "Standardform" (2015: 6).

61 E.g. in *In PA* 27,11–22. See two more examples (*In GA* 143,19–21 and *In DJ* 136,27–29) examined by Kalogeridou 2010: 191–192.

62 See the relevant discussion in Kalogeridou 2010: 82–91.

63 See Fazzo 2004. As Koch notices, this division between *dianoia* and *lexis* is rather fluid in Michael, since we find glossary explanations in parts where he examines Aristotle's *dianoia* (2015: 5).

appreciation of the philosophical merits of the *PA*[64] – which implies that the version of the *PA* that Michael and we possess was available to his readers/students. For this reason, he introduces a rather palatable way of commenting, prologizing his comments with a sort of summary or brief remark of what is to follow. When Michael chooses not to comment on a passage, he customarily informs the reader of his decision using the phrase "what follows is clear."[65]

As a sort of introduction to their commentaries, the Neoplatonic commentators describe the plan of their exegesis for the purpose of presenting with clarity the content of the original work. Topics or questions that such a plan often addresses include the purpose or topic of the treatise under discussion (*skopos*), its usefulness (*chrēsimon*), its place in the reading order of the treatises of the same author (*taksis*), the justification of its title (*aitia tēs epigraphēs*), the work's authenticity (*gnēsion*), and the chapters into which it is divided (*eis ta kephalaia diairesis*).[66] Since this scheme is not always to be found in (the introductory part of) Michael's commentaries, its appearance at the beginning of his *In GA* is impressive for an additional reason. It includes a parallel discussion of the relevant scheme of Michael's commentary on the *PA*. Specifically, he tells us that the *GA* is the natural sequel of the *PA*, its five books following 'naturally' (*kata phusin*) the books of the *PA*.[67] In the *PA*, Michael explains, Aristotle seeks the causes due to which the parts are as they are and defines the material and the final cause while omitting the efficient cause.[68] These phrases reveal Michael's understanding of the purpose of Aristotle's *PA* as well as of the reading order of its books in relation to the *GA*. Moreover, they make clear that Michael is not contesting the *PA*'s authenticity. In what comes after this passage, he proceeds to give some information regarding the topics of discussion that are included in the *PA*.

This description shows that Michael captured aptly the main topic and focus of Aristotle's *PA*.[69] So, why is it that he refrains from including this

[64] Cf. Kalogeridou 2010: 185–186. In general, he holds Aristotle in high esteem among the philosophers; cf. the use of the adjective *daimonios* ("marvelous") and the adverb *daimoniōs* ("marvelously") in *In PA* 5,18–19 and 16,13.

[65] *Saphē de ta hexēs*, *In PA* 90,19–22. Actually, this phrase is characteristic of him and it is not found in any other commentator. See Kalogeridou 2010: 130; Koch 2015: 4.

[66] See in detail Hadot 1990: 25, 34–47; Kalogeridou 2010: 167–174.

[67] συνεχής ἐστι τῇ Περὶ ζῴων μορίων ([Philoponus] *In GA* 14,3.1.3–4); συνεχῆ κατὰ φύσιν τῇ Περὶ ζῴων μορίων (14,3.3.5–6); ἔχεται γάρ, ὡς εἴρηται, καὶ συνήνωται κατὰ φύσιν τὰ Περὶ ζῴων μορίων (14,3.3.9–10).

[68] *In GA* 14,3.1.4–11.

[69] However, there might have been points at which Michael got confused. Cf. Koch 2015: 2 with reference to *In MA* 112,18.

introductory scheme at the beginning of his *In PA*, while still referring to this information here? One plausible explanation is that in doing so Michael establishes the continuity and coherence among Aristotle's works, which is one of his constant concerns. Interestingly, he follows the same practice elsewhere. Right after the opening of his *In MA*, he provides information about the purpose of Aristotle's *IA*, adding a comment on a direct thematic correspondence between the *IA* and *Phys.* Θ.[70] Additionally, we may observe, the idiosyncrasy of Aristotle's *PA*, that is, that *PA* II opens up the real discussion of the parts of animals and not *PA* I, may have led Michael to avoid putting forth an introduction in terms of the above-mentioned scheme. Indeed, we find an introductory comment at the beginning of his *In PA* II. He first summarizes what one reads in Book I and then announces that in what follows Aristotle will speak about the causes of the *phainomena* that concern animals.[71] By this brief and broad introductory description, Michael clearly refers to the rest of the three books and not merely to Book II. Finally, he draws a contextual link between the *PA* and the *HA*: the latter deals with the questions of *what* (*ek tinōn*) and *how many* (*posōn*) parts belong to each animal, concerning itself with the characteristics of all animals (*ta peri pantōn tōn zōōn dialambanei*); the former provides answers to the questions of *why* animals have these characteristics (*tas de aitias toutōn*) and *how* each one of them has been arranged by nature (*kath' hon tropon etachthē hekaston para tēs phuseōs*).[72]

After these preliminaries, I will now proceed to an assessment of Michael's comments on Aristotle's treatment of marrow and the brain, two terms that seem to have been commonly used interchangeably already from the fifth century BCE, despite Aristotle's silence on this subject. The straightforward manner in which Michael offers his comments makes us

[70] *In MA* 104,8–17. See also *In Eth.* X 619,27–29, where Michael claims that the *Pol.* should be read immediately after the *Ethics*, and the opening of his *In PA*, where he establishes a thematic link with Aristotle's *EN*, emphasizing that both works are addressed to (ἀκουσόμενον) the educated person (πεπαιδευμένον).

[71] *In PA* 25,3–9.

[72] *In PA* 25,9–12. In this introductory section, two things certainly deserve further attention, which, due to space limitations, cannot be detailed here: First, Michael's reference to the *HA* as a work comprising nine books – does he contest the authenticity of *HA* X, a topic that has divided scholars of Aristotle over the past decades? Second, Michael seems to have taken the *HA* as a broad and introductory study to zoology – this view might lie behind the phrase περὶ πάντων τῶν ζῴων διαλαμβάνει. At least from two other passages, it becomes clear that Michael puts the *HA* before the *PN* (*In PN* 123,23) and before the *PA* (*In PA* 77,11) in the order of examination of Aristotle's writings. How can this view be explained, especially in light of the fact that Michael did not write a commentary on the *HA*?

suspect that his readers would have been able to recognize this interchangeable use without further clarification.

In *PA* II.6, Aristotle discusses marrow, distinguishing the marrow found in bones (651b24) from the spinal cord, which he calls "spinal marrow" (ῥαχίτην) (651b32–33).[73] According to this chapter, marrow "is of the nature of blood," that is, results from blood concoction, and is "not, *as some think*, a seminal potential of the seed" (651b20–21). This latter doxographical remark suggests that some thinkers believed that marrow was the source of seed and hence the origin of life but, unfortunately, Aristotle does not offer any hint as to who these might be. Here Michael's text can provide assistance. At 37,2–3, Michael claims that "those belonging to the circle of Plato (*hoi peri Platōna*) took marrow to be the source of seed, just as they took the brain to be the source of the nerves."[74] Initially, we may think that Michael alludes here to some Neoplatonists. However, as James Wilberding has shown, Michael's interpretation of Plato's embryology is pretty far removed from the Neoplatonic core theory. Although problematic in some respects, Michael's interpretation, according to Wilberding, shows that he was aware of Plato's commitment to the encephalomyelogenic theory in the *Timaeus*, and the passage under consideration should be taken as an indication of this awareness.[75]

Indeed, in *Timaeus* 73b–d, Plato describes the formation of marrow, which he regards as the chief foundational cornerstone in the construction of flesh and bones.[76] Marrow is qualified as "a universal seed contrived for every mortal kind" (*panspermian panti thnētō genei*), resulting directly from the appropriate mixture of fire, water, air, and earth, and is said to have been divided into two main portions answering to the kinds of the soul implanted in it. That portion that was to contain the "divine seed" (*theion sperma*) was given the name *encephalon* (spinal cord).[77] "Each living thing," Plato stresses, "was at its completion to have a head to function as a container for this portion." *Encephalon*, thus, seems to have taken its name from its characteristic location within the most precious bodily part, that is, from its being kept firm and protected "within the head" (*en tē kephalē*). In a similar way, that portion that was to retain the rest of the soul, the mortal part, was divided into round and elongated shapes, which he called

[73] ῥαχίτην. Cf. Plato *Tim.* 91a6–b2. At *HA* III.2, 512b2, Aristotle uses the phrase νωτιαῖον μυελόν.

[74] Translations of Michael's text are my own. There is no translation in any modern language for Michael's *In PA*.

[75] Wilberding 2017: 177–183, especially 182–183.

[76] Cf. Taylor 1928: 518–521.

[77] ἐπωνόμασεν τοῦ μυελοῦ ταύτην τὴν μοῖραν ἐγκέφαλον; trans. Zeyl 1997.

marrow (*muelon de panta epephēmisen*). Living bodies were constructed around these figures, first and foremost the bones.[78]

An intriguing aspect of Plato's current account of marrow is his special emphasis on *naming* the bodily places under consideration – we actually encounter twice the idea that the god proceeded to name the bodily part: in the first case, it is a specific portion of the marrow that is designated as encephalon, while in the second case, the name *muelos* is bestowed upon other body parts. Aristotle makes no comment on the history of the names encephalon and marrow, and hence one might be led to think that Plato's stress on this issue is one more instance of his concern about the process of coining words. However, the way in which Michael introduces his comment on the section on marrow leaves little room for doubt; there seems to have been a discussion among the ancients concerning the names encephalon and marrow:

> After these Aristotle speaks also about the marrow, not the so-called encephalon [*ou tou kaloumenou encephalou*] (for he will talk about that next and will show that the brain is not marrow) but the marrow which is found in bones. (36,33–35)

According to Michael's comment, the name *encephalon* was attributed by many to spinal cord while the term *muelos* was the regular one to refer to the marrow that constructs the bones. One might propose here that through this phrasing Michael reformulates one of Aristotle's sentences found at the beginning of *PA* II.7, according to which "many people think that the encephalon is marrow."[79] However, Michael's text gives us something quite different on this point. While Aristotle says that many people regard encephalon and marrow as identical organs, Michael states that marrow *was called* encephalon by many people. Shall we take Michael's statement as indicating an existing controversy that is not mentioned by Aristotle? Or could it be a topic of discussion that arose in the years after Aristotle's time? Moreover, does the doxographical report of

[78] The main idea here lies, according to Taylor, in the indication by Timaeus of the "difference between the shape assumed by the μυελός in the head, where the immortal element of the ψυχή resides, and in the vertebral column, which is the home of the θνητὸν εἶδος ψυχῆς" (1928: 522). It is noteworthy that Aristotle criticizes as unusual and hence obscure expression the appellation ὀστεογενής ("produced in the bones"), 'used' by Plato to describe marrow (*Top.* 2, 140a5; cf. χρῆται as the implied verb in the sentence). However, this adjective is not to be found in the surviving corpus of Plato's writings.

[79] 652a24–25. This sentence is sometimes taken as indicating that Aristotle believed that encephalon *is* marrow; so Rocca 2019. However, Aristotle only says that marrow is continuous with encephalon (συνεχὴς τῷ ἐγκεφάλῳ, 652a30; see also *HA* IV.8, 533a12–13), although they are opposite in nature – the encephalon being the coldest among the parts while marrow is a hot part.

Michael on Plato a bit further down in his text (which we saw at the beginning of this section) refer solely to the Platonic view that marrow is the source of seed, or are Plato and his followers among those who were actually calling the marrow 'encephalon'?

In Diogenes Laertius, we find the Pythagorean view that "sperma is a drop dripping from the brain (*encephalou*)," an idea representing the so-called encephalogenetic or myelogenetic theory, which Alcmaeon of Croton and Hippo of Rhegium probably endorsed.[80] Traces of this theory can be found in Plato and Diocles.[81] As H. von Staden explains, "This encephalogenetic theory, which appears also in a myelogenetic version (i.e. with the spinal marrow instead of the brain as the ultimate source of seed), was perhaps originally inspired by the conviction of the supremacy of the brain in all vital functions (or by the related conviction of the 'divine' nature of the head and the 'divine' origin of seed)."[82] Two later sources can assure us not only that the terms marrow and encephalon were sometimes used interchangeably in everyday language but, perhaps more importantly, that, rather than being a matter of anatomical knowledge, the choice between the two terms depended upon the potential religious connotations that the context might have borne. The first text comes from Galen's *UP*:

> This source[83] *should not be spoken of as analogous to the encephalon*, as Aristotle is wont to do in such cases, being at times deceived by names which are derived not from the essence of the thing, but from accidental characteristics. This is also true of the term *engkephalos*, since this name is derived from its location [i.e. in the head: encephalon]. When Plato wished to make clear what he thought to be the true nature of its substance, *he called it myelos* (marrow, medulla). Even if it is medulla,[84] however, there is still something lacking in its appellation; for there is a spinal medulla [*nōtiaios tis esti muelos*] and another medulla in the bones, and these are not sources of all sensation and motion. Hence *many call by the same name, i.e. muelon,* both the encephalic medulla [*encephalitēn*] and the spinal medulla [*rhachitē*], and *others think fit to call muelon* not encephalic, but the medulla of the encephalon. But these designate the part not by a name but by a phrase [. . .] It is not simply medulla, since every medulla does not

[80] τὸ δὲ σπέρμα εἶναι σταγόνα ἐγκεφάλου, D.L. 8.28 = 58B1ADK. The examples are cited by von Staden 1989: 288–289.

[81] von Staden 1989: 289; Connell 2016: 107. [82] von Staden 1989: 289.

[83] I.e. the source of the senses and nerves. In animals such as crabs and whales, the source of the senses is located in the breast according to Galen and not in the head, as they have no head properly speaking.

[84] Cf. *UP* XII 11 (II, 216 H), where Galen claims that "the spinal medulla was formed to be like a second encephalon for the parts below the head."

> have this faculty, nor is it simply an encephalon (something in the head), since certainly it is no encephalon in animals that have no head [. . .] Hence, though the encephalon most particularly derives its name from its position (for it is called *engkephalos* because it lies *en tē kephalē*, in the head), I shall not say that it is something else and analogous to the encephalon when I find it situated in the parts of the thorax in animals that have no head. I shall say that it is the encephalon itself and that its antiquated name [*tēn archaian prosēgorian*] is not suitable.[85]

At first, Galen addresses the issue of the proper usage of the name encephalon by targeting and criticizing Aristotle's appeal to analogical similarity. Although he speaks about Aristotle's relevant practice in general, his observation probably takes its cue from Aristotle's discussion in *PA* II.7, to which we have already referred.[86] In Galen's view, animals without a head, properly speaking, cannot be described as possessing a part analogous to the encephalon in another place in their body. So the first thing we should observe here is Galen's insistence on needing to use the correct names in order to properly designate or identify something, and the fact that his remark is exemplified by the case study of encephalon (and marrow, in the second part of the text).

Next Galen mentions Plato and specifically his choice to name 'marrow' (and not 'encephalon') the *muelos* found in the area of the brain with the intention of avoiding the use of a term that would be descriptive in terms of location but not indicative with respect to the essence of the thing under question.[87] This example leads Galen to a necessary clarification regarding the two kinds of marrow and to a further comment on the variety of terms employed up to his time to describe these two kinds.[88] Galen's testimony actually provides us with a piece of information from which we draw the exact opposite conclusion of what we read in the *Timaeus*, that is, that Plato employed the term *muelos* where others used 'encephalon.' Besides

85 *UP* VIII 4 (627–628 H), translation modified, with emphasis added. In this passage, I have added the Greek text in brackets. The parentheses originate in May's translation.

86 May 1968 *ad loc.* cites *PA* II.7, 652b23–25 as Galen's point of reference along with *PA* IV.5, 681b12–17.

87 Elsewhere Galen implicitly criticizes the followers of Praxagoras and Philotimus "who maintain that the brain is nothing more than a *growth* (*blastēma*) of the spinal cord" (Rocca 2003: 164; cf. 165).

88 According to Ogle 1882: 164, "That the spinal cord is the marrow of the vertebrae is an error, the memory of which is still preserved in the popular term 'spinal marrow.' Although A[ristotle] observed that the cord was in substance quite unlike any other marrow, having indeed no other resemblances to this substance, than its position in a bone cavity and its light color, and though he knew that it was directly continuous with the brain, which latter he knew was not marrow. . . yet he held fast to the popular notion, because his theory required the cord to be made of some hot substance, such as he took fat to be."

anything else, this remark shows clearly that the two terms could substitute each other occasionally in Plato's time.[89] But was Plato the originator of this tendency? Here our second source becomes of great importance.

In *Deipn.* II, Athenaeus makes an interesting comment while discussing the pig's brain. The passage reads:

> None of the ancients, at any rate, ate pig's brains, because they contain almost all the senses. Apollodorus of Athens[90] denies that any ancient author (*tōn palaiōn*) even uses the word encephalon. Sophocles, for example, when he describes Heracles throwing Lichas into the sea in *Trachiniae*, does not use the word encephalon, but says "white marrow" (*leukon muelon*), avoiding a term that is not used:
>
> He made the white marrow ooze of his hair, as
> his head was split in two, and the blood along with it [. . .]
> So too Euripides brings Hecabe onstage mourning for Astyanax, who was thrown from the walls by the Greeks, and says:
> Poor thing, how cruelly your paternal walls,
> the battlements erected by Loxias, sheared from your head
> the locks of hair your mother often tended
> and kissed, whence from your shattered bones
> shines forth the gore – my purpose being to avoid shameful words.
> The interpretation of these verses involves some difficulty; because Philocles[91] does use the word encephalon:
> He wouldn't stop eating encephalon.
> Also Aristophanes:
> I'd be wasting two encephalon croquettes.
>
> And other authors as well. Sophocles must therefore have said "white marrow" for poetic effect (*poiētikōs*), while Euripides chose not to put an ugly, shameful sight on open display, but made the matter clear in a way that suited him. That they regarded the head as sacred is clear from the fact that they swore by it and treated the sneezes it produced as holy.[92]

This passage provides indisputable evidence in support of the view that thinkers already before Plato had reservations about using the terms encephalon and marrow in specific contexts. It is also enlightening with

[89] Cf. Galen, *AA* XV 1: "It makes no difference to the exposition of Anatomy which we have in mind whether . . . we name the spinal cord according to the customary usage of the Greek language 'vertebral marrow,' or simply 'marrow.' For Plato applies the term 'marrow' to the spinal cord, which he calls 'vertebral marrow,' and to the brain also, which he calls 'cranial marrow.' It is clear that he uses this term (marrow, I mean) of every soft delicate substance which bones envelop" (trans. Duckworth 1962: 223). The first part of this passage is cited by Rocca (2003: 163 n. 229) with further bibliographical references.

[90] *FGrH* 244 F 246. Apollodorus lived in the second century BCE.

[91] *TrGF* 24 F 5.

[92] 65f–66c; trans. Olson 2007, slightly modified.

respect to the actual replacement of the term encephalon with the term marrow when the topic touched upon sensitive issues of morality and religion. From the fact that Sophocles resorted to this interchangeable use in one of his plays, we must assume that he presupposed a common understanding among non-experts.

Thus, I would suggest that when Michael refers to the so-called encephalon, he need not have Plato specifically in mind but seems to be witnessing the common understanding of the term. The two passages from his commentary discussed earlier (37,2–3 and 36,33–35) do not refer to each other but serve distinct purposes:[93] the first, stating that those associated with Plato believed that marrow is the source of seed, illuminates Aristotle's remark at 651b20–21; the second, which is introductory to the whole discussion and stresses that the subject of Aristotle's treatment is "the marrow, not the so-called encephalon," refers to the common practice of using the terms marrow and encephalon interchangeably – a practice, as we saw a moment ago, with which already the spectators of Sophocles' plays were familiar.

On the other hand, the fact that Michael, unlike Aristotle, deems it necessary to refer to common opinion in the matter at hand shows that this view must have been of some significance for his current scholiastic venture, in particular for its informative side: he seeks, above all for clarificatory purposes, to draw his readers' attention to the fundamental difference between Aristotle's view and common opinion. His concise manner of expression leads us to think that this ancient interplay between the two terms is in all probability a familiar *topos* for Michael's contemporaries too and that his readers must have been aware of the distinction between the two kinds of marrow. This assumption is in fact confirmed by three Byzantine sources, possibly earlier than Michael. In the relevant discussion of the marrow found in Meletius' *On the Constitution of Man* – which in many respects follows Galen's account in the *UP* – Meletius uses a similar expression claiming that encephalon is called also *muelos* because it lies in the innermost part of the body (52,28–29 Cramer). According to E. Gielen, Meletius' work was intended for non-experts.[94] Similarly, Theophilus (seventh or ninth century) states that the common name of encephalon is *muelos*, whereas its special name is

[93] Besides, it would have sounded awkward in Greek to refer to an opinion held by a specific group of thinkers, as is the case with οἱ περὶ Πλάτωνα in Michael's text, by employing the participle *kaloumenos*.

[94] Gielen 2018: 154.

encephalon[95] and Leo the Physician (perhaps ninth century) that encephalon "is called also *muelos*, because it is shut up or lodges in a recess."[96]

3.4 Conclusion

With this chapter, I have tried to spotlight the different stages in the process of reception of Aristotle's *PA* until the period when its first Byzantine commentator took up interpreting it. The circulation of zoological compendia in Hellenistic times made it possible for a number of the *PA*'s doctrines to survive (especially those included also in the *HA* and *GA*) but at the same time seems to have put up barriers to the work's examination by those few thinkers who might have been potentially interested in it. Section 3.1 showed that certain views found in Hellenistic and later texts might indeed stem from the *PA* but that they seem to have been incorporated in these texts to accommodate their authors' line of argument rather than aiding them significantly to examine or adopt the research project of the *PA*. Section 3.2 considered a number of passages from Galen's *UP* that provide information about the status of the *PA* and its readership until Galen's time. Yet Galen's remarks about the recognition the *PA* had gained in the period before the composition of the *UP* can hardly be verified by the written sources that have come down to us. After Galen, the picture drawn seems equally disappointing until we come to the twelfth century and the biological reflections of Michael of Ephesus (Section 3.3). Michael's genuine notes throughout his *In PA* allow us not only to capture more accurately Aristotle's thought but also to raise anew questions regarding the content and the addressees of the *In PA*, as well as the level of experience in biological matters in Michael's time. Moreover, they show that sometimes, as in the case of marrow and the brain discussed above, Michael does not hesitate to create the conditions for further reflection on the topic under investigation, either by adducing small details on the conceptual overtones of some of Aristotle's terms or by adding information about views held by Aristotle's predecessors. Michael's scholiastic endeavor is an original interpretative contribution to the intellectual activity of his time, characterized by an arduous and meticulous approach to the ancient text for which Michael shows full respect. Such a

95 τῷ κοινῷ δὲ ὀνόματι ὀνομάζεται μυελός, ὥσπερ τῷ ἰδίῳ ἐγκέφαλος (*On the Constitution of Man* 168,14-15 Grimm-Stadelmann).

96 *Epitome On the Nature of Men* 26 Renehan; trans. Renehan. My thanks to Petros Bouras-Vallianatos for the references to Theophilus and Leo. For the Galenic influence on the works of these thinkers, Nemesius and Meletius, the reader may consult Bouras-Vallianatos 2019: 86-91.

treatment of Aristotle's biological material invites the crucial question of whether we are talking about a specific and unique case of a commentator or whether the environment in which Michael flourished called for new perspectives in philosophy. A question like this should remain open until further study on Michael's texts is carried out. Turning our deep attention to this neglected figure from the point of view of the reception of Aristotle's biological doctrines is indeed a desideratum, seeing that a hidden treasure lies in his still untranslated works, especially those dealing with zoological inquiries.

CHAPTER 4

Teleology and Understanding

Jessica Gelber

Introduction

Why does Aristotle insist that the natural scientist should give teleological explanations? One obvious reason is that Aristotle believes that ends (*tele*) are causes (*aitiai*), and explanations (*apodeixeis*) cite causes; a putative explanation that fails to cite the final causes of the phenomena it purports to explain is, consequently, a bad explanation. However, in addition to thinking that ends are causally operative in nature, Aristotle also has reasons, from the point of view of demands on explanation, for holding that teleological explanations of natural phenomena are indispensable. These specifically "epistemological" reasons have to do with the role he believes those teleological explanations play in facilitating our understanding of natural phenomena.[1] As I read it, Aristotle is appealing to such reasons in a passage in *PA* I.1, a chapter devoted to discussing various methodological questions that arise about any inquiry into nature. There he is arguing, given certain reasonable assumptions about explanation and understanding, *for* giving final cause explanations and *against* giving explanations in terms of "necessity" alone, as he describes the attempts of many of his predecessors.

> [A] (i) And in addition to these questions, since we see more than one cause of natural generation, e.g., both that for the sake of which and that from which the change originates, we need also to determine, about these causes, which sort is naturally first and which second. (ii) Now it is apparent that the first is the one we call "for the sake of which"; for this is the definition

[1] In what follows, when I speak of reasons as "epistemological" or say that teleological explanations are primary "in the order of exposition," I do not mean to suggest that these are merely pragmatic reasons or that the explanations are first "to us." The order of exposition of a completed science will, for Aristotle, track the causal structure of that which is in the domain of that science. What I am distinguishing here are considerations that appeal to what *makes things happen* from ones that appeal to what *makes things intelligible*. Although in Aristotle's view these mirror one another, they are nevertheless distinguishable.

> and the definition is the starting point alike in things composed according to art and in things composed by nature. (iii) For once the doctor has defined health, and the builder has defined house, either by thought or perception, they provide the accounts and the causes of each of the things they produce, and the reason why it must be produced in this way. Yet that for the sake of which and the good are present more in the works of nature than in those of art.
>
> [B] And what is of necessity is not present in all natural things in the same way; yet nearly everyone attempts to refer their accounts back to it without having distinguished in how many ways "necessary" is said. That which is necessary without qualification is present in what is eternal, while that which is necessary from an assumption is present also in all that is generated, as it is in artefacts such as a house or any other such thing. It is necessary that a certain sort of matter be present if there is to be a house or any other end, and this must come to be and be changed first, then that, and so on continuously up to the end and that for the sake of which each comes to be and is. It is the same way too with things that come to be by nature.
>
> [C] (i) But the manner of demonstration and of necessity is different in the case of both the natural and the theoretical sciences. (ii) This has been discussed elsewhere. (iii) For the starting point in some cases is *what is*, and in other cases is *what will be*. For, since health or human *is* such, it is necessary that this such and such *is* or *comes to be*. But it is not the case that since this such and such *is* or *has come to be*, that such and such from necessity *is* or *will be*. (iv) Nor is it possible to connect the necessity in such a demonstration into eternity, with the result that one says, "Since *this* is, therefore *that* is." (These matters too have been determined elsewhere – in what sorts of things necessity is present, what sort of necessity converts, and owing to what cause.) (639b11–640a9)[2]

While there are fairly typical ways of interpreting some parts of this passage, and notorious controversies about how to interpret others, there is no "standard" reading of this entire stretch of text. On the contrary, from the way these lines are discussed, it would be easy to form the impression of it as a collection of loosely connected ideas, many of them familiar and even repetitive, with the relevance of references to discussions in other works opaque. In what follows, I will propose instead that the entire passage should be read as a tightly unified discussion, organized into sections that follow naturally upon one another. In Section 4.1, I argue that Aristotle's point in [A] is that teleological explanations should precede non-teleological explanations in the order of exposition. His reasoning is

[2] Throughout this chapter, the translation of *PA* 1.1 largely follows Lennox 2001a, though it departs most substantively in what I have labeled section [C].

that the ends cited in teleological explanations are definitions, and definitions – which are not subject to further explanation – are appropriate starting points, insofar as definitions prevent explanations from going on *ad infinitum*. Moreover, I argue that Aristotle proceeds in [B] and [C] to criticize his predecessors' non-teleological accounts for being explanatorily defective: I take Aristotle's complaints to be that those accounts – unlike teleological explanations – neither begin from appropriate starting points (Section 4.2) nor entail the phenomena that they purport to explain (Section 4.3).

4.1 The End Is the Starting Point: [A]

The explicit aim of the first book of *Parts of Animals* is to set out and evaluate certain standards or "norms" of inquiry into nature.[3] Aristotle is particularly concerned to address procedural issues regarding inquiries into natural entities that are subject to generation, such as animals and other living beings. At 639b11, he raises the following question:

> And in addition to these questions, since we see more than one cause of natural generation, e.g., both that for the sake of which and that from which the change originates, we need also to determine, about these causes, which sort is naturally first and which second. [A](i) (639b11–14)

Understanding (*epistēmē*), for Aristotle, is knowing *why* something is the case. That is, "unqualified scientific knowledge" consists in grasping causal explanations.[4] As Aristotle here in *PA* points out, given that there are multiple causes of natural phenomena, there is a question about how the natural scientist ought to proceed.[5] Aristotle's answer is that teleological

[3] There is disagreement about whether these are norms for the presentation of a completed inquiry or norms for conducting one. For example, Kullmann 1974 and Balme 1992 take these as standards for evaluating explanations, while Lennox 2021 takes them as norms for engaging in an investigation. Here I am addressing Aristotle's argument for the priority of teleological explanations, and my concerns are orthogonal to that debate. The interpretation I offer of Aristotle's reasoning can be accepted regardless of whether one takes it to be about how to *give* causal explanations or about how to *judge* what has been presented.

[4] In a completed science, these explanations are given in the form of demonstrations, which are syllogisms whose middle terms refer to the causes of the phenomena being explained.

[5] Although Aristotle is famous for his doctrine of *four* causes, he mentions here only two: the final cause ("that for the sake of which") and the efficient cause ("from which the movement originates"). At 642a1–2 and a13–14, he again refers to "two" causes. There he calls them the cause "*to hou heneka*" and "*to ex anangkēs*." This division into two types of causal explanations, those that say why something is necessary and those that say what something is for, is again illustrated in the explanation given of respiration from 642a31 to the end of the chapter. Appeals to "the necessary and/or the

explanations – those that cite final causes – are naturally "first." In support of this he makes two claims:

> Now it is apparent that the first is the one we call "for the sake of which"; for this is the definition (*logos*) and the definition is the starting point (*archē*) alike in things composed according to art and in things composed by nature. [A](ii) (639b14–16)

As I read this, these two claims constitute an argument for teleological explanations being prior (in some sense here left unspecified) to other causal explanations. The first is that a teleological explanation cites the definition or account (*logos*)[6] of the end for which generation occurs. In explanations of natural generation, one specifies what it is "for the sake of" by citing the account of that which is coming to be. The second is that this definition cited is a starting point (*archē*). Although it is true, in general, that it is "absurd to begin explaining the processes of natural generation before we have established what it is that is being generated,"[7] when Aristotle says that the definition is the starting point, he is articulating a specific way in which final cause explanations are primary: the definition cited in a teleological explanation is an adequate starting point from which explanation can proceed, insofar as a definition can ground the explanation, preventing it from going on *ad infinitum*. When one has specified what something is, there is no further "Why?" question to be asked. This is what I propose Aristotle means to convey by saying that the *logos* is the *archē*: a definition is a starting point of explanation.[8]

There are stronger and weaker readings of this, depending on how much of the formal apparatus from *Posterior Analytics* one thinks is being assumed here.[9] But the core idea is simply that explanations, if they are to yield understanding, must not go on indefinitely. Explanations should

better" are common in Aristotle's biological works, e.g., when he (in *GA* II.1) distinguishes two explanations of sexual dimorphism, one that cites the "necessity and first mover and what sort of matter" (731b18) and the one that cites "the better" and "for the sake of something" (718b22–23). Aristotle also claims that the natural scientist needs to give "both" the "for the sake of which" and "of necessity" cause in *Physics* II.9, 200a32ff.

6 I intend for this discussion to remain neutral about both whether a *logos* here is a definition that signifies the essence of a biological kind or merely an account that says what something is, as well as whether these are *logoi* only of kinds, or of body parts, or both. See Henry 2021a for discussion.

7 Johnson 2005: 181

8 I think this is echoed at 640a33ff, when Aristotle says that "most of all one must say that since *this* is 'what it is to' be a human, therefore it has these things."

9 On its very strongest reading, Aristotle is relying on *APo's* particular conception of understanding (*epistēmē*) and demonstration (*apodeixis*). According to that conception, understanding involves knowing *why* something is the case and cannot be otherwise. A complete science begins with very general axioms (such as the Principle of Non-Contradiction) as well as special principles (such as the

ultimately proceed from starting points that are self-explanatory, that is, not subject to further explanation, and accounts saying what something is – those that cite the *logos* – are self-explanatory. Teleological explanations, because they cite *what* something is, thus provide a point past which no further explanations will be sought.

As I interpret him, Aristotle is giving a reason for teleological explanations being "naturally first" from the point of view of demands on explanation. This is not to suggest, of course, that *aitia* are merely types of explanation, as opposed to types of causal relations. I agree with the majority of interpreters who have argued that for Aristotle, there is some fact about the world – an "ontological basis," as it has been called[10] – that sanctions the use of teleological explanations. However, although arguments for the existence of final cause relations also provide reasons for giving teleological explanations, there can be complimentary arguments for teleological explanations as well. I propose that we read 639b11–16 as giving us one such argument, one that harnesses specifically epistemic considerations to support the priority of teleological explanations.

In contrast to the reading I am proposing, the passage is more often read as concerned with the *causal* priority of ends over efficient causal processes.[11] Aristotle's point is typically taken to be that the end of a process of generation is causally prior to the process of generation, being its causal origin (*archē*). This is in fact a point Aristotle makes a page later, at 640a10ff, where it is put in terms of the priority of being (*ousia*) over generation (*genesis*). In that later passage, Aristotle is asking whether the proper procedure is to try to explain how natural beings *come to be* or whether one should also focus on the *beings* that come to be. Aristotle opts

definitions of the objects in its domain) and derives from these starting points (*archai*), through a series of syllogisms called "demonstrations," all the other necessary facts. In order to qualify as starting points, the premises of these demonstrations must have certain features. In addition to being true, they must also be "primitive and immediate and better known than and prior to and causes of the conclusions" (*APo* I.2, 71b19–25). A premise is "primitive and immediate" (*prōtōn kai amesōn*) when there is no further explanation for it, no additional term that "mediates" the terms in the premise. (An assertion of the form "S is P" is mediated when there is some other term, R, that connects or mediates S and P. Immediate assertions, on the other hand, are such that no other term is needed to connect the terms occurring in them.) Definitions – the "indemonstrable posits" of the essence (94a9–11) – are, moreover, primitive and immediate. Since each complete branch of science must ultimately proceed from starting points that are primitive and immediate, and since definitions that signify the essence of the objects in the domain of any branch of science are primitive and immediate, those teleological explanations that cite definitions are naturally to be given first.

10 Gotthelf 2012: 69 contrasts "ontological" with "pragmatic" or "epistemological" bases.

11 For example, Lennox 2001a: 125 takes the *logos* identified as the *archē* to be not the definition that serves as the starting point of explanation but rather as the "content" of such a definition – the end – which is the causal origin of the process that will bring it about.

for the second: he thinks that one must investigate the beings – the ends that come to be – and not merely the processes whereby they are formed. Empedocles was wrong, Aristotle says there, to think that a vertebrate's spine is due to its twisting about during the process of generation; in fact, the process of generation involves twisting *in order to* produce that sort of backbone. In Aristotle's view, characteristics of generated beings causally determine features of their formation: generation is for the sake of being, not the reverse.

Although closely related, I do not think that Aristotle is merely anticipating that same point in section [A]. In this earlier passage, he is asking which of two different kinds of causes, the final or the efficient, is naturally "first." At 640a10, he turns to discuss the phenomena that those different kinds of causes are invoked to explain. That is, section [A] is about which type of *explanans* should be given first, whereas the later passage is about what the *explananda* should be, and these are different issues. After all, both types of causal explanations can be given for each of the two kinds of phenomena. Facts about the being (the *ousia*), such as an organism having certain parts or engaging in certain activities, are explained by citing both the purpose served (the final cause) and that which brings the part about (the efficient cause). Similarly, questions about generation (the *genesis*), such as why a process occurs in a specific way, are answered by citing both its final cause as well as its efficient cause.[12]

So, on my reading, Aristotle's concern here is with the order in which causal explanations are to be given, *if* one is to have unqualified understanding of living beings. Teleological explanations make understanding possible since answers to "for the sake of what?" questions are (or lead back to) statements of the essences – definitions – that are not subject to further explanation. The "for the sake of which" is the definition and the definition is the starting point.

This interpretation of the focus of the passage is supported by the analogy with artisans that he immediately proceeds to draw:

> For once the doctor has defined health, and the builder has defined house, either by thought or perception, they provide the accounts and the causes of each of the things they produce, and the reason why it must be produced in this way. Yet that for the sake of which and the good are present more in the works of nature than in those of art. [A](iii) (639b16–21)

[12] In *GA* II.6, 742a16–b17, Aristotle gives teleological explanations for the order of body part development, for instance.

Here Aristotle observes that artisans begin with a definition of the product, whether that be a house or health, and from that definition proceed to give the accounts and causes of each thing they do, explaining why certain things *must* be done. That is, the artisan is depicted as proceeding from a conception of the end product, either "in thought or perception," to the necessary means to achieving that end.[13] The same is true of explanations in natural science. The natural scientist, like the artisan, begins from a definition of the end and proceeds to explain what *must* occur in order for that end to be achieved.

If we read the analogy as supporting the priority of teleological explanations, as I have just done, the analogy is between how the artisan reasons and how the *natural scientist* proceeds. If instead we were to read the analogy as supporting the priority of final causation, this analogy would be between the artisan and *nature*. But on that reading, it is not clear how apt the analogy would be. For, in the example, the end is depicted as guiding the artisan's reasoning, and it is difficult to see what is meant to be playing an analogous role in the case of nature's operation. Nature is not a quasi-intentional agent, in Aristotle's view, and so there is no obvious candidate for something in natural objects corresponding to the artisan.[14] It is more straightforward, for this reason, to take Aristotle to be comparing the artisan and natural scientist here, his point being simply that definitions are appropriate starting points of a natural scientist's reasoning, just as they are for artisans.[15]

4.2 Necessity Is "on an Assumption" (*ex hupotheseōs*): [B]

After giving this argument for the priority of teleological explanations, Aristotle next raises a complaint about attempts to reduce explanations to "necessity." His remarks are directed at those who try to explain natural generation by identifying temporally prior states of matter as the causes of what results "of necessity" from them. He elsewhere describes these theorists

[13] This part of artistic production Aristotle elsewhere calls *noesis*: reasoning from a grasp of the end to be achieved (or from the form in the artist's soul) to the steps one can take to produce it. See, e.g., *Metaph.* VII.7 1032b15–17 and *passim.*

[14] This issue is noted by Lennox 2001a: 126 and Kullmann 2007: 282–283. Lennox 2001a: 132 proposes that the analogous element is the *dunamis* for producing an organism, present in the seed from the beginning. If we were compelled to identify something in natural beings that is analogous to the artist's operation, this would be an attractive proposal. On my reading, however, we are not so compelled.

[15] This is the explicit point of analogy between craft and nature in *Physics* II.9: the end (*telos*) is the starting point (*archē*) not of activity (*praxis*) but of the reasoning (*logismos*) (200a22–24).

as holding that "because the hot and the cold and each of those things is naturally such as it is, therefore certain things are and come to be of necessity" (*Phys.* II.8, 198b12–16). Usually (*malista*), he says, the search for the causes of generation involves finding "what comes to be after what, and what first acted or was affected, and so always in succession" (*Phys* II.7, 198a34–35).[16] In Aristotle's view, people "nowadays" incorrectly believe that because what is heavy naturally goes down and what is light naturally goes up, something else then results "of necessity" (cf. *Phys.* II.9, 199b35–200a5).

Although he thinks these people are mistaken about something, Aristotle also thinks that many natural changes are due to the nature of the matter involved and often describes changes that are due to matter as being "due to necessity."[17] Elsewhere he describes this way of being due to necessity – often called "material necessity" by interpreters – as being "in accordance with nature and impulse" (*APo* II.11, 94b37–95a1) or being "thus in respect of their character and nature" (*PA* I.1, 642a34–35). So, given his recognition of necessity as a cause, Aristotle is surely not objecting to their appeal to material necessity, as such. Nevertheless, there is something objectionable about their attempts to explain natural generation by appealing to it, which he proceeds to describe at 639b21:

> And what is of necessity is not present in all natural things in the same way; yet nearly everyone attempts to refer their accounts back to it without having distinguished in how many ways "necessary" is said. That which is necessary without qualification is present in what is eternal, while that which is necessary from an assumption is present also[18] in all that is

[16] Although it is unclear whether it is an explanation that anyone had actually put forward or not, in *GA* II.1, 734a25–33 Aristotle calls "absurd" an account of the production of animal body parts that would involve mistaking the coming to be after (*meta*) something for coming to be *because of* (*dia*) something. A body part, or anything else, can only come to be *because of* (*dia*) something that has the form (in actuality) of whatever is being produced.

[17] To forestall possible confusion: I do not mean to suggest that Aristotle thinks that temporally prior states of matter necessitate later ones. (As will be discussed in Section 4.3, this is at the heart of his subsequent criticism, beginning at 639a30.) For example, saying that fire "necessarily" goes up, or that it has a certain sort of effect when it interacts with other elemental powers "of necessity," is expressing the idea that the latter is something that follows – logically, so to speak, not temporally – from the *nature* of fire. Thus, Aristotle describes this way of being due to necessity as being in accordance with something's nature.

[18] A recent edition (Isépy 2025) of the manuscript retains this *kai*, though his research on the entire independent Greek, Latin and Arabic tradition determines that omission is also a stemmatical possibility. If omitted, Aristotle appears to be claiming that necessity in the realm of generated phenomena is restricted to necessity *ex hupotheseōs*. Balme 1972/92: 84 claims that the *kai* means "actually," yielding the same idea. With the *kai* retained and translated as I have here, the passage suggests that both types apply to what is generated. So read, on the proposed interpretation of this contrast that I will offer, a natural thought is that the *haplōs* necessities would be the definitions that serve as the indemonstrable principles of explanation.

> generated, as it is in artefacts such as a house or any other such thing. It is necessary that a certain sort of matter be present if there is to be a house or any other end, and this must come to be and be changed first, then that, and so on continuously up to the end and that for the sake of which each comes to be and is. It is the same way too with things that come to be by nature. [B] (639b21–30)

The criticism here is that "nearly everyone" overlooks a distinction between two ways something can be said to be "necessary." Some things are necessary "absolutely" or "unqualifiedly" (*haplōs*) and others are[19] necessary "from an assumption" (*ex hupotheseōs*).[20] Because of their failure to recognize this distinction, there is some mistake those he is criticizing are making.

The way it is typically read, Aristotle is criticizing his predecessors for their failure to appeal to the *right* kind of necessity. They ought to appeal to necessity *ex hupotheseōs*, but they do not. That reading presupposes that Aristotle's reference here to necessity *ex hupotheseōs* is to the type of necessity that Aristotle describes in passages such as the following:

> We call the necessary that without which, as a condition, a thing cannot live, e.g. breathing and food are necessary for an animal – for it is incapable of existing without these—and the conditions without which good cannot be or come to be, or without which we cannot get rid or be freed of evil, e.g. drinking the medicine is necessary in order that we may be cured of disease, and sailing to Aegina is necessary in order that we may get our money. (*Metaphysics* V.5,1015a20–26, Ross trans., modified)[21]

The type of necessity Aristotle describes here is usually called "hypothetical necessity"[22] or "conditional necessity" in the literature and is presumably what Aristotle has in mind when explaining phenomena that are necessary *for an end*.[23] This kind of necessity is, for this reason, sometimes even described as a "special kind of necessity operative in teleological

[19] Or "are also." See previous note.

[20] The apparent implication of this is that *only* necessity *haplōs* is applicable to what is eternal. I read this as restricted to eternal *natural* phenomena. On the interpretation I am going to propose, both types of necessity can apply to what is eternal but non-natural. For instance, mathematical theorems are necessary "from an assumption," while mathematical axioms are necessary without qualification. However, Aristotle argues in *GC* II.11 that eternal natural phenomena, such as the cyclical rotation of celestial bodies, occur of necessity *haplōs*. (I will briefly return to this point in Section 4.3.) I thank James Allen for urging me to clarify this.

[21] See also *PA* I.1, 642a7–8 and 642a32–4.

[22] Note that Aristotle does not use the phrase "*ex hupotheseōs*" when he describes this type of necessity.

[23] That Aristotle thinks ends necessitate what they are assumed to explain seems clear, but how that is supposed to work is notoriously difficult. See Stein 2016 for an illuminating discussion of these issues.

explanation."[24] According to a reading of [B], which assumes this is what necessity "*ex hupotheseōs*" refers to, Aristotle's complaint here is that his predecessors failed to recognize that there is a type of necessity associated with teleology that differs from the unqualified necessity applicable in other domains. As a result of their failure to recognize this other type of necessity, they use the wrong kind of necessity: they should be appealing to necessity *for an end* but do not. This amounts to the charge that his predecessors ignored the cause "for the sake of which" and are being faulted for that neglect.

Although Aristotle does think that their failure to recognize that ends are causes in nature is a mistake, it seems his point is different here. For, as I will argue, necessity *ex hupotheseōs* does not refer to a special type of "teleological" necessity, that is, being necessary *for an end.* Consequently, the objection in [B] – that people overlook a distinction between ways "necessary" is said – is not to their failure to appeal to that teleological necessity. Accordingly, the bad result of their overlooking the distinction between ways "necessary" is said is not that they use the wrong kind (or that fail to use the *right* kind) of necessity. Instead, I will propose an alternative understanding of the distinction that Aristotle is referring to in [B]. On my proposal, the criticism of his predecessors is that their explanations are not grounded in starting points that are not themselves subject to further explanation. This construal of the distinction and the reading it yields of [B] thus connects this part of the passage with section [A]: "nearly everyone's" explanations of generated natural phenomena that appeal to "necessity" alone fail to meet the demands on explanation that Aristotle has just claimed that final cause explanations *do* meet.

Before offering this alternative understanding of the phrase "necessity *ex hupotheseōs*," I want to emphasize that I am not denying that Aristotle thinks being necessary for an end is a way of being necessary. Although there are many questions about why being needed for the sake of an end should render something *necessary*, it seems clear that Aristotle thinks it does. The proposal I am making is merely about what Aristotle is referring to when he qualifies necessity with the phrase "on an assumption" (*ex hupotheseōs*) and about how that way of being necessary differs from being necessary "without qualification" (*haplōs*). (On the construal I will suggest, what is necessary for an end is rather a *subset* of what is necessary *ex hupotheseōs*.)

[24] Johnson 2005: 58

Instead of seeing these phrases as contrasting a teleological, "conditional," or "hypothetical" necessity with a non-teleological, "unconditional" necessity, I propose we understand them to be contrasting the two ways of being necessary that Aristotle distinguishes in *Metaphysics* V.5, several lines after describing being necessary for the sake of an end:

> Of some things the cause of being necessary is something different; of other things something different does not cause their necessity, but other things are necessary because of these things. (1015b9–11)

The difference expressed here has to do with *why* something is necessary. Some things, including generated phenomena, owe their being necessary to something else. These are necessary from an assumption (*ex hupotheseōs*), or as I will call it, "dependently" necessary. Geometrical theorems are a prime example of such dependent necessities: they owe their necessity to those definitions and axioms from which they follow. Other things, such as axioms and definitions, do not owe their necessity to anything else. These are simply (*haplōs*) necessary, or as I will call it, "independently" necessary.[25] On this view, the difference between the two ways of being necessary is not one of degree, and dependent necessities are no less necessary than independent necessities. Just as theorems of mathematics are necessary, though dependent for their being necessary on the self-grounded necessity of the axioms they are derived from, dependent necessities, in general, are dependent on independent necessities. The status of something's necessity is neither impugned by its being so "on an assumption" nor does it signal any demotion with respect to its being "incapable of being otherwise."

While a thorough defense of this construal of the distinction will require separate treatment, I will note here some of the reasons to doubt that necessity *ex hupotheseōs* is necessity for an end. There is nothing about the grammar of the phrase "*ex hupotheseōs*" that requires that we understand it to mean "for an end." On the contrary, as Jacob Rosen has argued, it is doubtful that the phrase *could* mean that.[26] More typically, Aristotle's use of the adverbial expressions "*ex hupotheseōs*" and "*haplōs*" tend to signify how something possesses a property and has nothing to do goals or ends. In *Politics* III.5, 1278a3–6, for instance, children are said to possess their

[25] Stein 2016 refers to these as "derived" and "underived" necessities, and (in my view) correctly claims that being necessary for an end is a type of derived necessity. However, Stein does not suggest, as I do here, that we take this contrast to line up with that alluded to in the six passages where necessity *ex hupotheseōs* is explicitly mentioned.

[26] See Rosen 2008.

citizenship *ex hupotheseōs*, unlike adult men who are citizens *haplōs*. Moreover, relative to the frequency of the expression "hypothetical necessity" in the literature about Aristotle's teleology, there are extremely few instances of *ex hupotheseōs* used to qualify "necessity" in Aristotle's extant corpus – only *six*, in fact.[27] And although he gives numerous explanations of phenomena that are necessary for an end in *PA*, Aristotle never uses the phrase *ex hupotheseōs* to qualify "necessity" after this first chapter.

Nevertheless, we might think that necessity *ex hupotheseōs* must be closely associated with teleology for Aristotle, since it appears in contexts where he is criticizing those predecessors who did not, as he did, appeal to ends in their explanations of natural phenomena. Moreover, the examples of dependent necessities that he gives are also often things necessary for ends. For instance, later in *PA* I.1 (642a2–13), Aristotle says that nourishment is necessary *ex hupotheseōs*, since it is not possible to live without it, using the same example as he does *in Metaphysics* V.5. This would appear on the surface to support the idea that necessity *ex hupotheseōs* is indeed necessity for an end.

However, we need not take this as conclusive evidence of that standard interpretation. For, on my proposed construal, it is natural to see why dependent necessity is related to Aristotle's belief in natural teleology: in the cases where something is necessary for an end, the ends are the upon which the necessity depends. So, it comes as no surprise that we find the majority of Aristotle's explicit references to a distinction between *haplōs* and *ex hupotheseōs* necessity in *Physics* II.9 and *PA* I.1. However, as I think, Aristotle's purposes in referring to this distinction are not the ones they are typically taken to be. For the same reason, it is not surprising that things necessary for ends are given as examples of things said to be necessary *ex hupotheseōs*, since these are obvious and salient instances of dependent necessities.

In fact, the passage (at 642a2–13) in which we find the second and only other occurrence in all of *PA* of the phrase "necessity *ex hupotheseōs*" is unclear in several respects, making it hard to determine the purpose for which it appears there at all. Here, we are told that there are two causes, the "for the sake of which" and "of necessity." (Given that contrast, "of necessity" here is unlikely to be necessity for an end.) Aristotle then turns

[27] The contrast between necessity *ex hupotheseōs* and *haplōs* occurs twice in *PA* I.1, twice in *Phys* II.9, once in *de Somno*, and once in *GC* II.11. Not all occurrences are universally considered to be referring to "hypothetical" necessity. Lennox ("Are Aristotelian Species Eternal?" in Lennox 2001b: 138), for instance, denies that this is what is meant in *GC* II.11.

to a "puzzle" that someone might raise (*an tis aporēseie*): what kind of necessity do they mean when they say "of necessity"? However, it is not clear whether "they" refers to Aristotle's own explanations, to Plato's discussion in the *Timaeus*, to Democritus, or someone else. Whoever's appeals to necessity he has in mind, Aristotle says it cannot be one of those ways of being necessary[28] from the "philosophical discussions" (*en tois kata philosophian*), though it is not clear whether he is referring to discussions found in *Posterior Analytics*, *Metaphysics*, or *Physics*, to "scientific treatises in general" (Louis 1957), or to something else, perhaps a lost treatise, *On Philosophy*. Aristotle next points out that in generated things, there is "the third" kind of necessity, which is then illustrated by the example (nourishment) that was also used in the *Metaphysics* V.5 discussion of necessity. It is at the point that Aristotle adds that "this" (*touto*) is "*hōsper*" a necessity *ex hupotheseōs*, though it is not certain whether "this" refers back only to the last example (nourishment) or to the third kind of necessity being introduced in this passage. (As Ogle [*ad loc*] rightly notes, read in the latter way, it is "strange" that he is introducing this as if it were something new, given that Aristotle has already spoken about this, in our passage [B].) Whatever "this" is, is Aristotle articulating what "necessity *ex hupotheseōs*" means or instead claiming that nourishment is an *instance* of it? That is, does *hōsper* indicate (as I think the Greek suggests) that something is an "example of" or "just like" a necessity *ex hupotheseōs* or that it "may be called" (Ogle) or is, "as it were" (Balme, Lennox) one? The passage deserves more scrutiny, since this is one of our key pieces of evidence, but it is far from certain that Aristotle means for this to be an "exposition of his theory" of necessity *ex hupotheseōs*.[29]

Further, whereas it would seem mysterious, as Rosen points out, why Aristotle would use the expression "*ex hupotheseōs*" to mean "for an end," I think there is a plausible explanation for Aristotle's use of it to indicate that something's status is "dependent" in the way I am suggesting. Although we most often call a proposition a "hypothesis" in English to indicate that it stands in need of proof or is itself something to be tested, that is neither the case for uses of *hupothesis* in Greek, generally, nor true,

[28] If one of the rejected kinds of necessity includes "material" necessity, Aristotle's statement here seems flatly contradicted by his practices throughout the remainder of *PA* (as discussed by Cooper 1987: 266ff and Lennox 2001: 150). This is yet another puzzling feature of this passage, which Balme 1992: 100 described as "one of Aristotle's clearest statements that all natural necessity is reducible to teleological necessity," a claim he later retracted (Balme 1987: 285) while nevertheless maintaining that the view is "suggested" (Balme 1987: 285 n. 33) by the passage.

[29] As Cooper 1987: 267 describes it.

in particular, of Aristotle's uses of it. For instance, he refers to the starting points of reasoning (*archai*) in theoretical sciences as *hypotheses* (*EE* II.11, 1227b29). Such starting points of science are not considered provisional or in need of confirmation.[30] In many cases, an *hupothesis* seems to refer to something "set down" in order to examine or establish something else, as for instance in the Greek mathematicians' practice of reasoning "from a hypothesis." Much about this practice is disputed, but we know that Plato was familiar with it, and Aristotle refers explicitly to procedures (such as "analysis") with which this practice was associated.[31] As Plato seems to use the phrase, a proposition is an *hupothesis* not because it is held only tentatively but because it stands in relation to some other proposition – one said to be "from a hypothesis" – as its grounds or support.[32] It is plausible that when Aristotle uses the qualifying phrase *ex hupotheseôs*, there is an expectation that it be heard as conveying this notion of being "dependent upon" something else.

This construal of necessity *ex hupotheseōs* and *haplōs* as dependent and independent necessity is also supported by *Physics* II.9, where we find Aristotle's most extended discussion of *ex hupotheseōs* necessity. Most of the literature on that chapter has focused on his examples, such as walls, saws, and houses, which are the products of goal-directed artificial processes. However, far less attention has been paid to the mathematical examples used in the second part of *Physics* II.9. These examples are mentioned just after Aristotle rejects the prevalent view that natural outcomes come to be as a result of the matter's natural propensities to behave in certain ways. That view, he says, is as absurd as believing that an artifact, for example, a wall, comes to be of necessity, in virtue of the fact that heavy things such as bricks and stones naturally move downward and light things such as wooden beams naturally go up. The comment Aristotle then makes, whose significance is not often registered, is that necessity in mathematics and in what is naturally generated is "in a certain way similar"

[30] In *NE* VII.8, 1151a17, *hai hupotheseis* in mathematics, just as the *to hou heneka* in actions, are the starting points (*archai*). And again, *hai hupotheseis* of *apodeixeis* are called the *archai* "from which the things (*pragma*) are first known" (*Metaph* V.1, 1013a14–16).

[31] For discussions see Robinson 1941; Huffman 1993a; Menn 2002; Morrison 2015; and Malink 2017.

[32] Robinson (chapters VII–X) provides a good entry into the variety of uses of "hypothesis" and the "hypothetical method" in Plato, arguing that Plato "regarded an hypothesis as a proposition posited in order to prove something else and not a proposition posited in order to be itself tested" (1941: 112). That idea is supported by the meaning of the verb *hupotithemai* from which it is derived, which according to Robinson is "to posit as a start, as a preparation for future activity, for something else" (112). One might also treat the proposition posited – the *hypothesis* – as the proposition to be tested, but that is not necessary.

(*tropon tina paraplēsiōs*) (200a16). Clearly, this does not mean that in both cases necessity is for an end, since Aristotle does not think that there are ends or goals in mathematics. So, what does this mean?

The similarity Aristotle goes on to describe has to do with a certain asymmetry exhibited in both domains. In natural and artificial generation, the ends are not necessary because of the presence of the matter and its changes, rather the reverse is true: the matter's presence and its changes are necessary on the assumption that certain ends will come to be. In other words, the matter is *dependently* necessary. That is true, moreover, despite the fact that *if* the matter and its changes were not present, those ends would not come to be. No saw or house would come to be, if not for the presence of iron and stones (200a28–29). Similarly, in mathematics, geometrical definitions and axioms are not necessary because of the truth of the theorems that follow from them, rather the reverse is the case: theorems are necessary on the assumption that the definitions and axioms from which they follow are true. Aristotle illustrates this point with the theorem stating that the internal angles of a triangle sum to "two rights." That theorem – 2R – owes its being necessary to the geometrical principles – the hypotheses, definitions, and axioms – from which it is derived, for example, that parallel lines never meet.

> For, since the straight is *this*, it is necessary that the angles of a triangle should equal two rights. (200a16–18)

As with other geometrical theorems, 2R is *dependently* necessary, for it is necessary on the assumption that the axioms and definitions hold.[33] That is true, moreover, despite the fact that those definitions, axioms, and hypotheses would not be true if the theorems derived from them were not also true (200a29–30).

On the construal of the distinction I am proposing, necessity is "in a certain way similar" in this respect: in mathematics, the *archai* make what follows from them necessary, though the *archai* are not made necessary by the conclusion; in generation, the ends make the matter and its changes necessary, but are not made necessary by the matter and its changes. On this reading, all mathematical theorems are necessary *ex hupotheseōs*,

[33] Although Aristotle does not call 2R "necessary *ex hupotheseōs*" in *Physics* II.9, in *de Caelo* I.12, 2R is given as an example of something impossible *ex hupotheseōs*: "It is impossible for a triangle to have two rights, if such and such (*ei tade*)" (281b5–6). Since, for Aristotle, "impossible" is equivalent to "necessarily not" (*APr.* I.13, 32a22–4; *De Int* 13, 22a19–20, 22b5–6), one might take this to suggest that theorems are necessary (or necessarily not) *ex hupotheseōs*.

but this does not impugn the necessity of those theorems.[34] It rather indicates a way in which theorems are distinct from the indemonstrable definitions and axioms on which they depend. And this is a point of similarity between mathematics and the subject of investigation by the natural scientist.[35]

This interpretation of the distinction requires more argument than I am giving here, but I hope to have shown that it is at least possible. It is also, I think, preferable. This interpretation provides a reading of [B] that is attractive in at least the following three respects: First, according to this interpretation, when Aristotle complains here that theorists do not appreciate that necessity is "from an assumption," the emphasis is not on *what* that assumption is that makes something necessary – ends or goals – but on the fact that *there is* an assumption that makes it necessary. Unlike the standard reading, this renders Aristotle's criticism in [B] compelling even to someone who does not believe that there are ends in nature, since he is not simply saying that those accounts have not cited the goal that the other things are only necessary *for*. That would not be heard as an objection at all by someone who does not think there are goals. Rather, Aristotle is merely assuming here that explanations – if they are to yield knowledge – cannot go on *ad infinitum*.

Second, if we take this to be the intended contrast between the ways of being necessary referred to in [B], Aristotle's first criticism is building on his preceding remarks. He just argued (in section [A]) that teleological explanations are naturally first because they proceed from self-explanatory starting points. Implicit there was the thought that explanation, if it is to yield understanding, must begin with something that is self-explanatory, something not subject to further explanation. Aristotle is now, in [B], criticizing those who treat all necessity as if it were alike. In particular, such theorists are faulted for overlooking the difference between independent and dependent necessity. Dependent necessities are not self-explanatory, since there is some *other* reason why they are necessary. Further, much of what occurs "of necessity" in generation – both of natural and of artificial products – does so only on an assumption. For example, foundations are laid of necessity but only on the assumption that a house will be constructed. So, since the "necessities" these theorists appeal to are dependent,

34 I have not found any evidence that Aristotle thinks all mathematical truths are necessary *haplōs*. He does say that all *haplōs* necessities are eternal (*NE* VI.2, 1139b23–24), though that is compatible with there being eternal truths (such as mathematical theorems) that are *not* necessary *haplōs*.

35 There is, however, an important respect in which things are "reversed" in mathematics and natural science, as I will mention in Section 4.3.

those who attempt to "refer their accounts back" to necessity wrongly posit as an *archē* something that is not an *archē*. What is necessary only dependently is not an adequate starting point for explanation, and thus their attempts at explanation cannot yield understanding.

Third, this reading of the complaint – that something that is not really an *archē* is being treated as if it were an *archē* – is one that Aristotle also makes elsewhere. On this proposal, Aristotle's claim in *PA* that his predecessors do not recognize that naturally generated phenomena are necessary *ex hupotheseōs* amounts to the charge that they begin their accounts from something that is not really an *archē*: Since there is some other cause for those putative starting points of explanation being "necessary," they are not indemonstrable or self-explanatory. In *GA* II.6, Aristotle criticizes people such as Democritus, who treat the fact that things are "always" a certain way as if it were an *archē* of explanation. However, as Aristotle points out, many things that are "always" a certain way are not thereby self-explanatory, but in fact are demonstrable. For instance, although geometrical theorems are always true, one can explain why they hold by showing that they follow from definitions and axioms:

> Yet on this line of argument, on the strength of which they undertake to dispense with trying to discover the cause, there will be no demonstration of any single one of the eternal things. It is obvious, however, that demonstrations of many of these (some of them things which always come to be, some things which always are) do exist. For instance, the angles of a triangle are always equal to two right angles, and the diagonal of a square is always incommensurate with the side; in both of these cases we have something eternal, yet there is a cause for them and they are demonstrable. (*GA* II.6 742b23–29, Peck trans.)

He also raises this criticism of Democritus' account of the existence of motion:

> And, in general, to think that this is a sufficient principle (*archē*), that something always is or comes to be in this way, is an erroneous assumption. It is to this that Democritus reduces natural explanation, saying that this is how things happened in the past also. He does not think it is requisite to seek a principle of what is always the case. It is right to say this for some cases, but wrong to say it for all. For instance, a triangle always has its interior angles equal to 180 degrees; but all the same there is some further cause of its being everlasting. (*Physics* VIII.1, 252a32–b4, Graham trans.)

On the alternative proposed, Aristotle is making the same criticism in *PA*. Just as saying that something is "always" the case is not thereby to provide an *archē*, neither is saying something is "necessary." For many natural phenomena, that which is necessary is so dependently, that is, it is dependent on some assumption that makes it necessary. In such cases, one will not have offered an explanation that begins from an *archē* without citing that assumption. In Aristotle's view, one does this by beginning ultimately from items such as definitions, as he claims that teleological explanations do. That is why teleological explanations are naturally to be given first.

4.3 The *tropos* of Demonstration and Necessity Is Different: [C]

As I have proposed we read [B], Aristotle criticized his predecessors' attempts to explain natural generation on the grounds that explanations, if they are to yield understanding, must not go on indefinitely. Those explanations that appeal only to necessity do not proceed from indemonstrable starting points, at least not when the necessity cited is "on an assumption." Next, in section [C], Aristotle raises a second criticism of accounts that appeal solely to necessity, again from the point of view of demands on explanation. In short, *explananda* are expected to be entailed by the *explanantia*,[36] and those explanations being criticized fail to fulfill this expectation. This section begins with a claim whose meaning is far from obvious:

> But[37] the manner (*tropos*) of demonstration and of necessity is different in the case of both the natural (*phusikê*)[38] and the theoretical sciences (*tōn theoretikōn epistemōn*). [C](i) (639b30–640a2)

[36] As with the assumption that explanations must begin from self-explanatory starting points, there are stronger and weaker readings of this idea, depending on how much of Aristotle's own formal theory of explanation one thinks is being imported here. In *Posterior Analytics*, the idea that the *explanandum* is necessitated by the *explanans* is exemplified by the fact that a demonstration is a type of syllogism, in which the conclusions are entailed by the premises. But Aristotle also treats as widely shared the idea that understanding involves knowing why something "is not possible to be otherwise" (*APo* I.2, 71b9–12). For instance, he takes for granted that explanations ought to entail what they purport to explain when he complains that theorists such as Democritus "do not state the necessity of the 'why' (*dia ti*)" (*GA* II.6, 742b18).

[37] It is unclear why this sentence begins with an adversative connective (*alla*). In line with the interpretation I will suggest, it could be intended to correct a misconception one may have formed, given Aristotle's immediately previous description of the continuous (*ephexēs*) causal sequence of matter's necessary changes. The thought would be that although it must be the case (*dei*) that various changes take place in a certain order (i.e. that something occurs first, then something else, and so on until the end is brought about), one should not mistakenly take this to imply that the direction of causation and explanation goes from what happens earlier to what happens later.

[38] The unexpressed noun being modified by *phusikē* might be *apodeixis* or *anangkē* instead of *epistēmē*. As far as I can tell, it would make no difference for my interpretation.

Interpreters of these lines disagree about nearly every point of detail, as well as its overall argumentative purpose and its relation to the surrounding context. There is general agreement that it concerns the "manner" or "mode" of explanation, and most interpreters have understood Aristotle's claim that this is "different" to mean that there are different models of explanation applicable to different domains of inquiry. But there is neither any consensus about what those domains are, for example, whether he is distinguishing natural science from a "theoretical" science such as mathematics, nor about what those models are or how they differ from one another.[39] Since natural science is elsewhere classified as a type of theoretical inquiry (e.g. *Metaphysics* 1025b24ff), some scholars think it is unlikely that Aristotle here is contrasting the two. Moreover, although this difference in "manner" of demonstration is assumed to be related to the comments about necessity made in [B], it is neither easy to see how a division between two kinds of inquiries aligns with the distinction between ways of being necessary (on anyone's construal of that distinction), nor clear how such a division coheres with the examples ("health" and "human") given in the lines that follow.

Rather than taking Aristotle to be contrasting the manner of explanation in natural science with the manner of explanation in theoretical sciences, it is possible to understand him to be contrasting these both with *something else*. In my view, that third thing from which the manner of "demonstration and necessity" in natural science and theoretical inquiries differs is what those predecessors mistakenly take it to be. That is, Aristotle is not contrasting different but equally acceptable types of explanations, but rather he is contrasting the *correct* ways of explaining natural generation with the *incorrect* attempts to do so. As I read [C], Aristotle is continuing the critique of certain non-teleological explanations offered by his predecessors that he began in [B]. Implicit here in [C] is the thought that successful explanations must necessitate what they explain, and those theorists who attempt to offer explanations that appeal to antecedent states of matter fail, for that reason, to give good explanations. For, in Aristotle's view, inferences cannot proceed in the temporal direction in which these other people think that they can.

Aristotle immediately goes on to remark, in [C](ii), that he has already discussed this point about the manner of "demonstration and necessity."

39 Some options can be found in Ogle 1912; Düring 1980; Pellegrin 1982; Grene 1985; Gotthelf 1987b; Charles 1988; Balme 1992; Lloyd 1996; Lennox 2001a, 2010d; Johnson 2005; Kullmann 2007; Leunissen 2010b.

I take him to be referring to his discussions in *Posterior Analytics* II.12[40] and *Generation and Corruption* II.11. In these chapters, Aristotle denies that what is temporally prior can entail what is temporally posterior.[41] In *APo* II.12, Aristotle argues that while there can be deductions from what is posterior (*apo tou husterou*), there cannot be deductions from what is prior (*apo tou proterou*): One cannot say that since this so and so (earlier) has come to be, this (later) so and so has come to be (*APo* 95a27–31).[42] In *GC* II.11, he brings up this idea in connection with a question concerning whether anything that comes to be (later) is made necessary by something earlier. Although both chapters are difficult, it is clear that Aristotle is preoccupied in these discussions with the illegitimacy of inferences from the temporally prior to posterior. Consequently, those discussions in *APo* and *GC* are relevant to Aristotle's purposes here in *PA*, where he is rejecting the idea that temporally antecedent material conditions can explain what occurs later in time.[43]

However, the relevance of the discussions in those chapters to Aristotle's treatment of methodological norms in *PA* I.1 is rarely acknowledged.

[40] Leunissen 2010b: 50–52 also takes this as a reference to *APo* II.12, though for different reasons than mine. In Leunissen's view, Aristotle is concerned in *APo* II.12 to extend his theory of demonstrations (a theory that treats mathematics as its paradigm) to explanations of changes, i.e., to explanations whose relata are not simultaneous, but occur in a temporal sequence. This extension is then relevant in *PA* 1.1, where Aristotle is establishing various methodological procedures for inquiry into the natural world, especially those inquiries about subjects that are generated. In Leunissen's view, the *APo* II.12 discussion provides the "blueprint" for further modifications to the theory of demonstrations that are needed for explaining such naturally generated subjects. (Although I see the relevance of *APo* II.12 differently, her reading provides a compelling response to Lloyd's charge that the theory as presented in *APo* leaves us unprepared for a "variety of modes of demonstration" (Lloyd 1996: 32) of the sort that many take Aristotle to be introducing in this passage.)

[41] Whether Aristotle is consistent on this point is disputed. For instance, Williams 1982: 200 claims that although this view is expressed here in *PA*, as well as in *APo* II.12, 95a27–31 and *Phys.* II.9, 199b34–200b8, Aristotle "seems to envisage" the opposite in *Metaph.* E.3, 1027a29–b16. According to Williams, that opposite view is "not without support from other passages in the corpus" (200), though he gives no reference to any passage.

[42] The reasoning there is not easy, and it is not clear how the point about inferences is connected to claims about causality. Kelsey (ms.), e.g., reads Aristotle as making a claim about causality (*viz.*, that what is in the past does not cause what is in the future) *by way of* a claim about deductions. Alternatively, in Leunissen's view, Aristotle countenances causation of later by earlier, but denies there are corresponding valid inferences from those (earlier) causes to their (later) effects. Such inferences are not valid, on her interpretation, because "there will always be a moment when the cause has occurred but not yet its effect, such that at that intermediate time inferences from cause to effect will be false" (Leunissen 2010b: 43).

[43] On this point, I disagree with Leunissen 2010b, who takes Aristotle to be allowing in *PA* I.1 that explanations of eternal, non-teleological phenomena can begin from temporally prior starting points. The lesson from *APo* II.12 being applied in *PA* I.1 does not seem to me to be that inferences from earlier to later are acceptable *in certain cases* but rather that they *never* are.

Although it is widely observed that the issue under consideration has something to do with the temporal order of the explanatory relata, the fact that he is rejecting altogether the legitimacy of inferences that go from earlier to later is not. According to some interpreters, for instance, the cross-reference at 640a2–3 is to *Physics* II.9, where Aristotle claims that things are "reversed" (200a19) in mathematics and natural science.[44] In that *Physics* passage, Aristotle refers to the changes that lead up to the end (and which are derived from it) as "things before" (*ta emprosthen*). This might suggest that the discussion there has to do with the temporal relation between *explanans* and *explanandum*. But there are no temporal relations in mathematics, and on closer examination, the fact that the conclusions in natural science are the "earlier things" is not relevant to the point he is making in *Physics* II.9. There, the sense in which natural science and mathematics are "reversed" is that in the former, the matter is the conclusion, and in the latter, the matter is the starting point. That is, in explanations in natural science, the matter is the conclusion derived from the *telos*,[45] whereas in mathematics, the definitions, axioms, and hypotheses are the "matter" of the geometrical theorems they entail.[46] But that difference between mathematics and natural science is not in view in *PA*. Rather, [C] is reflecting Aristotle's concerns about some particular issues that arise for explanations of nonsimultaneous happenings, such as natural generation.

> For the starting point in some cases is *what is*, and in other cases is *what will be*. For, since health or human *is* such, it is necessary that this such and such *is* or *comes to be*. But it is not the case that since this such and such *is* or *has come to be*, that such and such from necessity *is* or *will be*. [C](iii) (640a3–640a6)

Here, in *PA*, he points out that some explanations might begin from what *is* the case and proceed to legitimately infer that something else *is* the case, and some might begin from what *will be* in the future and proceed to legitimately infer that something else *has* or *will have* come to be (earlier).[47] In no case, however, can explanations of natural generation begin from what is or has come to be earlier and proceed to infer what will

[44] E.g. Balme 1972/92: 84–85; Cooper 1987: 269 n. 29, 1987b: 197.

[45] As already mentioned, Aristotle makes it explicit here (at 200a24) that the *telos* is the *archê* of the reasoning (*logismos*) in natural science.

[46] Aristotle says this explicitly in *Phys* II.3, 195a16–19. See Malink 2017 for illumination of this claim.

[47] I read the *tois men*/*tois de* clause as indefinite, rather than as corresponding to either the two *epistemai* just mentioned or to their corresponding *apodeixeis*. (In either case, as noted by Lloyd 1996: 31, one would expect feminine articles *tei* or *tais* instead of the neuter *tois*). These describe

be later.[48] That is, Aristotle is denying that inferences can proceed from the prior to the posterior, and registers this now as his second complaint: these theorists who cite the "necessity" of the matter and its changes to bring about certain effects get the temporal direction of inference all wrong. Aristotle is reminding us here that antecedent conditions do not necessitate later ones, as he discussed in *APo* and *GC*.

In fact, in [C] Aristotle immediately makes a second reference to *APo* II.12 and *GC* II.11, now because those passages also address certain exceptions to the rule that Aristotle has just been criticizing others for violating:

> Nor is it possible to connect the necessity in such a demonstration into eternity, with the result that one says, "Since *this* is, therefore *that* is." (These matters too have been determined elsewhere – in what sorts of things necessity is present, what sort of necessity converts, and owing to what cause.) [C](iv) (640a6–9)

On my interpretation, Aristotle has just censured explanations that proceed from what is temporally prior on the grounds that they do not entail what they purport to explain. For, as discussed in *APo* and *GC*, one cannot legitimately reason that something is or will be of necessity, given that something else is or has come to be. However, there are certain exceptional cases – namely, eternally recurring, cyclical processes – that are also discussed in those same chapters of *APo* and *GC* and to which he is alluding in [C]. In *GC*, sequences of happenings that occur eternally and cyclically, such as winter and summer solstices, are conceived of as importantly different from noncyclical processes, such as natural generation, that occur "in a straight line" (*GC* II.11, 338b11). In *APo*, Aristotle appears to envision explanations of these cyclical phenomena as taking the form of circular proofs; even if the necessity with which any member of a cyclical sequence occurs is dependent on something else occurring later, the necessity can be "connected to eternity" since it will "come round in a circle" (*APo* II.12, 96a6). Explanations of cyclical and eternal phenomena are exceptional in that they can involve, in a certain way, inferences from the prior to the posterior. In fact, in these specific cases, inferences can be viewed as proceeding in both directions, that is, from later to earlier as well as the reverse. But, as

two different but equally legitimate ways that explanations can proceed: in some cases (*tois men*), the starting point is "what is," and in some cases (*tois de*) the starting point is "what will be."

48 This would imply that Aristotle's own explanatory appeals to material and efficient causes should not be understood as inferences from what is earlier to what is later. Although beyond the scope of this chapter, that is a welcome result, in my view.

I understand Aristotle's reasoning, that is because any stage that is "later" than another stage in an eternal cycle is also going to be "earlier" than it, and so there is "conversion." Since, in demonstrations of cyclical phenomena, what is "earlier" is also "later" and vice versa, there is thus a sense in which it is possible for the later to be deduced from the earlier.

So, there are some cases, namely cyclical and eternal phenomena, in which an appeal to what is earlier are permissible. However – and this is the point he makes in 640a6–9 – natural generation is not among those cases. In explanations of natural generation, the earlier and later stages do not "convert," as they might in explanations of eternal and cyclical phenomena. So, as Aristotle argues here, those who try to account for generation by appealing to antecedent states of matter have not succeeded in giving explanations of the sort required for understanding.

4.4 Conclusion

This chapter advanced a reading of 639b11–640a9 that treats the passage as a unit, in which Aristotle is pursuing a single aim and is proceeding with a particular preoccupation that fits the local context. This passage begins by arguing that teleological explanations ought to be given first. As I understand his argument, it is that teleological explanations are to be given first because they proceed from self-explanatory starting points, namely, the definitions of the phenomena being explained. Consequently, teleological explanations serve to ground the whole body of causal explanations that ultimately constitute each branch of natural science. In addition, in this passage Aristotle identifies two specific epistemological defects of certain non-teleological explanations: First, those who appeal only to "necessity" are being charged with offering explanations that are not properly grounded. They appeal to "necessity" as if that were an adequate *archē* of explanation, that is, as if it were self-explanatory. But what is "necessary" is not always self-explanatory. Many cases of necessity, especially as it applies to naturally generated phenomena, are only necessary "on an assumption." Second, Aristotle thinks that an explanation should show why what is being explained cannot be otherwise: an *explanandum* ought to be entailed by the *explanans*. But those accounts of natural generation that appeal to material necessity alone are defective in this respect as well. For, these accounts cite antecedent conditions, such as the presence of a certain type of matter, as reasons for the occurrence of natural phenomena. However, antecedent conditions do not necessitate later ones and thus do not explain them.

On my reading, we can see criticisms of Aristotle's predecessors that are not often visible. Aristotle is not merely faulting others for not doing what *he* does. He is not criticizing them because they fail to give teleological explanations but instead arguing that their attempts to explain natural phenomena do not satisfy some rudimentary conditions on explanation and understanding. Understood in this light, the passage at 639b11–640a9 provides us with valuable insights into why Aristotle insists upon the indispensability of teleological explanations. In this passage, he is driven by concerns about the possibility of applying the ideals for scientific understanding to a world in which there is contingency and in which things do not occur simultaneously but unfold over time.[49]

[49] I want to acknowledge my gratitude to the editor of this volume, Sophia Connell, and to the other volume contributors in attendance at the workshop in Cambridge. I am grateful for valuable feedback from participants at Stephen Menn's McGill University seminar, especially Andrea Falcon, as well as audiences at MUSAPH, Beijing Normal (especially my commentator, Marilu Papandreou), and UC Berkeley. Many thanks to James Allen, Taylor Barinka, Ian Blecher, Kathleen Cook, Joseph Karbowski, Sean Kelsey, James Lennox, Marko Malink, Thomas Marré, Samuel Meister, Jessica Moss, and Nathaniel Stein for generously reading and discussing earlier versions.

CHAPTER 5

Hylomorphism in the De Partibus Animalium

Michail Peramatzis

Introduction

In Aristotle's hylomorphism, objects are compounds consisting of matter and form. For instance, a statue consists of its constituent bronze (the matter) and its perceptible shape (the form). In his view, particular living beings such as Socrates or Bucephalus are paradigmatic hylomorphic compounds in which form and matter are so intimately related as to constitute genuinely unified objects. Aristotle also thinks that there are universal compounds, species, or substance-kinds such as human or horse, which consist of matter and form "taken universally." Such items, however, are not substances (*Metaph.* Z.10, 1035b27–30). By contrast, particular compounds, form, and matter are substances of different types (Z.3, 1029a1–5; Z.10, 1035b11–27; H.1, 1042a24–31). Indeed, he argues that the form is identified with things' essence and so is their primary substance (Z.11, 1037a21–33). For, first, it makes particular compounds and matter what they are; second, it defines them; and, third, it explains their necessary features.

These three types of priority of the form over compounds and matter are not grounded simply in the idea that the form is a sort of shape, arrangement, or structure. Rather, in his metaphysical, physical, and, more importantly, biological treatises Aristotle contends that especially in the case of living beings the form is identified with the final cause of compounds and their matter and, more specifically, with their soul. In what follows, I shall assume without further discussion not only the idea that form and essence are the same but also the identification (in the case of living beings) of forms, formal causes, and formal features with final causes, functions, and telic features. Thus, the relation between Aristotle's hylomorphism and his teleology lies outside the scope of the present study.

Aristotle's application in *De Partibus Animalium* (*PA*) of hylomorphism to animals and their bodily parts raises several questions. The question I shall discuss is about the form of such living beings: what is essentially involved in such a form? I have argued elsewhere that there are three main approaches to this question:[1]

> [PURISM] A form is pure; it is essentially without matter; its definition does not mention any matter.[2]
> [COMPOUND-FIRST] Form and matter are not real-world, external, or mind-independent items but are abstracted in thought from the particular compound itself. The compound is a material (type of) object and is metaphysically basic.[3]
> [TRUE-GRIT] A form is essentially matter-involving; its own definition – not just that of the compound – mentions some appropriate sorts of material item. The latter are in some sense genuinely 'material'; they cannot be fully defined in terms of, or eliminated in favor of, formal, functional, or telic features.

I shall discuss some key passages of *PA* to examine what they reveal about Aristotle's view of the nature of form and its relation to the matter and the compound. More specifically:

1. Does [PURISM] or [TRUE-GRIT] describe more accurately Aristotle's view of form in *PA*?
2. Assuming that he subscribes to his standard view of form as ontologically, definitionally, and causally-explanatorily prior to the matter and the compound, how can the form of living compounds satisfy this priority requirement if *PA* favors [TRUE-GRIT]?

[1] Peramatzis 2015: 195–196.

[2] Frede and Patzig but also the so-called spiritualists favor [PURISM] (Frede and Patzig 1988; Frede 1990; Burnyeat 1992; Wedin 2000; Devereux 2011; for a reply to Devereux, see Code 2011). Interestingly, [PURISM] also seems to underlie the view of Aristotelian functionalists: they think that natural form is essentially defined simply in terms of functions, *telē*, potentialities for change, etc. Modern Aristotelian functionalists include Gill 1989, 1993, 2001, 2010; Lennox 2008; and Shields 2016. Some of these thinkers, however, also tend to subscribe to [COMPOUND-FIRST]: for they often take the compound as fundamental. In his view of substantial holism, Scaltsas 1994 also follows [COMPOUND-FIRST] as he conceives of the substantial form in actuality as being identical with the compound itself. By contrast, he thinks that the form is distinct from the compound only by thought abstraction. While Jennifer Whiting roughly follows the functionalist interpretation – e.g., by emphasizing the distinction between functional and non-functional matter – she takes the form as primary and as the principle of the compound's being and unity (Whiting 1986, 1991, 1992).

[3] In my view, [COMPOUND-FIRST] gets Aristotle's priority relations between form, matter, and compound the wrong way around. In what follows, I shall not discuss further [COMPOUND-FIRST].

3. If, by contrast, *PA* follows [PURISM], how can forms be the ontological, definitional, and causal-explanatory principles of *material* and *changeable* living compounds?

I shall argue that in *PA*, too, Aristotle subscribes to [TRUE-GRIT] and shall indicate (mainly in Sections 5.3–5.4) how this view makes better sense of certain aspects of his account in this treatise.

[TRUE-GRIT] faces a serious challenge, however, arising from the discussion of *PA* I.1. This chapter seems to argue for the following two theses. First, forms are prior to, more authoritative than, and indeed the final causes of matter and compounds. Second, matter is not essential to forms but is merely conditionally necessary for the existence of a form or telos in a complete compound: if the form or telos is to exist, necessarily it exists in a material compound. These two claims suggest, in line with [PURISM], that a form's defining account is "matter-less."

In response to these twin challenges, I shall argue (in Section 5.2) that Aristotle's theses about the priority of form over matter and about the matter's merely conditional necessity concern only what I shall call "cruder" types of matter. In his *Metaphysics*, Aristotle often draws a distinction not only between particular substance-compounds and universal compounds but also between *particular* material bits (or *token* materials) and *types* of matter. Thus, at *Metaph.* Z.10, 1035a4–17, he suggests that the form of (for instance) the morpheme (or 'syllable') BA – say what-it-is-to-be BA – essentially includes a quasi-material feature such as consisting of the phonemes (or "letters") B and A. He also contends, however, that the material items included in such a form are not particular phonemes enmattered in (for example) wax or air (*this* waxen or airen B or A). Nor are they the airen or waxen type of phonemes (*the* waxen or airen B or A).

As I have argued elsewhere,[4] this entails that token- or type-materials are not essential to a natural form. Such items are 'crude' types of matter, which are not mentioned in the definition of an essentially enmattered form. In the case of living beings, examples of such matter would be the particular material bits or bodily parts constituting the body of Socrates at any time or even excessively specific or narrow types of matter such as healthy muscles, blue eyes, or long bony fingers. By contrast, a natural form essentially contains, includes, or encodes *material features* or *ways of being material* such as *being made of flesh, bones, and tissues* or *having legs,*

[4] Peramatzis 2011: 39–137, especially 45–50.

feet, arms, and hands. It should be noted that this formulation and generally the [TRUE-GRIT] picture do not entail that (e.g.) the human form itself *has* flesh, bones, and tissues or hands, arms, legs, and feet. I use the jargon of "involving," "containing," "including," or "encoding" as opposed to "having," "exemplifying," or "instantiating" to invoke the relevant distinction made in modern discussions of (the content of) abstract objects, properties, and essences: an essence may *encode* certain features – perhaps by being partly identified with such features – but need not *exemplify* or *instantiate* them. For example, while the human form may encode the feature of having flesh, bones, hands, etc., it does not itself have flesh, bones, hands, etc.[5] Hence, certain material features can be, and indeed are, essential to the form. They are, as I shall label them, 'subtler' or 'more refined' than the inessential 'crude' types of matter just introduced. In my view of *PA* I.1, it is 'crude' matter that is merely conditionally necessary for the form's existence. Similarly, 'crude' matter is not essential to the form but is posterior to it. In Section 5.2, I shall argue, however, that *PA* I.1 is compatible with [TRUE-GRIT] as it allows that natural forms include in their very essence "subtler" ways of being material.

The distinction between 'crude' and 'subtle' matters is not an un-Aristotelian artificial subtlety. Nor is it a distinction that Aristotle invokes in his theoretical discussions but does not deploy in his scientific treatises. It is analogous to the distinction he draws in *De Generatione Animalium* (*GA*) V.1 between essential animal features and features that (if anything) are necessary merely by crude material and/or efficient causal necessity:

> **[T0]** We must now study the affections in respect of which the parts of animals differ. I mean such affections of the parts as the following: blue and dark color of the eyes, high and deep pitch of the voice, and differences of color and of hair or feathers .... About these [affections] and all others like them, we must not anymore suppose that the same sort of cause applies. For among those features which do not belong to nature in common or are not peculiar to this or that kind of animal none of these either is or comes-to-be such-and-such for the sake of anything. Thus, while an eye [is and comes-to-be] for the sake of something, it [is and comes-to-be] blue not for the sake of anything, unless this affection is peculiar to a certain kind. Nor, in some cases, do such affections contribute anything to the account of the animal's being but since they come-to-be by necessity we must refer their causes back to the matter and the moving principle .... Each thing, then, is for the sake of something, while as regards their coming-into-being all those features which inhere in the account, or are for the sake of something, or are

[5] For further discussion, see Zalta 2006.

that for the sake of which, these come-to-be on account of this cause as well as the remaining causes. But for those features which are not such, and of as many of them as there is coming-into-being, we must already look for their cause in their change and their coming-into-being since they acquire their differences in the very process of constitution. Thus, for example, an animal will of necessity have an eye (for it is posited to be such-and-such an animal), but it will have a particular type of eye of necessity, though in a different way, i.e. because it is by nature such as to act or be acted upon in this or that way. (*GA* V.1, 778a16–20; 29–b1; b10–19; Peck's translation adapted)

After introducing the attributes (778a16: *pathēmatōn*; 33–34: *pathos*) to be discussed using the examples of specific eye-color and particular types of hair or feather, Aristotle distinguishes between (i) features that are common to the natures of many animal-kinds or characteristic of just a specific animal-kind and (ii) features that do not fall under (i) and are not for the sake of anything (778a30–32). As an example of (i), he offers the feature of having eyes, to be contrasted with a case of (ii) having eyes of a specific type, say blue (32–34). He notes that, in at least some cases, type-(ii) features do not "tend towards" or "contribute to" an animal-kind's essential account (778a34–5: *oute... pros ton logon sunteinei ton tēs ousias*). The vague phrase "tending towards" or "contributing to" is neutral as between essential features and those which, though not essential, are conditionally necessary for an animal-kind and so may 'contribute to' its essence/form in the sense that they obtain *because of it* or that, *if the essence/form is to exist, necessarily they obtain.* Later in [To], however, Aristotle clarifies that type-(i) features, those which are and come-to-be for the sake of a telos or are themselves the telos, are parts of the essential account (778b12–13: *hosaper en tō(i) logō(i) enuparkhei tō(i) hekastou ē estin heneka tou ē hou heneka*). At 778b16–19, he provides the same example as that offered at 778a32–34: having eyes, a feature that embeds a bodily organ – a clear case of a material item. This sort of material feature – an analogue of my "subtle" matter – is encoded in the essence/form and mentioned in the definitional account. By contrast, having blue eyes – an instance of my 'crude' matter – is a not an essential (nor a conditionally necessary) feature, though it comes to belong to certain members of a given animal-kind by an absolute (not conditional) sort of material and/or efficient causal necessity (b13–16; cf. 778a34–b1).

In Section 5.3, on the basis of some important passages taken from *PA* II–IV, I shall seek to bolster [TRUE-GRIT]. I shall argue for the strong position that in these books Aristotle takes natural forms (partly) to encode "subtle" material features. I shall also offer evidence for the further idea that the material and formal features encoded in a form are essentially indissoluble: for they cannot be defined independently of each other. It should be stressed, however, that these two points do not eradicate the claim of form to being metaphysically, definitionally, or explanatorily prior to matter. Rather, as noted, matter-involving forms can be, and indeed are, prior in these ways to "cruder" types of matter. Section 5.4 will conclude by raising some questions about my overall view of Aristotle's hylomorphism in *PA*.

5.1 The Challenges of *PA* I.1

[TRUE-GRIT] faces two serious challenges, both of which grow out of the discussion of *PA* I.1:

1. While formal features are thought to be encoded in a natural form, matter is deemed merely conditionally necessary for (but not essential to) a form's or a compound's existence (639b27–640a1; 640a5–7). This idea is sometimes also framed as the claim that the matter is not part of the form or is not mentioned in the form's definition.
2. Natural form is generally considered to be prior to (more authoritative than, that because of which, the final cause of, more/rather a principle than, etc.) matter (640a16–19; 641a29–33; 642a7–18). If so, matter cannot be encoded in the form: for it is metaphysically determined by, defined in terms of, or explained by reference to, the form but not conversely.

My aim in the present section is to defuse these twin challenges. I shall argue that the passages from *PA* I.1 that give rise to (1) and (2) can be read as implying that the sorts of matter that are merely conditionally necessary (but not essential) and so are posterior to but not encoded in the form are the "cruder" types of matter (Section 5.1). It should be emphasized that my present argument is fairly moderate. It suggests just the following two points: (i) *PA* I.1 is compatible with [TRUE-GRIT]; and (ii) *PA* I.1 also contains some evidence for the distinction between 'cruder' and 'subtler' types of matter. In Section 5.3, I shall offer stronger positive evidence for [TRUE-GRIT].

In his comparison of artifacts and their form with natural forms such as what-it-is-to-be a human being, Aristotle remarks:

> **[T1]** The art is the account of the product without the matter (*hē de technē logos tou ergou ho aneu tēs hulēs estin*) . . .. Hence it would be best to say that, since this is what it is to be a human being (*tout' ēn to anthrōpō(i) einai*), on

> account of this it has these things (*tauta*); for it cannot be without these parts (*aneu moriōn toutōn*). If one cannot say this, one should say the next best thing, i.e. either that in general it cannot be otherwise, or that at least it is good thus. And these things follow. And since it is such (*toiouton*), its generation necessarily happens in this way and is such as it is (*hōdi kai toiautēn*). This is why this part (*tōn moriōn tode*) comes to be first, then that one (*tode*). And in like manner one should speak in precisely this way about all of the things constituted by nature. (640a32–b4)[6]

The first sentence of **[T1]** poses a serious difficulty for [TRUE-GRIT]. The comparison with art suggests that just as (the definition of) the art or the form of an artifact is the account of its product, goal, or function without the matter, so too (the definition of) a natural form is the account of the relevant natural function (e.g. seeing for eyesight and eyes) without the matter. But what sort of matter is to be discarded? In the case of the house-form, for instance, the definition correctly grasped in the housebuilder's mind does not of course literally contain in it these bricks, stones, etc. Nor does it mention such particular material bits. Moreover, this sort of definition need not, indeed should not, mention any specific types of matter such as bricks or stones generally, for the house-form may be realized in different types of material. Similarly, in the natural case, what is merely conditionally necessary (but not essential) are the particular, "thises," token, or type materials (*tauta, aneu moriōn toutōn, tōn moriōn tode, tode*). To be sure, a form or telos being essentially thus-and-so (*toiouton*) – for instance, what-it-is-to-be a human (*tout' ēn to anthrōpō(i) einai*) – determines the way a process of coming-to-be unfolds, the type of process it is, and presumably the nature of the incomplete, matter-like entities at earlier stages of the process (hence the *toiautēn* used of *genesis* at 640b2). Yet the material items that are conditionally necessary are particular tokens or types as suggested by the use of the more specific deictic phrases at 640a34 (*taut'*), 35 (*toutōn*), and b3–4 (*tode... eita tode*) in contrast with the more generic, qualitative place-holder 'such as it is' or 'such-and-such' (*toiouton*) to pick out the form and its content.[7] **[T1]** is still compatible with [TRUE-GRIT].

It is important, at this point, to compare briefly the claims about the matter's conditional necessity and about the form's teleological causal

6 All translations of *PA* are taken from Lennox 2001a with minor changes.

7 For similar uses of deictic phrases to pick out narrowly specific token- or type-matters, see 639b27–640a1; 640a5–7; 640a16–19. The only exceptions are found at 639b27, where matter is described as 'such,' and at 640b2, where coming-to-be (not matter) is said to obtain not only 'in this here way' (*hōdi*) but also 'such as it is' (*toiautēn*).

priority made in **[T1]** and elsewhere with Aristotle's account of conditional necessity in *Physics* II.9. There too he insists that while artifactual or natural ends cannot exist without certain types of matter, yet it is because, and for the sake, of the form/telos being as it is that matter has its characteristic nature (200a5–11). For instance, it is because or for the sake of cutting or dividing in a certain way (the form/telos of a saw) that the saw's matter is iron or some other appropriate type of metal. The converse does not hold good (200a11–13; 24–27; cf. *PA* I.1, 642a8–14). Toward the end of that chapter, though, Aristotle contends that certain types of matter are within the definitional account of the form or telos (and not just of the compound; 200b4–5). Thus, he argues that the very function of a saw, identified with its form, say such-and-such a division, cannot be defined properly without mentioning material items such as having a serrated blade or being made of appropriate metallic materials (200b5–7). Indeed, he closes the chapter by holding that certain material items are mentioned in the form's definitional account (200b7–8). This surprising conclusion comes immediately after he reaffirms the claim that the form/telos is the (final) cause of matter but not conversely (200a33–34). Further, the form/telos is the definitional principle referred to in a proper account (200a34–35). Matter, by contrast, is only conditionally necessary for the existence of the form/telos in a complete compound (200b1–4). These claims are equivalent to the ideas outlined in **[T1]** about the matter's conditional necessity and the form's (teleological) causal priority. If they are compatible with the thesis sketched at the end of *Physics* II.9 that the form's definition mentions certain material items, Aristotle ought to draw the distinction I am suggesting between 'cruder' tokens/types of matter and 'subtler' material features. While the former are only conditionally necessary and so are not encoded in the essence, the latter are integral to a natural form's content.[8]

The discussion of *PA* I.1 continues at 640b4ff. with a criticism of Aristotle's predecessors, who were crude materialists. This criticism need not imply that offering definitions that mention matter (or indeed its efficient cause and its motions; 640b7–8; 13–16; cf. *Phys.* II.9, 200a30–32) is as such a misguided approach to natural science. Rather, the mistake is to think that matter by itself, without any reference to the form/telos, can undergird the whole explanatory task. This idea is confirmed in the following passage:

[8] For further discussion, see Peramatzis 2011: 133–134, 283–285; 2018: 24–25.

[T2] But if human beings, animals, and their parts are by nature, one should speak about flesh, bone, blood, and all the uniform parts. Likewise too, about the non-uniform parts such as face, hand, and foot, one should say in virtue of what each of them is such as it is, and in respect of what sort of potential (*hē(i) te toiouton hekaston estin autōn kai kata poian dunamin*). For it is not enough to say from what things they are constituted (*ou gar hikanon to ek toutōn estin*), e.g. from fire or earth. It is just as if we were speaking about a bed or any other such thing; we would attempt to define its form rather than its matter, e.g. the bronze or the wood (*epeirōmetha mallon an dihorizein to eidos autēs ē tēn hulēn, hoion ton chalkon ē to xulon*). And if we could not do this, we would at least attempt to define the matter of the composite (*ei de mē, tēn ge tou sunholou*); for a bed is a 'this-in-that' or 'this-such', so that we would have to mention its shape as well, and what its visible character is (*klinē gar tode en tō(i)de ē tode toionde, hōste kan peri tou schēmatos eiē lekteon, kai poion tēn idean*). For the nature in respect of shape is more important than the material nature (*hē gar kata tēn morphēn phusis kuriōtera tēs hulikēs phuseōs*). (640b18–29)

Against the crude materialists, Aristotle maintains that living beings such as humans, as well as their parts, are by nature. Presumably, this entails that they possess not only material but also formal natures (*Physics* II.1). Hence, he suggests that their definitional accounts should also mention their homogeneous and non-homogeneous bodily parts. However, such accounts should always describe in virtue of what features and capacities bodily parts of this sort are as they are. That is, they should characterize them by reference to their very essence or form. To provide such characterizations it is not sufficient to specify merely their material constituents or even the material origins of their coming-to-be. It does not follow from this, however, that a scientific definition ought to disregard matter altogether. Rather, it ought to consider it together with, and as being because or for the sake of, the form.

In his discussion of the example of a bed, Aristotle contends that we ought to specify the form rather (or more?) than the matter. The way he develops this example, however, suggests that he ostracizes from definitional accounts not matter *tout-court* but highly specific type- and token-matters: the (type) bronze or wood; this (token) bronze or wood (see the use of the articles at 640b25: *ton* ... *to* ...). It is still left open that material features at an appropriately higher level of generality are encoded in the form. More importantly, at 640b26 he holds that if we must characterize the matter too, we ought to describe the compound's fully enformed matter (*tēn ge tou sunholou* [*sc. hulēn*]). That it is necessary to include matter in our definitions will become clearer from some of the scientific accounts given in *PA* II–IV; see Section 5.3). The claim made at 640b26 indicates that a definition may, indeed should, include a certain sort of matter, that which is determinate and functional, the one constituting the complete compound. Indeed, at 640b26–28, Aristotle argues that the bed's essence or form is a compound: this (form) in this (matter; *tode en tō(i)de*) or this (matter) being such-and-such, that is, enformed (*tode toionde*). That this claim applies not only to the compound bed but also, and primarily, to the bed's form becomes clear from the use of the very same example a few lines later, where Aristotle speaks of the bed's form (641a17–18: *tou eidous tēs klinēs*).

There is a strong affinity between this point and the argument developed in *Metaphysics* Z.11 in Aristotle's criticism of Socrates the Younger. In my reading of this controversial passage (Peramatzis 2011: 56–59; 91–99; 312–316), Aristotle rejects Socrates the Younger's model in which the human form is defined without mentioning any material features or parts. Instead, he argues that there is a definitional dependence of such a form – a form of a natural substance-kind – upon being perceptible, being changeable, and having bodily parts in a certain condition (1036b24–30). Thus, a bodily part such as a hand is essentially characterized by reference to its ability to perform its appropriate function and so is essentially in a certain state (1036b30–32). In fact, as is clear from other passages, Aristotle also defends the converse dependence of form on material items. This is why he introduces his criticism of Socrates the Younger by insisting that forms of natural substance-kinds (as well as these substance-kinds themselves) are essentially 'this-in-this' or 'these-items-in-this-state' (1036b23–24: *tod' en tō(i)d' estin ē hōdi tadi echonta*), expressions that are virtually identical with the characterization of the bed-form in **[T2]**. These twin phrases describe natural forms and their bearers as certain forms in certain matter or as materials enformed in a certain way. Further, in both **[T2]** and *Metaph.* Z.11, the alternative formulations, with their indifference as to the order in which form and matter are picked out, suggest not only that both formal and material features are encoded in the form but also that they are neither prior to nor independent of each other. Elsewhere, and with the help of further passages, I have argued that such material and formal features are essentially interdependent.[9]

Aristotle's injunctions in **[T2]**, then, are, first, that our definitions ought to include not just matter (for matter by itself is indeterminate and explanatorily insufficient; 640b23: *ou gar hikanon to ek tinōn estin*)[10]

[9] Peramatzis 2011: 110–132; 2013–2014; 2015: 210–13. [10] See Lennox 2001a: 125.

and, second, that formal and material items encoded in a form are not metaphysically, definitionally, or explanatorily prior to each other. At the same time, however, this formal-cum-material defining essence is 'more important,' 'authoritative,' or 'dominant' than (*kuriōtera*), and so prior to, matter. In my view, this implies that an essentially matter-involving form is still metaphysically, definitionally, and explanatorily prior to 'cruder' types of matter, such as this (token) or the (type) bronze or wood. For, without any reference to the form, which determines the nature of the compound's matter (*tēn tou sunholou hulēn*), such 'cruder' types of matter by themselves are indeterminate.[11]

5.2 Essentially Matter-Involving Form in *PA* II–IV

In Section 5.2, I compared some central passages of *PA* I.1 with the discussion of conditional necessity in *Physics* II.9. I pointed out that in the closing section of *Physics* II.9 there is an apparent tension between the following two claims:

1. Matter is only conditionally necessary for the existence of the form/telos. It is not an essential part of the latter. The form is metaphysically, definitionally, and causally-explanatorily prior to matter (as the matter's final cause).
2. Matter is encoded in the form/telos and is mentioned in its essence-defining account.

If (2) is the case, the form cannot be prior to matter in the ways specified in (1). Conversely, if (1) obtains, matter cannot be encoded in the form, as required in (2). I suggested that to mitigate this tension Aristotle ought to distinguish between different types of matter. 'Cruder' matters are the sorts of matter excluded from scientific definitions in (1). By contrast, 'subtler' matters are the material items mentioned in scientific definitions, as (2) suggests. In Section 5.2, I argued that even *PA* I.1 allows for this distinction and so is not incompatible with [TRUE-GRIT]. In the present section, I shall seek to show that *PA* II–IV supports the stronger claim that there are specific, 'subtler' types of matter that are included in the forms of certain living beings. The obvious way to bring *PA* I.1 in line with II–IV to avoid the apparent clash between (1) and (2), then, is to think that Aristotle distinguishes (implicitly or explicitly) between different sorts

[11] Similar claims about matter are made in *Metaph.* Z.10, 1036a8–9; Z.11, 1037a24–7; cf. *Meteor.* IV.12, 389b29–30; 390a2–b2. For further discussion, see Peramatzis 2018.

of matter in the manner just outlined. Indeed, his considered view is (or so I shall argue) that forms of living beings are not only essentially matter-involving but also prior to "cruder" matters.

There are two groups of passages occurring in *PA* II–IV that could support the matter's basic role in natural science. In the first group, we encounter passages in which matter is invoked as a fundamental causal item on the basis of which certain important features of an animal are explained.[12] Hence, in his discussion of blood in II.4, Aristotle suggests that the fineness and purity of blood is responsible for the subtler intelligence or higher-level cognition of some animal-kinds (e.g. bees and ants): for example, their sharper, more mobile, or swifter sense perception (650b20–28). Animals whose blood is too watery, however, tend to be more cowardly (b28–29). By contrast, animals such as bulls and boars that have blood of a rather earthy constitution (for instance, blood with many fibers in it) are excessively spirited (b33–651a5). Aristotle concludes that blood, a clearly material item, is the cause for the presence of important animal characteristics over and above bodily constitution, such as character or cognition (*PA* II.4, 651a12–15).[13]

It may be thought, however, that such explanations do not go far enough. For they describe matter as a basic causal-explanatory item in virtue of which certain other, less basic features obtain. But this is just to affirm the role of matter as one among Aristotle's four types of cause. Furthermore, it is to understand matter as conditionally necessary for the existence of the form/telos and as figuring in the relevant types of explanation: if the form/telos obtains, necessarily the matter obtains, but it is because of the form/telos being as it is, that the matter is such-and-such. Such explanations, however, do not support [TRUE-GRIT]. To vindicate [TRUE-GRIT] we have to turn to a second group of passages, in which Aristotle clearly holds that material features are encoded in the form and/or are mentioned in scientific definitions. Here I am indebted to Allan Gotthelf, who has collected and discussed the relevant sections of *PA* II–IV (Gotthelf 2012a, 221–236). I shall not, however, examine all the passages he studies. Moreover, I shall highlight some details that may have gone unnoticed but are crucial for [TRUE-GRIT].

[12] Lennox ([1996b] 2001b: 183, 186–187, 196, 202 n. 11) and Gotthelf (1985: 219 n. 4; 1987b: 161–162, 179; 2012: 187–1878) argue convincingly that in Aristotle's biological treatises matter seems to be not only part of the essence of animals but also causally and explanatorily fundamental.

[13] Gotthelf [1987b] 2012: 161–162. Similar points are made in the study of eye-protectors (*PA* II.13, 657a31, 657b5ff., 657b30ff.) and horns (*PA* III.2, 663b13–15; 663b22–25).

In his discussion of flesh, Aristotle remarks:

> **[T3]** For flesh is in virtue of itself a principle and a body of animals. This is clear even from our account (*kata ton logon*); for we define (*horizometha*) animal by the possession of perception (*tō(i) echein aisthēsin*), and primary in this account is the primary mode of perception. This is touch, and it is of touch that such a part is the sense-receptor – either the primary one, just as the pupil is of vision, or it taken with the medium, as if one were to add the entire diaphanous medium to the pupil. (II.8, 653b21–27)

That having flesh is an important part of the essence is suggested by the use of *horizometha*: it is not any old account (*logos*) but an essence-defining account that mentions this sort of material feature (Gotthelf [1985] 2012a: 222). But what is the definiendum: the whole compound or the form? It would not be controversial to think that the defining account of natural compounds mentions matter: for, after all, they are essentially enmattered objects and so their definition ought to refer to matter. This would not support [TRUE-GRIT]. There are two textual points outside the context of **[T3]**, however, which imply that the definiendum is not just the compound but also, and primarily, its form. First, in *PA* I.1 Aristotle employs the example of a bed twice, at 640b23ff (my **[T2]**) and 641a17–18. While in the first case he writes that *the bed* is essentially a 'this-in-this' or a 'this-such-and-such,' in the second he clarifies that he equally speaks of *the bed-form* (*peri tou eidous tēs klinēs*). Second, in *Metaphysics* Z.11, especially the criticism of Socrates the Younger, he formulates his claim in a general manner – *some things* are 'this-in-this' or 'these-in-this-state' – and offers the example of a human or a specific type of animal (1036b23–32). While these formulations may suggest that he refers to the compound alone, the broader context of Z.10–11 clearly entails that his question is primarily about the form (Z.10, 1034b20–24, b4–12; Z.11, 1036a26–9, b2–5). More generally, it would be odd if Aristotle held this sort of separatist thesis, in which the compound's definition mentions matter, whereas the form's definition does not. For a form just is the compound's essence or at least its basic essence. If so, how could the compound's fundamental essence fail to be matter-involving, while at the same time the compound is, allegedly, *essentially* enmattered? I shall return to this question in Section 5.4.

Aristotle's argument in **[T3]**, however, has a further similarity to *Metaphysics* Z.11. As I pointed out in Section 5.2, *Metaphysics* Z.11 invokes a chain of definitional dependencies in which a specific type of animal-form, say a human-form, cannot be defined without mentioning changeability or bodily parts in a certain condition. Similarly, in **[T3]** an animal-form is defined in terms of (inter alia) having perception (after all, this is what distinguishes animals from plants). More specifically, having the primary type of perception, touch, and ultimately having the relevant perceptual organ (or quasi-organ), flesh, seem to be essential features encoded in an animal-form. It is important to highlight the formulation of the features encoded in the animal-form: Aristotle uses the phrases 'having (primary) perception' (653b23–24: *tō̆(i) echein aisthēsin*; cf. 653b20–21: *tois echousi sarkas*). A matter-involving form, such as the human-form, does not itself *have* touch or flesh; it is particular compound humans that have touch and flesh. Rather, such a form partly encodes features such as *having touch/to have touch* and *having flesh/to have flesh*. Similarly, its definition mentions such items. A matter-involving form is not an object or a compound that instantiates or exemplifies features but is a complex essence that encodes or is identified with features.

A similar cautious formulation is used in the discussion of lung-possessors (III.6, 669b8–12). The very essence/form of lung-possessors or, perhaps more plausibly, of specific (e.g. human) lung-possessors contains having a lung (*huparkhei to pleumona echein*).[14] Clearly, this does not entail that the form, being a specific lung-possessor, itself has lungs; rather, this form encodes an essential material feature such as *having lungs/being lunged*.[15] Moreover, in his discussion of some bloodless animals in IV.5, Aristotle does not argue simply that bloodless animals are constituted from materials other than blood or that the viscera of blooded animals are made of blood. Rather, while the nature of blooded animals' viscera (partly?) consists in (being made of) blood, the essence of bloodless animals (partly?) consists in (having) material constituents other than blood (678a32–35). This is clear from the closing sentence of this passage, where he specifies that he is dealing with the essence-defining account of certain animals (*en tō̆(i) logō̆(i) enhuparchei tō̆(i) horizonti tēn ousian autōn*). Furthermore, the phrase deployed at 678a33–34, "because some such affection ... is of their substance" (*dia to tēs ousias autōn einai ti toiouton pathos*), is singularly revealing: it emphatically suggests that a feature, such

[14] That Aristotle is focusing on the essence/form and not just the compound is clear from the comparison with the bird-kind, which has a name, 'bird': for he speaks of being a bird (669b11–12: *to ornithi einai*), the kind's essence/form, not this (particular compound) bird or the (kind) bird.

[15] In the present case, the relevant common kind is unnamed. It may even be argued that it does not exist as such but only in its specific manifestations: human lung-possessors, bovine lung-possessors, etc. Yet having lungs seems to be an essential part of the form of each of those specific kinds. Thus, the phrase "in the essence of those" (*ekeinōn en tē̆(i) ousia(i)*) can be read distributively: "For each specific lung-possessing kind, having lungs or being lunged is part of its specific essence or form."

as being blooded or being bloodless, is a constitutive part of the relevant animals' essence. By no means does this formulation entail that the essence/form has, or does not have, blood in the manner in which blooded or bloodless compounds do. Rather, being blooded or being bloodless, the corresponding material feature, is encoded in the form. Similarly, at 678a33–35, Aristotle holds that what is part of the form is *that* certain animals are blooded or bloodless – their essential feature of being blooded or being bloodless. All such formulations are to be strongly contrasted with the articulate nouns "the/this wood/bronze . . ." or other similar expressions used in **[T1–2]** and elsewhere (639b27–640a1, 640a5–7, 640a16–19), presumably to pick out 'crude' token- or type-matters.

Aristotle's discussion of birds gives rise to an important clue about the relation between material and formal features encoded in a form:

> **[T4]** They are two-footed of necessity (*dipoun de ex anankēs estin*); for the substance of the bird is that of the blooded animals, but at the same time that of the winged animals (*tōn gar enhaimōn hē tou ornithos ousia, hama de kai pterugōtos*),[16] and blooded animals do not move by more than four points. Accordingly, the attached parts are four –as in the other locomotive land-dwellers, so too in the birds. But four arms and legs are present in the one group, while in the birds, instead of forelimbs or arms, wings are the common feature (*pterugion koinon estin*);[17] and in virtue of these they are able to stretch out, and the ability to fly is in the substance of the bird (*kata tautas gar tonikoi*[18] *eisi, tō(i) d' ornithi en tē(i) ousia(i) to ptētikon einai*). So it remains for them to be, of necessity, two-footed (*ex anankēs diposin einai*). (PA IV.12, 693b6–13)

Birds' essence is thought to include the clearly material features of being blooded and having wings. Moreover, at 693b13–14, it is added that a formal or function-related feature such as being capable of flying is part of their essence. The main question concerns the relation between having wings and being capable of flying: are these two distinct essential features, are they simply different ways to latch onto the very same feature, or are they essentially interdependent? The claim made at 693b11–12 suggests that having wings is that 'by which' birds are able to stretch out. It seems

[16] There is a textual variant at 693b6–8: instead of the *tōn gar enhaimōn hē tou ornithos ousia* found in most MSS, Y reads *tōn men gar enhaimōn tēn ousian ho ornis*. While there is no need to adopt the latter reading, Y's text, especially the nominative *ho ornis*, seems to agree better with the rest of the clause: *hama de kai pterugōtos*. Moreover, the phrase *tēn ousian ho ornis* suggests even more emphatically that birds *essentially* or *in their essence* are blooded and winged.

[17] At 693b12 most MSS have *pterugion koinon estin*, as opposed to *pteruges to idion estin*.

[18] Some commentators conjecture *ptētikoi* instead of *tonikoi* at 693b12 but I see no compelling reason to emend the text.

plausible to think that the ability to stretch out is a feature to be explained by having wings (as the prepositional phrase *kata tautas* at 693b12 suggests). But, at 693b13, in the very same clause, and after the occurrence of the explanandum being capable of stretching out, Aristotle also invokes being capable of flying as part of birds' essence. Presumably, being capable of flying is also a basic explanans for the same explanandum, being capable of stretching out. This strongly suggests that having wings and being capable of flying constitute equivalent parts of the same explanans-essence. It is difficult to ascertain whether the corresponding descriptions are different representations of one and the same essential feature. However, the fact that being winged is more akin to matter, while being capable of flying is cognate with a formal/telic feature suggests that there is some difference between them. The least demanding reading would be that they are mutually incomplete and essentially interdependent. Thus, it would be difficult to see how birds' ability to fly could be essentially characterized without the material feature of having wings. Conversely, too, having ("organic" or "instrumental," i.e. properly functional) wings is essentially inseparable from being able to fly.

A similar argument can be advanced even if we think that being capable of stretching out is not an explanandum feature but is a part of birds' essence that is intimately interwoven with being winged. In this reading, "by which" (693b12: *kata tautas*) would indicate that the ability to stretch out is part of birds' essence alongside having wings. But if so, how could this ability be defined without reference to having wings? Being able to stretch out immediately raises the question "stretch out what?," the answer to which seems to require (in the case of birds) the presence of wings. My conclusion is that **[T4]** suggests not only that material features such as being blooded and being winged are essential parts of birds' form. It also implies that formal/telic features such as being capable of flying (or even being capable to stretch out, if indeed this is an essential feature included in birds' form) and material features such as being winged are inextricably interdependent.

There is an additional important claim made in **[T4]**. Both in the beginning and at the end of the passage Aristotle emphasizes that being biped is a necessary feature of birds (693b5–6; 14: *ex anankēs*). Moreover, this necessary feature seems to be an explanandum that is grounded in birds' essential features such as being blooded, having wings, and being capable of flying.[19] **[T4]** clearly indicates that such necessary explananda

[19] Depending on how we read 693b11–13, being capable of stretching out may be a further necessary explanandum feature of a similar sort.

are to be explained on the basis of not only formal/telic features, such as being capable of flying, but also material features such as having wings and being blooded. If being biped and other such features are merely necessary explananda, they could be understood as matter-like items, analogous to the 'cruder' matter, which is merely conditionally necessary and posterior to the form but is not encoded in it. That such features may be matter-like is clearly suggested by the example of being biped, which embeds bodily parts such as feet or legs.

Finally, and related to **[T4]**, in his explanation of why fish do not have feet but rather have fins Aristotle remarks:

> **[T5]** The fish do not have distinct limbs, owing to the fact that the nature of fish, according to the account of their substance, is to be able to swim (*dia to neustikēn einai tēn phusin autōn kata ton tēs ousias logon*), and since nature makes nothing either superfluous or pointless. And since they are blooded in virtue of their substance (*enhaima esti kata tēn ousian*), it is on account of being swimmers that they have fins, and on account of not being land-dwellers that they do not have feet (*dia men to neustika einai pterugia echei, dia de to mē pezeuein ouk echei podas*); for the addition of feet is useful in relation to movement on land. And they are not able to have four fins and at the same time feet, nor any other such limb; for they are blooded. (*PA* IV.13, 695b17–25)

Fish's nature encodes not only a formal/telic feature such as being capable to swim but also a clearly material feature such as being blooded: for they are said to be blooded in respect of their essence (695b20: *enhaima esti kata tēn ousian*; cf. b25: *enhaima gar*). Moreover, both features seem to explain, in some sort of causal collaboration, the necessary feature of having fins: being swimmers and having blood grounds their being finned. Put negatively, their non-terrestrial nature – presumably, a feature that is implied by their being essentially swimmers – explains the lack of feet/legs. In this reading, having fins in the case of fish is analogous to birds' necessary feature of being biped discussed in **[T4]**. But this is not the only available or the most plausible reading. It may be doubted whether having the capacity to swim (in the case of fish) could be essentially specified without invoking the material feature of having fins. If this is correct, having fins would be analogous not to the merely necessary feature of being biped in **[T4]** but to birds' essential feature of being winged. This would make an intuitively and indeed morphologically strong parallel: being capable of swimming, a formal/telic feature, would be like being capable of flying, while having fins, a material feature, would be like having wings. Moreover, both being capable of swimming and having fins would be essential parts of fish's form.

If this is correct, however, why does having fins appear to be *outside* the scope of the essentialist explanatory clauses "owing to their (essential) ability to swim" (695b18–19: *dia to neustikēn einai tēn phusin autōn*; b20–1: *dia men to neustika einai*) or "because they are (essentially) blooded" (b20: *epei d' enhaima esti*)? Indeed, why does having fins appear to be explained by reference to such essential features if it is itself part of the very same essence? There are two possible lines of response to these twin questions. The first is deflationary. The explanandum nexus (or 'fact') *that fish have fins* is not itself an essentialist truth but is a merely necessary claim, conditional upon there being fish that realize their form/telos as blooded swimmers. What is part of this latter form/telos is rather the very material feature *having fins/being finned*. It is because fish's form is essentially *to be blooded-finned-swimmers*, that, if such a form exists in fully enformed fish, it is necessary *that they have fins*. This would be a robust (if not satisfactorily informative) way in which the essentially matter-involving form of fish would ground the conditionally necessary explanandum that they have fins.

The second type of response would deny that an explanans-essential feature such as being capable of swimming is more fundamental than the feature of having fins. Indeed, it would contend that part of Aristotle's aim in **[T5]** is to show that while material features such as being finned are essential, they cannot be specified without mentioning formal/telic features such as having the capacity to swim. But were he to thematize the question of how to characterize essentially the capacity of fish to swim, he would equally invoke material features such as having fins. In this second view, Aristotle conceives of formal/telic and material features as essentially interdependent and incomplete without reference to each other. The two strategies are not incompatible. Thus, it could be argued that not only is it impossible to define the items encoded in the form (being a blooded-finned-swimmer) independently of each other. It is also the case that such essentially interdependent items ground merely necessary features of the kind: for instance, the explanandum of (fish's) having fins.

5.3 Conclusion

Irrespective of the complications arising from **[T4–5]**, the passages examined from *PA* II–IV provide strong evidence in favor of [TRUE-GRIT]. The contrasting claim made in *PA* I.1, that matter is just conditionally necessary but not essential to natural forms, can be accommodated by being restricted to just 'cruder' matters, as I understood them in Sections 5.1. *PA* II–IV, by contrast, would require "subtler" matters as part of the form's essence. In effect, the distinction between 'cruder' and "subtler" matters would eradicate the tension between these two different parts of Aristotle's study of animal nature. Let me conclude by addressing four questions that could be raised about my view of Aristotelian hylomorphism in *PA*.

(1) The [TRUE-GRIT] reading of *PA* implies that not only the compound but also, and primarily, the form includes in its own essence material features. A weaker view would maintain that while the compound's definition mentions matter, the form's definition is understood in the manner of [PURISM]. This view could also alleviate the tension between the claim that matter is only conditionally necessary and the explanations provided in *PA* II–IV, where the form itself seems to involve matter. For, in this view, it is only the compound's essence that contains matter. By contrast, matter is conditionally necessary for the existence of the form but is not essentially involved in it.

There are two sorts of response to this view. The first is exegetical. I argued that the claims made in *PA* seem to be not only about the compound but also about the form. The example of a bed provided at *PA* I.1, 640b26–27 and 641a17–18, clearly shows that Aristotle takes not only the compound object (a bed) but also the bed's form as a 'this-in-this' or a 'this-such' in the manner of [TRUE-GRIT].[20] The second line of reply is philosophical. Is it plausible to separate so sharply the form from the compound's essence? Aristotle holds that the form just is the essence, or the fundamental essence, of the compound. The separatist view agrees that the compound *essentially*, or *in its basic essence*, is matter-involving. But how can this obtain if the compound's form, its very essence, is matter-less? Is the separatist contending that the material items mentioned in the compound's definition are merely contingent or accidental accretions to its basic essence, the form? But if so, why are such nonessential items mentioned in an essence-defining account? If, by contrast, the separatist holds that there are two sharply distinct essences, one impure of the compound and one pure of its form, how could such disparate essences be of one and the same (kind of) object, the compound?

Alternatively, the separatist might argue that the material items mentioned in the compound's definition are not essential but are still necessary. Hence, while a pure form constitutes what is strictly essential to the compound, the matter mentioned in its definition is just necessary. This gives rise to a further challenge to the separatist. A form is supposed to be

[20] For further discussion, see Peramatzis 2011: 39–137.

the basic cause – in most cases the final cause – that grounds the compound's being. Not only does it explain why the compound has its characteristic necessary features, it also undergirds the necessity of such features. Let us take up necessity. How can a pure form, one in which there is no place for material features, ground the necessity of the compound's material features: for instance, birds' *necessarily* being biped or fish's *necessarily* having fins? Moreover, how can a pure form entail specific features *with overtly material content*, such as those just mentioned, *without itself encoding any material content at all*? It would be a miracle of almost cosmic dimensions if a completely immaterial form entailed any determinate material features. It would be even more miraculous if such a pure form also grounded the necessity of gritty features of this sort. By contrast, [TRUE-GRIT] seems to make good sense of passages such as **[T3]** in which the form of animal-kinds involves not just a sanitized version of a formal feature such as having touch but rather *having-touch-in-the-flesh-involving-manner*.

(2) In [TRUE-GRIT], do forms turn out to be material objects or particular compounds such as Socrates or Bucephalus? And, if so, given that compounds have such forms in them as their essences, would this imply that Socrates and Bucephalus have (for example) double their weight? Further, to take advantage of the example of Goliath and Lumpl offered in modern discussions of identity and temporary intrinsics, would [TRUE-GRIT] entail that there are two distinct but spatiotemporally coincident objects, Socrates and his human-form or Bucephalus and his equine-form?[21] Would it also imply an intolerable causal overdetermination, in which when (for instance) Socrates is being ironic toward me, there are two objects that are responsible for my feeling that I am being the subject of Socratic irony?

[TRUE-GRIT] is not vulnerable to such paradoxical implications. The distinction between 'cruder' and 'subtler' matters suggests that a form's essence does not contain any particular material bits, bodily parts, or token materials. Only such 'cruder' matters constitute particular material compounds, have mass or weight of their own, and causally interact with other material objects. But such matter is not encoded in the form. Nor are any inessential, overly specific types of matter included in the form. Rather, only 'subtler' matters are essentially involved in a form. The first key point about such types of matter concerns logical grammar and ontology. The

[21] For a recent discussion, see Koslicki 2018: 104–124; for a criticism of Koslicki's view, see Peramatzis 2019.

material items encoded in a form are not token- or type-matters such as, respectively, this particular bronze or wood or the type bronze or wood. Rather, they are material features or, perhaps better, ways of being material that are essential to *how* or *why* the compound is as it is: for example, being fleshy, having hands, having wings, etc. As I pointed out, such items are not had, instantiated, or exemplified by the form in the way in which a particular animal has flesh, a human has hands, or a bird has wings.[22] Rather, material features are encoded in the form: for instance, the human-form's content partly consists in having hands. The second question about 'subtler' matter is how a scientific definition of an animal-form ought to characterize it. This is not a question that can be settled by mere armchair speculation. It requires actual scientific investigation and discovery. Intuitively, while having hairy, freckled, and pale flesh is an inessential feature, having flesh seems to be an essential material feature for humans.

(3) Is this sort of characterization of the matter encoded in a form too rarefied or abstract to pick out properly physical matter? After all, physical matter is just the sort of matter found in ordinary material objects, whereas material features are not such objects, as my discussion of question (2) just suggested. Moreover, is there a danger that this refined essential matter is just the modern functionalist's proximate or functional matter?[23] If so, does 'subtler' matter end up being fully reducible to purely formal and telic features, without any involvement of 'bad,' constituent, or realizer matter?[24] If this were correct, [TRUE-GRIT] would collapse into functionalism.

In response, [TRUE-GRIT] ought to achieve a 'balancing act' between latching onto types of matter that are not excessively 'crude' and avoiding material features that are insufficiently robust.[25] The rebuttal to the charge that [TRUE-GRIT] collapses into functionalism is that the material

[22] Lennox 2001b, 153–154 argues for a claim that is congenial to my view: forms (of kinds) are not eternal, as they are neither individuals (or numerically one) nor particular objects that could exist eternally either "by persisting at any and every time or by being a member of a continuous series of individuals which are the same in form such that for each moment of time at least one member of that series exists at that moment."

[23] Irwin 1988: 245–247, 250–252; Gill 1989, 1993, 2001, 2010; Whiting 1992: 79–81, 84–87; see also her 1986 and 1991; Shields 2016: xxvi–xxviii. Nussbaum and Putnam 1992: 32 add that the matter must be in the right condition for the form. Whiting 1986, 372 also argues that functional matter is part of the form as it is defined exclusively in terms of the form.

[24] Nussbaum and Putnam 1992: 37, 39–40, 40–45, 55 use the language of "realizer-matter." Whiting 1992: 86–87 distinguishes between the 'thin' and the 'thick' material object: the former is a compound of form and functional matter, while the latter a compound of form and constitutive matter (at a time).

[25] See Lennox 2001a: 138; [1996b] 2001b: 183; Gotthelf [1985] 2012: 238–239.

features encoded in a form are appropriately robust or 'bulky' and so cannot be eliminated in favor of purely formal, functional, or telic features. They are characterized at the correct level of generality and so are not particular bodily bits or parts, mere disjunctions of narrowly specific material features, or inessential types of matter.[26] Nor are they abstract types of matter such as geometrical extension, purely functional matter such as "what is capable of perceiving" or items characterized in vacuous terms such as "material features appropriate/suitable for the relevant form." Thus, for example, having flesh is not just being able to touch but is a bona fide way of being material; the same seems to apply to features such as being winged and being finned.

(4) In [TRUE-GRIT], then, the form encodes not only formal, functional, or telic features but also robust or 'bulky' material features, such as those just mentioned. Moreover, such material features are not fully definable in formal, functional, or telic terms. Nor should formal, functional, or telic features encoded in the form be reducible to material features. For, if this were the case, Aristotle's hylomorphism would degenerate into a strand of materialism, in which the form could not serve its role as a primary substance and cause. In this picture, however, form proves to be a plurality of formal, functional, or telic features taken together with several material features. Indeed, not only does a form encode both formal and material features, it also includes *many* formal and *many* material features. But how can such a manifold of essential features constitute a unified essence? Moreover, how can such a highly complex essence be the basic unifier of the relevant substance-kind and its members?[27] Aristotle clearly holds that an essence or form, as the primary substance of things, is itself genuinely one. Moreover, he thinks that in virtue of this sort of unity a form unifies substance-kinds and particular compounds, thereby privileging them over

[26] For a different view, see Balme 1987: 304–305, who argues that an animal-kind's definition involves disjunctions of material characteristics that the relevant animal-kind might possess. He seems to be including items such as "either blue or green or brown eyes."

[27] Charles 2000: 333–335 raises a similar, unity-related worry about the biological works and their relation to Aristotle's view of scientific inquiry as set out in the *Posterior Analytics* and the *Metaphysics*. Moreover, he points out that several of Aristotle's biological explanations do not refer to a formal feature, such as (e.g.) being a swimmer in the case of fish, but to features of bodily composition. Nor do such explanations show why a certain type of bodily composition is necessary or good for survival, as Aristotle's teleology seems to require. [TRUE-GRIT] can overcome such worries: for, if formal, functional, and telic features are essentially and inextricably matter-involving, just as the whole form is, they cannot, by themselves, carry through the basic explanatory task. Rather, in invoking them one already introduces material features even if one abstracts away, or does not explicitly mention, the latter. In Aristotle's view, the essence/form was never purely formal, so it is unsurprising that essentialist explanations resort to material features too.

mere heaps and accidental compounds such as, for instance, a pale human (*Metaphysics* Z.12; Z.17; H.6).

In my view, such unity-related worries are not well-founded. For, as I argued in Section 5.3, the formal and material features encoded in a form are not independent conjuncts. Nor are they parts in any strong sense of this term. For, according to [TRUE-GRIT], they are essentially interdependent in an inextricable manner. To be a flier of the relevant sort is essentially to be winged (and perhaps also blooded) in a specific way. To be a swimmer, in the case of fish, is essentially to be finned in a particular way. How could we characterize essentially the formal or telic feature of being capable of swimming or being capable of flying without mentioning, respectively, having fins or having wings? How could we distinguish the relevant flying ability from the sort of flying that airplanes or even CGI birds are able to perform, without mentioning having (avian) wings? How could we define fish's ability to swim without mentioning having fins of a certain type if we wanted to differentiate it from the swimming capacity of a human or even a sperm? Conversely too, it seems difficult to define having this particular kind of wings without mentioning birds' ability to fly so as to rule out chicken wings or jet wings. Similarly, it would be difficult to specify what having fins essentially consists in without invoking fishes' capacity to swim, for otherwise how could we rule out diver's fins or sailboat fin keels? This sort of interdependence and non-decomposability suggests that the formal and material features encoded in the form are not a disparate plurality but are essentially unified. This is how [TRUE-GRIT] understands Aristotle's unity-related requirement that primary substance or form should "not be defined by one thing being said of an essentially different thing" (*Metaph.* Z.4, 1030a6–14; Z.6, 1031a28–31; b13–14; 1032a4–6; Z.11, 1037b3–4).[28]

The pressing question, however, is whether we can specify such interdependent formal and material features encoded in the form without referring back to the definiendum substance-kind. Can we characterize the relevant type of being a winged-flier or being a finned-swimmer without smuggling in being a bird or being a fish, respectively? If we cannot, we could not specify a substance-kind's essence or form non-circularly. While this is a crucial question in the foundations of Aristotelian hylomorphism and generally essentialism, I shall postpone it for a future study.

[28] For further discussion, see Peramatzis 2015: 210–213; 2011: 122–32. Charles 2000: 343–346, proposes a similar solution.

CHAPTER 6

Animal Species in Parts of Animals *Book I*

Devin Henry

Introduction

In this chapter, I argue that Aristotle was a realist about animal species.[1] My thesis consists of two related claims. First, I argue that indivisible species constitute the *ousiai* of Aristotle's zoology. Second, I show that the aim of Aristotelian zoological division is to identify and organize the features specified in the definition of those species. The latter (epistemological) claim is explicit in the discussion of division in *PA* I 2–3, while the former (ontological) claim is advanced in *PA* I 4.[2]

The claim that Aristotle was a realist about animal species may seem obvious to those who come at the zoological works through the lens of the *Metaphysics*. However, the suggestion that Aristotle's zoology even has a concept of species has fallen out of fashion among specialists who study the biological works.[3] Balme (1987b: 297), for example, argues that Aristotle "does not need a technical word for species, since he does not hypostatize it into an entity or absolute; its status is merely that of a universal." Pellegrin (1985: 112), in his aptly titled paper "Aristotle: A Zoology without Species," goes even further and argues that "the idea of animal species is only an empirical notion and has no theoretical function in Aristotle's biology." Instead, he argues that "all programatic passages of the zoological

[1] I will take it for granted here that "form" and "species" pick out distinct entities within Aristotle's hylomorphic theory (Loux 1979). In *Metaphysics* VII.10–11, for example, Aristotle treats particular species (e.g. human, horse) as composite-kinds (1035b27–30, 1037a5–7), which he distinguishes from both the generated composite (e.g. Socrates) and the form (1033b25–26). See also Kullmann 2007: 345, who takes the species of *PA* to be species-composites.

[2] My focus in this chapter is restricted to the evidence from *PA* Book I. I shall set to one side the question of whether Aristotle's views concerning species there are consistent with what he says in the rest of zoological works, including *PA* II–IV. For a suggestion about how the epistemological claim fits with Aristotle's remarks on inquiry in *PA* I 1 (e.g. 644a23–b1, discussed below), see Henry (2021b). For a reply to Pellegrin's 1985: 99 objection that *PA* II–IV does not contain a single definition of any animal species see Henry 2021a: 778–779.

[3] One notable exception to this is Kullmann 2007.

works agree on this point: the true objects of the science of living things are the parts, their properties, and their variations" (106). In his Clarendon edition of the *PA*, Lennox (2001a: 122) goes so far as to avoid translating *eidos* as "species," even when the context seems to demand it, in part because he thinks that that translation is "embedded in modern taxonomic theory and practice in ways that systematically mislead the modern reader of Aristotle."[4]

6.1 Species in the Relevant Sense

Let me start by clarifying what I mean when I attribute to Aristotle a concept of "species" in the relevant sense. For, as we shall see, in the zoological works Aristotle uses *eidos* to designate animal groups higher up in a system of classification. For example, at *HA* IV.2, 525a30–35 crayfish and crabs (very extensive kinds) are called "species" (*eidē*) of crustacean. But this is not what I mean by species in the relevant sense. Rather, I am using the concept to pick out those lowest groups that Aristotle calls "atomic species" (*ta atoma eidē*), "indivisible species" (*ta adiaphora eidē*), or "last species" (*ta eschata eidē*), what Balme (1962: 84) calls "the most immediately visible type . . . whose members are particular individuals distinguished only by differences in *hulē*" (e.g. human, lion, ox). The following passages can serve as examples of what I have in mind:

> While there are many species (*eidē*) within the genus (*genos*) of live-bearing four-footed animals, they do not have a common name. These species are only named individually, as it were, in the way that humans are (e.g. lion, stag, horse, dog, etc.) – although there is actually a single genus in the case of the so-called bushy-tailed animals such as the horse, ass, mule, jennet, as well as the animals that are called mules in Syria based on their resemblance to them. However, they are not the same species strictly speaking; for they copulate and generate from one another. For this reason we have to try and

[4] Lennox's view is part of an intellectual tradition stretching back to the arguments of Balme 1962 and Pellegrin 1982, who were in turn reacting to an earlier tradition of interpreting Aristotle's *History of Animals* through the lens of a post-linnaean world (Leunissen 2021: 67). This is the historical context of Lennox's claim that the conventional translation of *eidos* as 'species' is "embedded in modern taxonomic theory and practice in ways that systematically mislead the modern reader of Aristotle." This is a complicated issue. At this point, I will simply note that, because the modern reader of Aristotle's biology has changed (thanks in large part to the work of Balme, Pellegrin, Lennox, and Gotthelf), we should no longer continue to assume that those same presuppositions are operative. In light of that, I shall continue to use the traditional translations of "genus" and "species" for *genos* and *eidos* with the qualification that these do not refer to the linnaean ranks of Genus and Species (see below).

> grasp each species and study their nature (*phusin*) separately. (HA I.6, 490b31–491a6; cf. 505b25–32)
>
> In all animals that are capable of progressive motion, the sexes are separated, one individual is male while another is female, though both are the same in species, as in the case of human and horse. (*GA* I.23, 730b32–731a1; see also *GA* II.8, 747b28–748b7)
>
> But our present inquiry, as I have said before, is concerned with the causes of why some living things are long-lived while others are short-lived. The difference occurs not only in relation to whole kinds (*genê*) but also between individuals falling under a species (*huph' eidos*). By 'differences in kind' I mean, for example, between humans and horses (for the human kind [*to tōn anthropōn genos*] is longer-lived than the horse kind); by 'difference according to species' (*kat' eidos*) I mean between one human and another. (*Juv.* 1, 465a1–8)

The fact that in the last passage humans (a paradigmatic species) are called both a *genos* and an *eidos* will be important in Section 6.2. For now, I only mean to draw attention to the fact that, when Aristotle speaks of "individuals falling under a species" or of one human being longer lived than another human as a "difference according to species," I take him to be referring to species in the relevant sense.

While it is true that Aristotle never rigorously defends any particular species concept in the zoological works let alone one of the many concepts currently in use by modern biologists,[5] he does employ several operational or diagnostic criteria for identifying them.[6] In the *Analytics*, for example, Aristotle often characterizes species as groups of individuals that are "alike and undifferentiated" (*ta homoia kai adiaphora*). This remains a key feature of species in the zoological works. There species are routinely characterized as groups of animals that are (a) *morphologically* alike and (b) indivisible *in account*. Consider, the following passage from *HA* I 1:

> Of animals, some resemble one another in all their parts while others have parts wherein they differ. Sometimes the parts are the same in form, as, for instance, one human's nose or eye resembles another human's nose or eye, flesh flesh, and bone bone, and in like manner with a horse, and with all other animals that we reckon to be of one and the same species. For as the

[5] When Pellegrin (1985: 108) claims that "the very concept of species is not rigorously construed by Aristotle," he seems to have in mind a sort of hybrid between the modern Biological Species Concept (which defines species in terms of interbreeding and reproductive isolation) and the Genetic Species Concept (which emphasizes a shared genotype). The concept of species that I attribute to Aristotle in this chapter is much less theory-laden than either of those modern concepts.

[6] Contrast Pellegrin (1985: 98, 104–108) who argues that the zoological works do not provide any means for identifying species.

> whole is to the whole, so each part is to each part separately. (*HA* I.1, 486a15–20, Thompson trans. in Barnes 1984, with modifications)

We can concede that *eidos* at a16 does not (or at least need not) mean "species" in the relevant sense. Instead, *eidos* is likely being used there in its morphological sense to refer to the overall shape and form of an animal's parts (e.g. the shape of one human's nose is similar the shape of another human's nose). But Aristotle is clearly using morphological resemblance here as a diagnostic feature of species in the relevant sense (= *eidos* at a19, b18).[7] This reflects Aristotle's tendency in the zoological works to use morphology as a criterion for individuating kinds more generally, following the lead of common sense (cf. *PA* I.1, 639b31–33, 641b23–25). For, as he tells us in *HA* I.6, it is "first and foremost" in relation to the overall shape and structure of the parts and the whole body "that animals differ one from another" (*HA* 491a14–16). The *HA* I.1 passage says that species are those groups of animals that are *homoion* in the overall shape and form of their parts. However, Aristotle mentions other types of resemblances that are relevant to animal groupings, including shared patterns of behavior (*praxis*), shared ways of making a living (*bios*), as well as common character traits (*ethē*) (*HA* I.1, 487a11–13). Presumably members of the same species are *homoion* in these ways as well.

Aristotle also characterizes species as groups that are *adiaphora* in account, as he does in the following passage from *PA* I.5:

> Therefore, one should first discuss actions (*praxeis*): those common to all; those 'according to genus (*genos*)'; and those 'according to species (*eidos*)'. I call 'common to all' those that belong to all animals, and 'according to genus' those whose differences from one another are a matter of degree. For example, I speak of bird 'according to genus,' but I speak of human and every other group that does not exhibit any difference in its general account 'according to species.' For what is common, in some cases they are common by analogy, in others according to genus, and in others according to species. (*PA* I.5, 645b20–27)

While Aristotle does not elaborate here on what it means for a species to be "without any difference in its general account," he seems to be thinking about it in relation to definitions reached through a process of division. According to *Metaphysics* VII.12, the definition of a species *S* is "composed from the differentiae (*ek tōn diaphorōn*), or, strictly speaking from the final

[7] Kullmann 2007: 341.

one (*teleutaias*)" (1038a28–30). By the 'final' differentia Aristotle means the one located at the end of a single branch of division that does not exhibit any further differentiation (e.g. two-footedness). This indivisible difference is said to pick out "the form and substance" of *S* that individuates it from other species of the same kind (1038a25–30).[8] In this context, we say that some species *S* is "without any difference in its general account" if the differentia *D* specified in its essence-defining account does not admit of any further differentiation; *D* is a *final* differentia located at the end of its respective branch of division. I will return to Aristotle's theory of division in *PA* I.2–3. As we shall see, while *PA* I.2–3 continues to employ the theory of definition by genus and differentia, Aristotle insists (against the Platonists) that each species must be defined by *multiple* final differentiae obtained through *multiple* lines of division.

As the *De Juv.* passage above makes it clear, Aristotle does not mean species members do not exhibit any differences. For example, humans differ in terms of their longevity. Likewise, animal species are divisible into male and female individuals (*GA* I.23, 730b32–731a1). However, at least in the latter case, Aristotle denies that sex divides animals *kat' eidos* since it is a material affection and not part of a species' essence (*Metaphysics* X.9). Not all differences, then, count as specific differentiae in the technical sense. This shows that Aristotle privileges a certain kind of indivisibility as a mark of species in the relevant sense (though it is unclear what exactly that is or how it is determined).[9]

All of this suggests that Aristotle not only recognizes species in the relevant sense but that he employs several means of identifying them. In light of this, I suggest the following working definition of an Aristotelian (zoological) species:

> A species is a group of animals that are morphologically (behaviorally, ecologically, etc.) the same, separated into males and females, who all share an indivisible account in common.

In this chapter, I argue that Aristotle privileges these species as the true objects of zoological science. In order to clear the way for this interpretation, however, we need to take a brief look at the Balme-Pellegrin thesis, which concerns the zoological sense of Aristotle's twin concepts of *genos* and *eidos*. As far as I can tell, this has been the main impetus behind the

[8] The example in *Metaphysics* VII 12 is two-footedness in the case of humans.

[9] For one suggestion, see Lennox [1985a] 2001b: 144.

anti-realist interpretation favored by specialists, which claims that Aristotle's zoology does not recognize species.

6.2 The Balme-Pellegrin Thesis

It is indisputable that, in the *Organon* and *Metaphysics* at least, we find what Balme and Pellegrin call a "logical" or "technical" use of *genos*, *eidos*, and *diaphora*, which forms the basis of Aristotle's so-called language of classification.[10] According to this usage, to call something a *genos* is to designate it as a wider class that is divisible into *eidē*, while calling something an *eidos* is to designate it as one of those sub-classes. Within this theory, the *genos* necessarily has a wider extension than the *eidos* (*Topics* IV.1, 121b11), while the *eidos* in turn is subordinated to, and implies, the *genos* above it (Balme 1962: 83). What Aristotle calls *ta atoma eidē* correspond to the lowest groups whose members are "alike and undifferentiated" in the ways discussed above. While this technical vocabulary is itself domain-neutral – it can apply equally to colors, numbers, meteorological phenomena, etc. – Aristotle often uses it in connection with animals. For example, at *Cats.* 5, 2a16–18, he says that the individual human, Socrates, is in the *eidos* Human, which is part of the wider *genos* Animal.[11] Because in this context *genos* and *eidos* appear to form the basis of a theory of classification, scholars have tended to translate them (not unreasonably) as "genus" and "species."

Historically, however, it was common to assume that this technical theory from the *Organon* provided Aristotle with the conceptual machinery for constructing a taxonomy of animals organized around the linnaean categories of Genus and Species, which pick out the two lowest ranks in a fixed system of binomial classification aimed at distributing living things into a hierarchy of mutually exclusive groups (or taxa). It was this assumption, specifically the claim that *genos* and *eidos* correspond to Genus and

[10] Balme 1962, e.g. 83; Pellegrin 1982, e.g. 53, 69; Lennox 2001a: 122. I use "classification" in a loose sense to refer to any system of organization that aims at identifying and arranging the kinds of a given domain with a nested hierarchy. A "taxonomy" (as I shall use the term) is a specific method of classification designed to locate things into a nested hierarchy of *fixed* and *exhaustive* taxa (e.g. Linnaeus' binomial system of classification: see below). Pellegrin 1982: 68–69 seems to have this distinction in mind when he says that *genos* and *eidos* have a "classificatory, though not taxonomic, function." By contrast, Gotthelf 2012: 264 uses "systematic classification" for what I am calling a taxonomy.

[11] See also *Metaphysics* VII.12 and X.8–9. This use of *genos* and *eidos* is widespread in the *Topics*.

Species in a linnaean taxonomy, that the Balme-Pellegrin thesis was intended to challenge.[12]

One of Balme and Pellegrin's main arguments was that, in the zoological works, *genos* and *eidos* – unlike Genus and Species – are level-neutral: they can apply to groups at any level of zoological classification, including one and the same taxon.[13] For example, crabs are both an *eidos* of crustaceans and a *genos* containing many *eidē* of its own (*HA* I.6, 490b10–16, *PA* IV.8, 683b25–30). By contrast, the linnaean categories of Genus and Species are supposed to pick out fixed ranks (the lowest two) that are also mutually exclusive (the same taxon cannot be both a Genus and a Species in a linnaean taxonomy). I see this as conclusive evidence that Aristotle's use of *genos* and *eidos* in the zoology do not correspond to Genus and Species in the linnaean (taxonomical) sense. Of course, the level-neutrality of *genos* and *eidos* in the zoological works is not incompatible with Aristotle's own technical usage in the *Organon* and *Metaphysics*. For even in the *Topics*, Aristotle recognizes that these are not fixed designations but can be applied at any level of classification. In light of this, it is reasonable to adopt the traditional translations of "genus" and "species" for *genos* and *eidos* with the understanding that these terms are not being used in the fixed linnaean sense but in Aristotle's own technical sense.

Now Balme (1962) agreed that *genos* and *eidos* do have this technical sense of genus and species in the *Topics*. But he denied that the zoological works ever use the terms in that way. Instead, he argued, the zoology uses them exclusively in a nontechnical sense. There *genos* simply means a certain *kind* or *sort* of thing, as when we say that schooners are a kind (though not a genus) of sailing ship, while *eidos* refers exclusively to the observable shape and form of a particular animal and is interchangeable with other nontechnical terms like *morphē* or *schēmata*. While Balme admits that there are some passages in the zoological works where it is possible to translate *genos* and *eidos* in their technical (classificatory) sense of "genus" and "species" (e.g. *PA* IV.8, 683b26–27), he argues that in each case the context only requires their nontechnical sense.

I think we can agree with main thrust of the Balme-Pellegrin thesis without claiming that Aristotle's zoology lacks a concept of species in the

[12] For a historical survey of these interpretations, see Pellegrin (1982: 53, 58–59, 75ff.). While the assumption that Aristotle used *genos* and *eidos* in this linnaean sense is mostly a relic of past scholarship on the *HA*, it continues to linger among nonspecialists and historians of zoology.

[13] Pellegrin 1982: 69; Lennox 2001a: 298. D'Arcy Thompson had already recognized this feature of *genos* and *eidos* in his translation of the *HA* (see his note on *HA* I.6, 490b17ff.).

relevant sense. For the claim is not that *eidos* always means "species." Obviously, it also bears its hylomorphic sense of "form" (as opposed to matter). And even when it is used in the technical (classificatory) sense, we have seen that *eidos* can be used for groups at any level of classification, including very extensive kinds (e.g. crabs, trumpet-shells). But it doesn't follow from this that *eidos* is never used to pick out species in the relevant sense (as it seems to do in the passages surveyed above). This is especially clear in cases where *eidos* is qualified by the word *atomon* (*adiaphoron*, *eschata*, etc.). The expression *to atomon eidos*, when applied to animals, remains the best candidate for serving as Aristotle's technical phrase for species in the relevant sense. That expression is not level-neutral (it does not apply to animal groupings at any level of zoological classification) but has a fixed designation that applies exclusively to so-called infima species (e.g. humans, lions, horses).[14]

Having said that, my argument in this chapter does not turn on the meanings of the words *eidos* and *genos*. Even Balme concedes that in *PA* I Aristotle employs the technical distinction between "genus" and "species" from the logico-metaphysical works. However, he insists that *PA* I is not part of Aristotle's "zoology" in the strict sense, and so he does not see this fact as undermining his central thesis.[15] While I am less inclined to treat this technical use of *genos* and *eidos* in *PA* I as markedly different from the rest of the zoology, I do not need to take a stand on this since my focus is on *PA* I itself. And it is evident from the way Aristotle characterizes certain animal groups there that he has a concept of species in the relevant sense.

In the remainder of this chapter, I will argue for two claims:

1. The epistemological claim. The aim of Aristotelian zoological division is to identify and organize the essential features specified in the definition of each indivisible species of animal.

[14] One might point to *Juv.* 1, 465a1–8 as a counter-example. There human (a paradigmatic *atomon eidos*) is called both a *genos* and an *eidos*. This would seem to confirm Balme's claim that *eidos* "represents no group that *genos* does not also represent" (Balme 1962: 87). However, if we grant (along with Balme) that *genos* has both a technical/classificatory sense meaning "genus" and a nontechnical sense meaning a "kind" of thing, then humans cannot be both an *atomon eidos* (atomic species) and a *genos* in the technical sense. For a *genos* in the technical sense is divisible into *eidē*, while an *atomon eidos* is indivisible by definition. For this reason, when Aristotle speaks of *to tōn anthropōn genos* in the *De Juv.* passage, I would contest that *genos* is not being used in its technical sense: Aristotle does not mean to designate humans as a genus that is divisible into different species of human (pace Ross: see Balme 1962: 95–96). Rather, by calling humans and horses *genē*, Aristotle only means to say they are two *kinds* or *sorts* of animal without implying anything about their place in a system of classification.

[15] Balme 1962: 84. For an argument against severing *PA* I from the rest of the treatise, see Henry 2021a.

2. The ontological claim. Those indivisible species constitute the *ousiai* of Aristotle's zoology.

As we shall see, the first claim is put forward in *PA* I.2–3, while its corollary is advanced in *PA* I.4. I begin with the former.[16]

6.3 *PA* I.2–3: Species as Objects of Definition

In Section 6.2, we explored Aristotle's technical terms *genos*, *eidos*, and *diaphora*, which I have translated as "genus," "species," and "differentia." The method within which those terms operate is division, which is the subject of *PA* I.2–3. Since the target of Aristotle's criticisms in those chapters is Plato's method of dichotomous division, I will start there.[17]

At *Phaedrus* 265d–266b, (Plato's) Socrates tells the young Phaedrus about a method for 'defining each thing,' which he separates into two parts. The first part 'consists in seeing together the many scattered things and drawing them into a single kind,' while the second part 'consists in the ability to cut things up according to their kinds along joints that are natural, trying not to break any part into pieces as a bad butcher might do.' Socrates refers to these twin procedures as collection (*sunagōgē*) and division (*diaresis*), respectively, though for convenience I will refer to them together simply as "the method of division." In several places, Plato makes it clear that division is a tool of definition (*Phaedrus* 265d4, *Politicus* 285d8–9) where the various terms obtained by division correspond to

[16] Two caveats are necessary here. First, in what follows I am concerned only with definitions obtained through a process of division (*Metaphysics* VII.12, 1037b29: *tōn kata tas diaireseis horismōn*). I will set aside the question of how these definitions relate to other types of definition recognized by Aristotle (e.g. *APo.* II.8–10) or how the method of division in *PA* I.2–3 relates to Aristotle's project in *PA* II–IV. Code 2010 denies that definitions obtained through division are essence-defining accounts. However, Aristotle is explicit in *PA* I.3 that "one ought to divide by what is in the essence (*ousia*) and not by incidental properties" (643a28). This is consistent with his view in *APo.* II.13 that division is useful for "hunting down" the attributes "predicated in the *ti esti*" of a thing (96a23–24), which is Aristotle's expression of choice for essence in the scientific works. The issue is complicated by the fact that Aristotle distinguishes between definitions of the form and definitions of the composite (e.g. *Metaphysics* VII.10–11). In the latter case, the essence is said to include properties of a thing's formal and material nature (*Metaphysics* VI.1). Since the *PA* includes material parts in the *logos tēs ousias* of animal kinds (e.g. 685b15; see Henry 2021a: 767–768), I take it that the object of definition in *PA* is the species-composite and not its form. Kullmann 2007: 345; contrast Peramatzis (Chapter 5 in this volume). Whether these count as "definitions" in some strict metaphysical sense (Frede 1990) is not relevant to my argument here. Second, while I shall have something to say about how the ontological claim relates to the doctrine of substance in the *Categories* and *Metaphysics* (see pp. 150–151), I do not have the space to pursue the full implications of that claim here.

[17] For a more detailed discussion, see Henry 2011 and references therein.

features in a thing's essence (*ti estin*, *Sophist* 218c1). The definition itself is constructed by "weaving together" (*sumplokē*) all of the terms located along a single line of division (*Sophist* 221a–c; cf. *PA* I.3, 643b13–16), the last of which – the final differentia – marks off the object from other objects of the same kind.

When done properly, division should help ensure that our definitions pick out natural kinds.[18] This goal is achieved, Plato thinks, by employing *dichotomous* divisions that involve partitioning each class into precisely two sub-classes by means of a single bifurcated differentia (e.g. footed/footless, wild/tame). Plato has an ontological reason for this. In the *Phaedrus*, Socrates asserts that the parts of reality naturally come in pairs, a right-hand and a left-hand part (265e3–266b1). And so dividing a kind into two equal parts ensures that we cut it along its natural joints. This is reinforced in the *Politicus* where the Eleatic Stranger tells us that when we "go along cutting through the middle of things" we will be more likely "to hit upon real kinds" (262b).

One of the central aims of *PA* I.2–3 is to expose the failures of dichotomous division in order to clear the way for Aristotle's own revised method of division by multiple differentiae. Aristotle's method is complicated and examining all of its details would take us too far afield. Instead, I will focus on its most relevant features here.

Like its predecessor, Aristotelian division is a means for identifying and organizing the features included in the definition of a thing. However, the way Aristotle's method accomplishes this aim is significantly different from Plato's. Condensing as much as possible, Aristotle recommends that we begin by dividing some wider kind *K* straightaway by means of multiple differentiae (rather than dichotomously) where each differentia picks out some generic feature in the universal account of *K* (*PA* I.3, 643b10–12). Suppose *K* is *bird*. We start by dividing the kind according to multiple differentiae that correspond to those features that belong to all birds *qua* bird (e.g. beak, wing, crop). We then divide each of those generic differentiae along multiple axes according to what Aristotle calls differences of "the more-and-less."[19] For example, different birds will have beaks whose lengths fall somewhere along a continuum between *longer* and *shorter* (one axis of division), whose curvature falls somewhere between *curved* and *flat* (another axis of division), whose hue falls somewhere between *black* and *white* (another axis of division), and so forth. We continue dividing along these multiple lines of division, making more and more fined-grained

[18] Deslauriers 1991; Henry 2011: 235. [19] Lennox [1987b] 2001b: 160–181.

differentiations within each feature, until we reach a level where the feature in question cannot be further differentiated (e.g. the determinate curvature of an osprey beak). At this point, we have reached the final differentia. Each species is then defined by collecting together all of those multiple final differentiae into a single essence-defining account.[20] With this sketch in hand, let's look at *PA* I.2–3.

The opening sentence of *PA* I.2 indicates that the aim of division is to "grasp the particular (*to kath' hekaston*)." Given what Aristotle says about division elsewhere (e.g. *APo.* 95a22–23; *Metaphysics* 1037b29), it is reasonable to assume that the sort of grasp he has in mind is one that comes from having a definition of it. But what are the objects of those definitions? What are "the particulars" whose essence we are attempting to grasp by means of division?

According to Pellegrin, *PA* I.2–3 supports the view that division aims at grasping animal *parts* and not the animals to which those parts belong:

> Aristotelian division cannot give us as "last species" the definition of a whole animal but is necessarily limited to the definition of organic functions or functional parts. Division, therefore, gives us species of blood, uterus, foot, but not animal species, or else it would fall into the errors of Platonic methodology. (Pellegrin 1985: 100)

[20] See Balme 1987a: 76–77. This revised theory of definition by division raises two issues that are worth mentioning here. First, one of the casualties of Aristotle's attack on Platonic division in *PA* I.3 seems to be his own account of definition in *Metaphysics* VII.12, which also identifies the essence of the species with a single final differentia (e.g. two-footed) obtained by pursuing a single line of division (1038a25–30). Charles 2000, e.g. 313, 330, argues that we can reconcile Aristotle's two accounts by taking one of the multiple differentiae specified in a *PA* I.3–style account to be the cause of all the others; that single differentia is the essence of the species. However, in that case, Aristotle's objection to Platonic division at *PA* II.3, 644a1–11 (see below) would seem to lose its force. For the Platonist could concede that the account of the species must include multiple differentiae but insist that the essence itself corresponds to a single differentia falling under a single branch of division. On my reading, the upshot of *PA* I.3 is that species are defined by multiple (essential) features and that this is why Aristotle concludes at 644a11 that "it is impossible for those who divide dichotomously to grasp any of the particular animals" (since dichotomy ends in a single differentia only). I return to this below. The other point worth mentioning is that Aristotle's revised theory of division generates a new problem of "horizontal" unity that he does not address in *PA* I.2–3 (Lennox 2001a: 166). In *Metaphysics* VII.12, Aristotle introduced the rule "divide each differentia by its own proper differentiae" (1038a10, repeated at *PA* 643b17). This rule was introduced as a way to ensure the "vertical" unity of the terms along a single branch of division. For example, in the division footed > four-footed > cloven-footed the first two terms are entailed by the last one. And so, the final account of the species only needs to mention that last term. According to the revised theory in *PA* I.2–3, however, species are defined by collecting together multiple final differentiae into a single account, which are located at the ends of multiple branches of division. The question (which Aristotle does not address) is how those multiple differentiae are unified "horizontally" so that the definition picks out one thing.

Since the object of definition is always a substance, it would seem to follow that in the zoological works animal parts constitute the substances "in the primary and strongest sense" (106). Indeed (as the title of his paper suggests), Pellegrin argues that Aristotle's zoology has no need for animal species. I will return to Pellegrin's ontological thesis (that animal parts constitute the real substances of the living world) in Section 6.4. In this section, I want to focus on his epistemological claim that, according to *PA* I.2–3 at least, the true objects of definition are "the parts, their properties, and their variations" *and not* the animal species themselves. By contrast, I shall argue that, on the most natural reading of *PA* I.2–3, the ultimate objects of definition are the indivisible species of animal to which those parts belong.

We can start by returning to Aristotle's opening remarks in *PA* I.2:

> Some people attempt to grasp the particular by dividing the genus into two differentiae. On the one hand, this is not easy to do. On the other, it is impossible. (*PA* 642b5–7)

Now the "some people" here are clearly the Platonists. And, as the closing lines of *PA* I.3 make clear, "the particulars" that their method fails to grasp are *particular animals* (644a10–11: *tōn kath' hekaston zōō[i]n*), for example, "humans, horses, cattle, Indian dogs, pigs, goats, and sheep" (643b5–6, cf. 644a1–10). The parts of those animals (e.g. feet) are among the multiple differentiae that Aristotle says must be included in a proper definition of those animals (643b27–644a9).

This is reinforced by several things Aristotle says elsewhere in *PA* I.2–3. Consider the following objection at *PA* I.3, 643a7–20:

> Now the differentiae will be equal in number to the indivisible animals, if we assume both that the animals and the differentiae are indivisible, and that no differentia belongs in common [sc. to more than one indivisible animal]. But, if it is possible for a differentia to be both indivisible and common, then clearly animals that differ in species would end up in the same species in virtue of that common differentia. Therefore, if the differentiae under which all the indivisible animals fall are unique (*idioi*), then none of them can be common. Otherwise, things that are different will be in the same division. But it is necessary (a) that none of the things that are the same and indivisible should go at one time under one differentia of the division and then at another time under another differentia, (b) that things that are other should not go under the same differentia, and (c) that all things should go under these. Clearly, then, on cannot grasp the indivisible species (*atoma eidē*) by dividing animals (or indeed any other kind of thing) in the way the dichotomists do. For, according to them, the number of

> ultimate differentiae must necessarily be equal to the total number of indivisible species of animals (*tois zō[i]ois pasi tois atomois tō[i] eidei*). (*PA* I.3, 643a7–20)

Balme (1992) argues that Aristotle's criticisms of Platonic division in *PA* I.2–3 are aimed, not so much at division into *two* classes but at division by a *single* differentia. That appears to be Aristotle's target here. One of the central commitments of Platonic division is that each indivisible species should be defined by a single unique differentia that is not shared in common by other species of the same kind. For example, in the *Politicus*, the Eleatic Stranger rejects their first division on the grounds that it ended up defining the statesman as a shepherd of the human herd, a function that is shared in common by all those other crafts that claim to look after various aspects of human life: "So how will our account of the king appear right and complete to us, when we posit him as sole herdsman and rearer of the human herd, singling him out on his own from among tens of thousands of others who dispute the title with him?" (*Politicus* 268c). The assumption throughout the *Politicus* is that the last differentia, which is obtained by following a single line of dichotomous division, should identify that one feature that belongs exclusively to the definiendum (267c5–d1). This constraint on definition can be traced back to the *Theaetetus* where Socrates says in connection with the third type of *logos*: "If you get a hold of the differentia that distinguishes a thing from everything else, then (so some people say) you will have gotten an account of it; on the other hand, so long as it is something common that you grasp, your account of it will be about those things which have this in common" (*Theaetetus* 208d5–9). It is this aspect of Platonic division that constitutes Aristotle's focus in *PA* 643a7–20.

Now, on one reasonable interpretation of the passage, Aristotle objects that this commitment leads the Platonists into a two-horn dilemma. If each indivisible animal is defined by a unique (*idion*) differentia, and no final differentia can be common to more than one indivisible animal, then the number of final differentiae must match the number of indivisible species out there in the world, which Aristotle sees as an unrealistic Platonic ideal. Yet, if we allow some (final) differentiae to be common to more than one species, then, since *ex hypothesis* each species is defined by a single differentia, we end up with a kind of cross-division where "things that are different will be in the same division." For example, both cattle and pigs will fall under the same differentia *cloven-footed*. But (Aristotle argues) in an exhaustive division of animals "things that are other should not go under the same differentia." Aristotle's revised theory of division avoids both horns of this dilemma since, on his account, each species is defined by multiple final differentiae, some of which may be common to other species of the same kind. What is unique to each species is the entire complex of final differentiae obtained through multiple axes of division.[21]

Whether or not this is the right way to understand Aristotle's objection in *PA* I.2, 643a7–20, the closing lines of the passage make it sufficiently clear that the particulars whose essence we are attempting to grasp by means of division are particular indivisible species of animal (*tois zō[i]ois pasi tois atomois tō[i] eidei*).

Finally, consider the closing section of *PA* I.3. There Aristotle presents a series of arguments to reinforce his claim that "it is impossible to grasp any of the particular species (*tōn kath' hekastōn eidōn*) by dividing the genus into two" (643b26–27).[22] His final argument reads:

> If human were split-footed alone, then by proceeding in this way [sc. dichotomously] one might arrive at this single differentia. But because human is not merely split-footed alone, it is necessary for there to be multiple differentiae that do not all fall under a single branch of division. But there cannot be multiple differentiae falling under a single branch of *dichotomous* division (at least not of the same thing); rather [sc. on the Platonic method], one must end with *one* differentiae according to *one* such division. For this reason, it is impossible for those who divide dichotomously to grasp any of the particular animals. (*PA* I.3, 644a1–11)

When Aristotle says that human is not merely split-footed alone, he must be talking about features present in the essence. For he has already said back at *PA* 643a26–27 that one should try to divide by a thing's essential features and not by its so-called necessary accidents (cf. 645b1–3). As Aristotle makes clear

[21] This is Aristotle's position in *APo.* II.13: "We should take items of this type up to the point where we have first taken just so many that, while *each* extends further [sc. than the species being defined], *all of them together* do not. And this must be the essence of the object" (96a32–35). For a different interpretation of *PA* 643a7–20, see Balme 1992: 110.

[22] Given the opposition with *genos* here, it is reasonable to translate *eidos* in its technical sense to mean "species." So Balme 1991: "It is clear from the following that none of the particular species can be obtained by dividing the genus dichotomously as some have thought." Likewise, the Loeb translation reads: "Here are further considerations to show that it is impossible to come at any of the particular species by the method of dividing the group into two, as some people have imagined." By contrast, Ogle (in the Barnes translation) and Lennox (2001a) translate *eidos* as "form." However, as noted, Lennox 2001a: 122 takes *genos* and *eidos* in their technical (classificatory) sense within the context of division, and so his "form (of a kind)" is conceptually equivalent to Balme's "species (of a genus)." But even if we do take *eidos* here to mean "form" in the purely hylomorphic sense (Peramatzis, Chapter 5 in this volume), Aristotle is still clearly talking about the forms of animal species and not the forms of their parts.

here, the reason this cannot be achieved using dichotomous division is the fact that the essence of each indivisible species is constituted by a complex of features, not by a single unique feature as the Platonists hold. For this reason, Aristotle concludes, "it is impossible for those who divide dichotomously to grasp any of the particular animals."[23]

According to *PA* I.2–3, then, definitions reached through a process of division have as their ultimate object particular indivisible species of animal. Those definitions are composed from multiple final differentiae (including the parts) located at the ends of multiple branches of division. To be sure, *PA* II–IV does have definitions of parts among its *proximate* aims (see Henry 2021a). But, according to *PA* I.2–3 at least, the account of those parts should be included in the account of the animal species to which they belong (cf. *Metaphysics* VII.10–11).

One might object here that Aristotle's project in *PA* I.2–3 is entirely negative and so cannot be taken as evidence for his positive views on definitions reached through a process of division. However, the overall goal of *PA* I is to establish certain norms governing a natural science of living things (*PA* 639a1–14). And it is easy to see how *PA* I.2–3 fits into this project. For, in addition to criticizing dichotomous division, Aristotle also identifies several norms governing the proper use of zoological division (e.g. "one ought to divide by what is in the essence (*ousia*) and not by incidental properties," 643a28; "one ought to divide by opposites," a33; "one ought not divide by the common functions of body and soul," a35; "one should divide the kind straightaway by many differentiae," b24).[24] To treat *PA* I.2–3 merely as a polemic against the dichotomists is to miss these more positive contributions those chapters make to Aristotle's zoology.

6.4 *PA* I 4: Species as Substances

I want to turn finally to the other part of Pellegrin's thesis, namely, that in the zoological works Aristotle treats the parts of animals as substances "in the primary and strongest sense." As far as I can tell, Aristotle never makes such a radical claim anywhere in the *PA* (or the zoological writings more broadly). What he does say is this:

> Since the last species are substances (*ousia men ta eschata eidē*), while within these are individuals that are undifferentiated in species (e.g. Socrates,

[23] Pellegrin 1985: 100 claims, without argument, that *PA* 1.3, 644a1–11 is not talking about how to define humans (the species) but the human way of having feet (the part). I find this reading to be forced especially given the concluding sentence of the passage.

[24] See also Lennox 2021: 170 n. 49.

> Coriscus), we must either first speak about the things that belong to them universally or say the same thing many times, as we mentioned above. But the universal attributes are common; for it is those that belong to more than one thing that we call universal. The question is which of these should be our subject. For, on the one hand, insofar as what is specifically indivisible is the substance, it would be superior, if possible, to study separately what is particular and specifically indivisible – just as one studies human, so too about bird (for this genus contains species), except our study would be about any one of the indivisible species of bird (*tōn atomōn*), e.g. about sparrows, cranes, or something of that sort. On the other hand, insofar as this will result in saying the same thing many times about the same attribute because it belongs in common to many things, speaking separately about each [sc. of the indivisible species] is somewhat ridiculous and lengthy. (*PA* I.4, 644a23–b1)

As the examples in this passage make clear, the "last species" (*ta eschata eidē*, a23) that Aristotle identifies as the *ousiai* of zoological science are particular indivisible species of animal, for example, humans, sparrows, cranes. This is the ontological side of his view from *PA* I.2–3 that those species constitute the objects whose essence we are attempting to grasp by means of division.[25]

To see what Aristotle has in mind in *PA* I.4, 644a23–b1, we have to read it in the context of a puzzle about how to conduct zoological inquiry originally formulated back in *PA* I.1:

> Should one take up each substance (*ousia*) itself and define it independently, for example, taking up one-by-one the nature of human, lion, ox, and any other animal as well; or should one first establish the attributes common to all in relation to something common? (639a15–19)

[25] Here I follow Kullmann 2007: 344–345 in taking *ta eschata eidē* to refer to those same indivisible species that were objects of definition in *PA* I.2–3. This was Balme's original view in his 1972 Clarendon text, but he later changed his mind and took *ta eschata eidē* to refer to the concrete individuals Socrates and Coriscus (a24) (or their particular forms?); species (along with genera) became "the things that belong to them universally" (a27–28) (Balme 1987b: 297; 1990: 121). I do not have the space to provide a full rebuttal of Balme's revised interpretation. Even on Balme's 1990 own translation, Aristotle uses similar expressions throughout *PA* I.2–3 to refer to indivisible species of animals (e.g. 643a7–20). And so it stands to reason that *ta eschata eidē* at 644a23 carries the same sense. More importantly, as I go on to show, the immediate context for Aristotle's remarks here is a puzzle about the starting-points of inquiry. Particular individuals like Socrates and Coriscus are not on the table as a possible answer to the question, "Which of these should be our subject?" (a28–29). Admittedly, *ta to eidos adiaphora* at a24 does refer to Socrates and Coriscus. But, given the context, those must be functioning as examples of things that are "undifferentiated in species" (form?) rather than examples of "the last species" (*ta eschata eidē*) themselves.

Aristotle repeats the two approaches to inquiry again at *PA* 639b3–5:

> Accordingly, the following question about how one is to carry out an investigation of animals should not be overlooked: should one first study animals in common within a genus (*genos*) and then go on to study their distinctive features, or should one study the particular straightaway (*kath' hekaston euthus*).[26]

The two alternative methods that Aristotle considers here are: [A] begin straightaway at the level of indivisible species (e.g. human, lion, ox) and study them one-by-one; or [B] begin by studying animals first at the level of extensive genera (e.g. bird, fish: *PA* 643b10–12, 644b1–6) before going on to study their specific differences.[27]

This is the puzzle Aristotle is recalling in our passage when he asks, "Which of these should be our subject?" (644a28). Understood in this context, it is clear that when he says, "insofar as *what is specifically indivisible* is the substance, it would be superior, if possible, to study separately what is particular and specifically indivisible," Aristotle is talking about [A] studying each indivisible species of animal straightaway:

> Just as one studies human, so too about bird (for this genus contains species), except our study would be about any one of the indivisible species of bird (*tōn atomōn*), e.g. about sparrows, cranes, or something of that sort.

What Aristotle is saying here is that the fact that these indivisible species constitute the *ousiai* of the living world provides us with an ontological reason for adopting the species-by-species approach. The problem with this approach, as he goes on to say, is that it can be needlessly repetitive (639a23–28, 644a26). Since species of the same kind share many features in common, it would be inefficient to give separate accounts of those features for each one (644a33–37). The virtue of the alternative method [B], then, is that it avoids this problem since we start with what belongs in common to animals as members of some wider kind and then (using the revised method of division) work our way down to those features that are unique to them as members of atomic species. However, as *PA* I.2–3 makes clear, those atomic species remain the primary objects whose essences we ultimately seek to define by means of division.

There is a question here, which I shall not pursue, about how Aristotle's claim in *PA* I.4 that animal species are *ousiai* can be reconciled with the

[26] As the original formulation makes clear *to kath' hekaston* stands for indivisible species of animal, just as it does in *PA* I.2–3. Lennox 2001a: 153.

[27] I discuss this in detail in Henry 2021b. The former approach [A] seems to be what Aristotle has in mind at the end of the *HA* I.6 passage quoted earlier when he says: "For this reason we have to try and grasp each species and study their nature separately."

doctrine of substance found in the logico-metaphysical works. In the *Metaphysics*, for example, Aristotle denies that species-composites (e.g. human, horse) are substances (e.g. 1035b27–30), while in the *Categories* he famously downgrades species (along with genera) to secondary substances (universals). One possibility, of course, is that Aristotle changed his mind in the zoological works and came to see atomic species as substances most of all. However, a less radical answer would be that, while particular matter-form composites (e.g. Socrates) are primary substances in the unqualified sense, species-composites are primary substances from the perspective of zoology.[28]

6.5 Conclusion

In this chapter, I have focused on the status of animal species in Aristotle's zoological works with an eye to the discussion in *PA* I.2–4. I argued that, whatever Aristotle actually gets up to in the rest of the treatise, *PA* I.2–4 is clear that the aim of Aristotelian zoological division is to identify and organize the features specified in the definition of each indivisible species of animal and that these constitute the primary substances of the science of zoology. This fits well with the orthodox interpretation that treats Aristotelian species as part of "the eternal fabric" of the sublunary world kept perpetually in existence by the mechanism of generation (*GA* IV 3, 767b8–10).[29] This naturally leads us into questions about the role of species in the *Generation of Animals*. But those questions will have to wait for another occasion.

[28] This may be what Kullmann 2007: 344–345 has in mind when he argues that "substance" in *PA* I.4 is being used in an epistemological rather than an ontological sense.

[29] Henry 2019: ch. 9.

CHAPTER 7

Parts and Wholes in the Parts of Animals

Thomas K. Johansen

Introduction

PA I.5 (645a35–36) sends a strong message that the parts of the animal body are to be studied for the sake of the animal as a whole.[1] This means that the material parts are primarily of interest for their contribution to the form of the whole living being. It is clear that this contribution lies in the bodily parts' instrumentality, their function in the performance of the psychological functions that characterize the various animals. Yet Aristotle's expository method in the *PA* may seem to run counter to his holism. If, as Aristotle suggests (644a24), it is the lowest or 'indivisible' species that are the substances, then (in accordance with holism) we should study the parts of animals at this level (644a32). Accounting for the parts of animals at a higher level would run the risk of abstracting from the explanatory dependencies between body and soul that obtain at the level of the actually existing species. Yet many of the parts of animals are common to several species, so explaining them for each species, Aristotle says (644a36), would be repetitive and tiresome. We find thus in the *PA* two opposed explanatory tendencies: one 'upwards' toward the more common and greater simplicity and another 'downwards' toward the ultimate species and greater complexity. His proposed solution is to account for the various bodily parts at a general level and to descend to the species only when the parts differ significantly. In this chapter, I address some of the difficulties that have to be resolved before one can accept Aristotle's solution.

Essences and Parts

In the first chapter of the *PA*, Aristotle recommends that we explain the parts of an animal with reference to the animal's essence:

[1] I am most grateful to Sophia Connell and James Lennox for their advice and criticism.

> **[T1]** Hence it would be best to say that, since this is what it is to be a human being, on account of this it has these things, for it cannot be without these parts. If one cannot say this, one should say the next best thing, i.e. either that in general it cannot be otherwise, or that at least it is good thus. And these things follow. And since it is such, its generation necessarily happens in this way and is such as it is. (This is why this part comes to be first, then that one.) And in like manner one should speak in precisely this way about all of the things constituted by nature. (*PA* I.1 640a33–b4)[2]

The essence is the object of the definition. Here, the essence, "what it is to be a human being," explains the parts of the animals. As the passage is often read,[3] this may be immediately because the part is contained in the definition, as "wing" might be part of a definition of "bird" as "an animal with wings." However, it may also be because the part is necessary for the essence to be realized, even if it is not itself part of the essence. Here the notion of hypothetical necessity captures the relationship between essence and part. So, we might say that feathers are hypothetically necessary for the bird if it is to have wings. In either case, however, the essence would be the starting point for explaining the parts of the animal.

A form of essentialism is, then, a presupposition of Aristotle's accounts of the parts. Importantly, the essence is not itself the target of explanation, the part is. We should therefore not expect Aristotle in the *PA* to spell out definitions of animals. Rather what is called for is to mention those features of the definition that are necessary for explanation. But this may only be some features of the animal's essence. His essentialism is therefore also compatible with the absence from the *PA* of any full definitions of animal kinds.[4]

Essence, Soul, and Body

The way in which the essence is the starting point for explaining the parts of the animals is explicated in two complementary ways in *PA* I.1. The first is in terms of the distinction between form and matter. Animals, like all natural beings, are composites of form and matter, with the soul playing

[2] Translations of the *PA* here and throughout by Lennox (2001a).

[3] In line with the reading of the passage advocated by Balme, Cooper and Gotthelf, cf. Lennox 2001a: 134–135. The point in **[T1]** about the order of generation may support this reading: the parts that will develop gradually toward the creature's essence will clearly include parts that are essential to it as a certain kind of animal, including nutritive, perceptual, and, sometimes, locomotive parts.

[4] Cf. Gotthelf 2012: 155: "It has often been claimed, and not without reason, that there are no definitions of animal kinds in *PA*, or anywhere in the biological treatises."

the role of the form. The nature of an animal is both soul and body. The natural philosopher thus studies both soul and body.

However, priority goes to the soul as Aristotle says in this programmatic passage:

> **[T2]** It will be up to the natural philosophers to speak and know about the soul; if not all of it, about that very part in virtue of which the animal is such as it is. He will state both what the soul or that very part of it is, and speak about the attributes it has in virtue of the sort of substantial being it is, especially since the nature of something is spoken of and is in two ways: as matter and as substantial being. And nature as substantial being is both nature as mover and nature as end. And it is the soul – either all of it or some part of it – that is such in the animal's case. So in this way too it will be requisite for the person studying nature to speak about soul more than the matter, inasmuch as it is more that the matter is nature because of soul than the reverse. (*PA* I.1, 641a22–32)

The natural philosopher will treat of both form and matter, soul and body, since both are nature. However, the starting point will be the soul, since the soul is the final cause of the body,[5] that for the sake of which the body is, and so prior in explanation to the body.

Secondly, the priority of the form as the final cause is explicated also in terms of necessity. The body is necessary, hypothetically necessary, for the functions of the animal, just as certain materials are necessary for the axe if it is going to be able to split (*PA* I.1, 642a9–10). Again, Aristotle gives priority to the form over the matter in this relationship: the matter is necessary *for* the form (I.1, 640b27–29, I.5 645a30–36, 645b15–20, II.1, 646b2–3) so the matter is to be explained by its contribution to this form.[6]

There is a debate, explored in this volume by D. Henry (Chapter 6) and M. Peramatzis (Chapter 5), as to whether the essence that provides the starting point of the explanation of the parts is purely formal or whether the form itself involves reference to matter. I shall not engage in this debate directly here.[7] But I shall suggest that it is important to see the way the essence explains the body in terms of levels of explanation. Here are two

[5] *De Anima* II.4 provides the subtext to this passage.

[6] Cf. "Nature is of two sorts, nature as matter and nature as form, and the form is the end, and since everything else is for the end, the form must be what things are for" (*Phys.* II.8, 199a30). "There are four causes: first, the final cause, that for the sake of which a thing exists; secondly, the formal cause, the definition of its essence (and these two we may regard pretty much as one and the same)" (*GA* I.1, 715a4–6).

[7] I argued in *The Powers of Aristotle's Soul* (Oxford University Press, 2012) that when defining the soul, the form should be specified by the natural philosopher in such a way that it would support an account of the body by hypothetical necessity.

examples, nutrition and perception, to clarify the point. Nutrition serves the function of maintaining the animal. In *DA* II.4, we were told that the soul was the efficient and formal cause of nutrition, understood as the process whereby the animal cooks up the food to provide the final nutriment for the body. The account of nutrition was functional, making no reference to any particular living being, nor therefore to any particular kind of body or food or any particular kind of final nutriment. The account of nutrition here is 'purely' formal: there is nothing in the account of nutrition that corresponds to the material account of anger as "boiling of the blood around the heart" (*DA* I.1).[8] It is in the *PA* and the *GA* that we learn that the final nutriment for some animals is blood, for others something analogous. We learn how the cooking works in terms of heat and we learn how blood works in relation to the flesh. A different story applies to plants, though this falls outside the brief of the *PA*. We may say that the account of the *DA* was formal and shied away from any physiological detail because Aristotle wanted here to articulate an account of nutrition at its most general level, an account that would apply to all living beings. In the *PA*, meanwhile, we are interested in the bodily parts of different kinds of animals so we shall now account for the psychic functions as realized in particular kinds of matter. So, we explain blood by its role in nutrition. Talk of a blooded animal will inherit, however, the functional understanding of nutrition from the *DA*: blood is final nutriment of the sort that will support the animal's continuation as a certain kind of *ousia.* Whatever we say about the different material parts of different animals will be said against the background of the general formal account of nutrition given in the *DA*.

Now when we proceed to use blood as a differentia in accounting for other parts of the animal, we should not think that the explanatory connection with the formal account of nutrition has been severed: it is just that what is of immediate relevance to the part that we are trying to explain may be that the animal has already been determined as blooded. So having blood will also explain why animals have fat, the result of a rich diet not having been absorbed by the flesh from the blood (*PA* II.5, 651a20–27). Fat in turn may explain infertility in some animals (651a12–17). Bodily differences with an explanatory origin in basic psychological functions may help explain other bodily differences.[9]

[8] This is not to say that it is not implied that concoction is a materially realized process but rather that there is no explicit mention of what kind of matter realizes the process.

[9] I focus here on blood, but much more could be said about the other aspects of nutrition, for example concoction. For a detailed analysis along the lines I am suggesting, see Gotthelf 2012: 165–171. He sums up the explanatory principles involved as follows: "The fact about animals

A parallel case is presented by flesh. All animals *qua* animals must have perception, of which touch is the most basic modality. The flesh is composed as the primary organ of touch. Being composed of the four basic tangibles, the hot and the cold, the wet and the dry, the flesh bestows on the animal the ability to be affected by these qualities. Again, we may say that flesh (or an analogue) is part of the nature of an animal as such, and the natural philosopher will offer an account of this part both in terms of the psychological function, perception, and the flesh. The flesh may in this way stand as a differentia of animals from non-animals. The flesh in turn works as an explanation of the animal's other parts. So being soft the flesh requires supportive structures, typically bone, which in turn will help explain the sinews. However, the fact that flesh explains bone does not mean (though it does not exclude either) that bone or sinews, in turn, is part of the definition of the animal kind in question. Again, it is important to keep in mind the explanandum: the various animal parts. To explain them we require differentiae that lead us back to the essence of the animal, but we are not thereby using these differentiae to define the animal. This is not the target of the exercise.

It should come as no surprise therefore that matter is referred to in the differentiae explaining the various parts. The starting point will ultimately be a differentia in function or end, like nutrition or perception, but such functions require bodily parts that themselves are explanatory of further differentiae. However, as we shall see, the differentiae that are referred to in the accounts of the parts are of a wide variety, much wider than just a material composition, like flesh or blood. Some refer to a function, others to a shape or size, yet others to a life-style, yet others again to the material composition.

7.1 Levels of Explanation

Now the explanatory priority of an animal's essence in relation to parts might suggest that the parts of animals should be accounted for at the level of the indivisible or ultimate species, cat or dog, rather than higher kinds,

which distinguishes them from plants, and is central to their mode of nutrition (their 'threptic' soul), is that they concoct their food inside themselves. The argument thus starts with a basic fact about all living things, then adds a basic fact about (in effect) animal soul, then derives the need for certain organ-systems. All this is a general framework within which existence of the stomach and associated organs is explained, as the organs which perform these necessary functions" (169).

mammal or animal.[10] The following argument, based on statements from *PA* I.4–5, would support this suggestion:

1. the matter of an animal should be explained in relation to the form that most properly defines the animal, for example, what it is to be a human being (I.5, 645b–36);
2. parts out of which an animal is composed are its matter (I.5, 645a25–36);
3. the form that defines the animal is the form of the indivisible species, for example, human being (I.4, 644a29–32);
4. the parts of the animals should be explained in relation to the form of the indivisible species.

So, the parts of animals should be discussed at the level of generality where the animal is most properly defined, which would mean the level of the lowest species, crow, sparrow or crane, say, rather than at a more general level such as bird or oviparous.

However, Aristotle also raises a procedural difficulty:

> **[T3]** Since, however, it is the last forms that are substantial beings, and these, e.g. Socrates and Coriscus, are undifferentiated in respect of form, it is necessary either to state what belongs generally first, or to say the same thing many times. And things that belong generally are common; for things that belong to many we call general. There is, however, a puzzle about which of these two should be our subject. On the one hand, in so far as

[10] *Atoma eidē*. This can also be translated "final species"; for this notion, see D. Henry (Chapter 6) in this volume. He defends two claims:

1. The epistemological claim. The aim of Aristotelian zoological division is to identify and organize the essential features specified in the definition of each indivisible species of animal.

2. The ontological claim. Those indivisible species constitute the *ousiai* of Aristotle's zoology. While I agree with his defense of (2), I am less convinced of (1). Henry argues that "Since the object of definition is always a substance, it would seem that in the zoological works animal parts constitute the substances 'in the primary and strongest sense'" (145). Indeed (as the title of his chapter suggests) Pellegrin argues that Aristotle's zoology has no need for animal species. I will return to Pellegrin's ontological thesis in Section 7.2 (animal parts constitute the real substances of the living world). In this section, I want to focus on the epistemological claim that, according to *PA* I.2–3 at least, the true objects of definition are "the parts, their properties, and their variations" *and not* the animal species themselves (145). By contrast, I shall argue that, on the most natural reading of *PA* I.2–3, the ultimate objects of definition are the indivisible species of animal to which those parts belong. However, the fact that *PA* I.2–3 presents the method of division of animal species by parts as differentiae does not mean that the parts are accounted for to define the species. Rather, Aristotle may mean to justify why the differentiae are the basis from which to account for the parts: they define the species. The essence is the explanatory starting points for the parts in that its differentiae will be referred to in the account of the parts. This alternative explanation is more readily compatible with two features of the *PA*: both that other differentiae are referred to frequently and that not all the differences are defining differentiae.

> what is indivisible in form is a substantial being, it would be best, if one could, to study separately the things that are particular and undivided in form – just as one studies mankind, so too bird; for this kind has forms. But the study would be of any one of the indivisible birds, e.g. sparrow or crane or something of this sort. On the other hand, in so far as this will result in speaking many times about the same affection because it belongs in common to many things, in this respect speaking separately about each one is somewhat silly (or 'absurd,' *hupatopon*) and tedious. (*PA* I.4, 644a23–b1)

Aristotle sees a tension here between two explanatory approaches to the parts of animals.

On the one hand, there is a preference, as we have seen, for an account of the parts at the level of the indivisible species, the so-called lowest or *infima* species. Aristotle's general view is that these species are ontologically prior to more common kinds. So, in *DA* II.3 (414b20–28), Aristotle says that "it is absurd to seek a common account (*logos*) in these and other cases (sc. like geometrical figures and souls) which is not a proper account of any of the things that are and is not in accordance with a proper and indivisible kind (*eidos*), while leaving out such an account" (414b25–28). While Aristotle is not saying in *DA* II.3 that there should not be a common account, the suggestion is that a common account is secondary in relation to an account of the indivisible species since this is the level of things that are.

To illustrate the point, imagine zooming in with your binoculars on a bird in the distance: at first you see features common to all birds, a wing, a beak, little by little features common to corvids, until a clear picture emerges of a crow. The other features were there in virtue of this crow. It is birds at the level of specificity of a crow that exist, the more undifferentiated features you saw at first are only there because there are crows, or jays, or sparrows. Dividing is like zooming in: only by homing in on the lowest species do we reach a point where the universal corresponds to real-world entities. If we insist that our accounts of parts should be real, then the accounts at the level the *infima* species would seem to have the strongest claim.

The countervailing consideration suggested by **[T3]** is this: if we begin from the level of individuals, like Socrates and Coriscus, it is clear that they will have most things in common, indeed everything that qualifies them as human beings. There is no need then describe these features relatively to each individual. But by the same reasoning, one might think that, since human beings share many features with other land-animals or blooded animals, those features should also be described at a higher level of generality than that of human being. Considerations of simplicity would seem then push us higher up toward the more general accounts.

Add to this that Aristotle says not only that it would be boring to go through the lowest species one by one but also that it would be "somewhat absurd", that is, I take it, lacking in rationality. A likely reason for this may be that there are features of the lowest species that belong to them not insofar as they are that species but insofar as they belong to a larger kind. So a human being has a heart not *qua* human being but *qua* blooded animal, as all blooded animals have a heart. "Blooded" is the term that explains why humans have a heart. Compare on this point Aristotle *APo* I.5, 74a26–33: "If you tried to demonstrate separately for each kind of triangle that each has internal angles equal to two right angles, you would not know [this], except sophistically . . . for you do not know [that it has two right angles] as a triangle, nor do you know it of every triangle, except in a numerical sense." The explanatory middle term is known as the "commensurate universal" in *APo* I.4. Similarly, to explain why the lowest species of animals have a certain part you may need to ascend to a feature of a more common kind. It would be less rational, and "somewhat absurd," to give the account here at the level of the ultimate species.[11]

We have then considerations for and against accounting for the parts of animals at the level of the indivisible species. On the one hand, it seems that the parts of animals should be explained at the level of the ultimate species since this is the most proper level of definition, and so the level at which we can best see how the parts contribute to the substantial being of an animal. On the other hand, many parts are common to many species of animal and are realized in them in virtue of an animal's more generic attributes and not because it is this particular species of animal.

In *PA* I.4 Aristotle moves to resolve the issue:

> **[T4]** Perhaps, then, the right course is this. In some cases – whenever kinds are spoken of by people in a clearly defined manner and have both a single common nature and forms in them not too distant – we should speak in

[11] The point is made by Lennox 2001a: 170 and developed in Johansen 2012: 64–66. See also Lennox [1987a] 2001b: 29–30. Henry 2021 is skeptical of this connection with the *APo*: "(Aristotle) need not have anything more complicated in mind here than the simple point that it is more efficient to give one explanation for why all birds have split wings than to repeat that explanation over and over" (95 n. 27). I take it, in contrast, that it is "somewhat absurd," "in the sense of less than rational to speak of individual cases when the part 'belongs in common to many things.'" But this point seems exactly to trade on the question of the level of generality at which an attribute properly belongs.

> common according to kinds, like bird and fish and any other there may be that, though it is unnamed, embraces, like a kind, the forms within it. But whenever they are *not* such as this, we should speak one by one, e.g. about mankind and any other such kind. (644b1–7)

We should speak of the parts together where they share a single common nature as do species of birds, but where the distance is greater, as between birds and fish, it seems that separate accounts are appropriate. The key point is whether there is a single explanatory feature, a common nature, that applies to the species. Where the species are too distant in not sharing a common nature, it is reasonable to think that the cause of their parts would not be of one kind or nature. Here the only way of bringing these different parts together in one explanation is by analogy. But, as we know from the *APo* I.10 (76a38–42), analogy has a lower explanatory value because the *explananda* do not fall under a shared genus.[12] So wings and fins may have similar functions in birds and fish, but you cannot demonstrate facts about birds from facts about fish.

Even if wings and fins are common by analogy why should there not be common explanations of them? They both perform the job of propelling the animal forward in their respective media. Admittedly, unlike perception, forward locomotion is not a function that birds and fish share *qua* animals, as shown by the fact that not all animals move. Yet they still share a highest genus, animal, so there is no reason to think that forward motion in the two cases would be ambiguous.[13] Why then insist on mere explanation by analogy? One answer might be that the shared function is specified at such a high level of generality that it has no specific consequences for the sort of material part that would serve it. We are after all, as emphasized in *PA* I.1 (640a33–b4), interested in explaining the different bodily parts of animals from the point of view of their forms. The point, as we saw, is developed by reference to the notion of hypothetical necessity. Aristotle envisages the formal-final cause to provide hypothetically necessary explanations of the material parts (642a32–b4). But, if so, I am suggesting, we may need to descend to the level of fish or bird form before

[12] See Lloyd 1996: 142–148 on the relationship between analogy and Aristotle's demonstrative method in the *APo*.

[13] Lloyd 1996: 142 says that "The feathers of birds and the scales of fish are not the same in genos in one sense, but they both fall perfectly satisfactorily within the purview of an inquiry into animals, where animal provides the determinate genos to meet the *APo* condition." The question is whether it is sufficient to have a shared genos 'animal' to provide an unambiguous explanatory term for the protective covering of fish and birds. Aristotle's emphasis on the distance of the two kinds – to be sure within the genos 'animal' – suggests that it is insufficient.

we can see why this form requires a certain bodily realization. The description of functions at too high a level of generality would not support such entailments.

To give one example, for an account of function to support an explanation of the bodily part as hypothetically necessary, we may need to factor in facts about how the animal lives, for example, whether it lives in water or air.[14] Simply saying that an animal is able to move forward will have only vague implications for the nature of its organs of propulsion. In contrast, forward propulsion in water will require a specific kind of body part, fins, as will progression in air, wings. But to bring in the environment is already to descend to a notion of function specific to these kinds of animals living in water or air.[15] Put differently, the level of specificity at which an account of the function ties up with an explanation of the animals' bodily parts is lower than the level at which the account applies indifferently to water and airborne animals, that is, we need to descend from the level at which they are one by analogy.[16]

There seems, on this interpretation, to be a justification for not giving common accounts of parts belonging to distant animals.[17] We can treat kinds that are relatively close as one, such as various birds that have a number of joint features as birds, while treating those separately that are most distant from each other, like birds and fish, or man, who in some ways occupies a special position in relation to all the other animals.[18] The distance in kind between the animals is reflected in the explanatory stance we take to their parts. Whether they differ at a high level of generality and

[14] Cf. *PA* III.4, 665b2–5: "Just as, with the external parts, there is not *a* use provided to all animals, but rather a distinctive provision has been made for each of them related to their ways of life and movements, so is it natural that the internal parts are also different in different animals." See further Lennox 2010c, who argues plausibly that "an animal's many parts and corresponding activities are provided with a unified explanation by identifying their contributions to the one bios, or 'way of life,' of that animal" (333). From the point of view of my argument here, the relevance would be that information about the way of life is crucial to generating the hypothetical necessities that explain the particular parts of an animal.

[15] Aristotle refers to the seal having rear feet that are "entirely fish-like," like fins (IV.13, 697b4–5), although the seal is a "dualizer," in some respects a water-animal, in others a land-dweller.

[16] Aristotle in general takes it that natural philosophy (as opposed, e.g., to mathematics) should account for the form in a manner that supports hypothetical necessitation of the matter, see *Phys.* II.2 and *DA* I.1 (cf. Johansen 2012: 147–57).

[17] It is worth noting that functional analogies for parts of the body sometimes extend outside living beings, for example, to irrigation systems (the blood vessels), anchors or nails (the viscera, III.7, 670a10, 14) and a hearth (the heart, III.7 670a25).

[18] See, e.g., *PA* I.1, 645b5–10; Lennox 1999.

are similar by analogy or whether they are the same kind and differ only in terms of degree,[19] 'kind' is here indexed to kind of living being.

> **[T5]** It is necessary first to divide the attributes associated with each kind that belong in themselves to all the animals, and next to try to divide their causes. Now it has been said before that many common features belong to many of the animals, some without qualification (such as feet, wings, and scales, and affections too in the same way), and others analogously. By analogously I mean that while some have a lung, others have, not a lung, but instead something different which is to them what a lung is to those that have one; and some have blood, while others have its analogue, with the same potential that blood has for the blooded. To speak separately about each of these animals as particulars, as we also said before, will result in saying the same things many times, whenever we speak about all the attributes; the same attribute belongs to many animals. (I.5, 645b1–14)

[T5] reiterates the message of **[T4]** that we can speak generally of parts that belong to many animals to avoid repetition.

However, one might still wonder on this account why the distance between kinds should by itself translate into a distance between the *parts* of animals of those different kinds. So one might in principle think that a bird and a fish could have some fundamental difference, say that one flies and the other swims, which would determine one kind having wings and the other fins, while taking it that they also share a large number of very similar parts. So while wings and fins might be the same only by the weaker relation of analogy, the other parts could differ by the stronger relation of the more and the less. In other words, there is no reason why a particular difference between two kinds, basic as it might be to differentiating them as kinds of animals, should translate into an equally basic *universal* difference between all the parts of those two kinds. One could point to the variety of differentiae that Aristotle uses to characterize the parts of animals. So, a fish is a water animal, as well as blooded and oviparous, while a bird is a flying, blooded, two-legged, oviparous animal. This in turn would suggest that some features of fish and birds will be common *qua* blooded, and so may differ by the more or less, while the means of propulsion, reflecting a more fundamental difference, will only be the same by analogy.[20]

[19] For some cases where Aristotle does not seem to apply the distinction consistently, see Lloyd 1996: 150–152.

[20] As J. Lennox points out to me, the fact that fish and birds are both oviparous also leads Aristotle in the *GA* to some explanations that are common to both, while others are different, cf. *GA* III.I–3.

The concern is then that the relative difference in kinds of animals does not allow us consistently to distinguish when their parts are the same merely by analogy or by the more or the less. Now there are at least two things to say in reply to this worry. One is that Aristotle sees the differentiae as closely interrelated. It is not the case that a fish operates as a blooded animal in isolation from its nature as a water animal. As Aristotle says,

> **[T6]** Yet if it is impossible for some indivisible and unitary form of substantial being to belong to animals that differ in form – rather, the form will always have a difference, as bird differs from mankind (for their two-footedness is other and different) – then even if they are blooded, either their blood is different, or blood should be reckoned as no part of their substantial being. (I.3, 643a1–5)

So, the options are either to treat "blooded" as different in the two animals that differ in form insofar as "blooded" is a differentia or to say that "blooded" is the same but does not enter into the definition of what the thing is. The latter alternative seems less attractive because the contrast between "blooded" and "unblooded" is a fundamental difference between kinds across the *PA*. Yet the former option might seem to open the door to homonymy in the terms that differentiate the kinds. If "blooded" differs when it is a differentia of dog and sparrow, representing different kinds of being for dog and sparrow, what it is to be "blooded" in these two cases must surely be something different. But if the being of "blooded" differs we seem to have a case of homonymy, that is, one term with two different definitions.

However, there is no need to draw this conclusion, as we can see from Aristotle's account of two-footedness in humans and birds. Certainly, the two kinds are two-footed in different ways: "[Birds'] legs are two, as with mankind, thought bent inward, as with four-footed animals, and not outward, as with mankind" (IV.12, 693b2–3). However, there is no ambiguity in terms of their being *animals with two feet*. In the same way, one may say that the two kinds are blooded in different ways, but there is no ambiguity in terms of their being blooded animals. The blood of these animals is in both cases a concoction produced by heat providing nutriment to the body, just as the two feet of both birds and humans support them and propel them forward in fundamentally the same manner. These are not cases of homonymy such as the living *man* and the *man* in a portrait or the living *body* and the dead *body*, where the difference does not arise from a specification of a shared functional property. The more general point here seems to be that differentiae allow for modification at the level of the lower kinds. As long as we allow for the different ways of being blooded as a specification of a more general way of being blooded, there is no need to worry about homonymy.

Aristotle's point that the parts of animals differ by degree presupposes that they are said without homonymy: it makes no sense to speak of one part being more less than another – say one animal having harder or softer skin – unless one is comparing them in the same respect. For the same reason that we may differentiate the parts by their degrees, we may also distinguish without homonymy the kinds by the degrees to which their parts are qualified by a property: some animals are more warm-blooded than others, for example.[21]

7.2 Parts and Holism

We began by asking whether Aristotle's procedure of discussing parts at a general level above the lowest species was consistent with the claim that it is the lowest species that are real and the fundamental objects of definition. So far this procedure has appeared justified. But it is worth considering the issue in more detail from the point of view of Aristotle's holism. It may seem here that the various explanatory factors are so interwoven that the appropriate level of accounting for living beings must generally be that of the ultimate species. This would be because the parts of the organism are so closely related to each other that it seems that they can only be understood if we consider the full range of explanatory factors, and this would inevitably bring the explanation down to the level of the indivisible species.

We start by observing the functional interconnectedness of the parts. The functions of the parts of animals form a hierarchical structure with a common point at the top, the essence of the animal. We see this clearly in *PA* II.8–9, where Aristotle treats the flesh and the bones. Flesh is a particularly central example since all animals must have it, or an analogue, to possess the basic capacity that defines them as animals, perception. Flesh is the organ of touch, without which there can be no perception. Moreover, the flesh is basic for the composition of the whole body in the sense that all the other uniform parts – bone, skin, sinews, blood vessels, hair, etc. – are for the sake of the flesh (653b30–33). These parts serve to support or protect the flesh. Aristotle has then a general functional

[21] On the importance of "the more and the less" in differentiating the animal kinds, see Lennox [1987b] 2001b: 166–167.

conception of flesh, as an instrument of touch, which applies to all animals, and he has a corresponding general functional conception of the other uniform parts as preserving the flesh. This function allows us, unambiguously, to fix a single reference not only for "flesh" and "its analogue" but also for the subservient parts. So, bone is a hard structure for "the preservation of soft tissue," with fish-spine or cartilage as the analogue parts in other animals (653b32–35). From this common starting point, the flesh, Aristotle now differentiates between the parts on the basis of how they preserve the flesh or its analogue in different animals. So, some animals have bones inside, others (like crabs) on the outside. The explanation for this difference is that the latter animals have relatively little internal heat and therefore need the covering on the outside to preserve it, like a lid on a saucepan. Other animals, like insects, have neither bone nor soft flesh but a substance of a nature in between flesh and sinew, which provides sufficient structural integrity for them not to require the additional support of bone (654a8–19).

Different animals have the function of touch, then, but have different parts to support this function. We might say that while their parts are compositionally different, their composition presents different solutions to the same problem of how to protect the animal's basic perceptive part. While there is no reason in principle why one function should not be realized by a part with the same constitution in two kinds of animals, we find that the organs vary according to a range of other constraints to do with their aims, ways of living, environment, size, shape, and weight, available material, etc. Explaining how a part is actually composed requires us not just to consider how perception in general might be best realized but how it might be best realized for a creature that is subject to all of these constraints.[22] Given the complicated matrix of factors involved, such an explanation, we might expect, will naturally tend 'downwards,' away from general facts about animals or their most common kinds, toward the lowest species. For each factor will introduce further complexities that bring us closer to the sort of animal that actually exists, the ultimate species.[23]

[22] As Lennox 2001a: 335 observes, "The organs are constructed out of the various uniform parts, which are constructed out of the blood; what is optional is the precise choice of organs to be constructed, and thus the shapes, sizes, and combinations of uniform parts to be made in particular locations. It is the functions required by the organism's way of life that put constrains on this process, constraints which define the actions of the formal nature." I take it that Lennox by 'optional' here means something like 'variable.'

[23] See further Henry 2021: 86, who rightly emphasizes the point.

A good illustration is when the part serves two or more functions. Here we often need to look at the ultimate species to understand how the functions are combined. So, Aristotle says in *PA* III.1:

> **[T7]** All those animals that breathe and cool themselves from outside have the nature of the mouth both for the sake of these functions and for respiration besides. For nature, in virtue of itself, as we just said, puts the parts common to all animals to many distinctive uses; for example, in the case of the mouth nourishment is common to all, while strength is distinctive to some and speech to others, and again breathing is not common to all. But nature has collected all these uses together in one, producing a differentiation of this part for the differences of its operation. That is why some mouths are narrower, some wider: those which are for the sake of nourishment, respiration, and speech are narrower, while of mouths which are for the sake of protection, all that are sawtoothed open wide. For since their strength lies in their biting, it is useful for the opening of the mouth to be large; for the mouth bites with more teeth and over a larger area to the extent that it opens more widely. (662a18–23)

Many body parts serve several functions. From the point of view of ingesting food, there may be no difference between the sort of mouth two species need, but combine ingestion with speech, and together the two functions will demand that the human mouth be narrow, while a dog, in contrast, needs a mouth that opens wide when it bites. Where a part has more than one use, Aristotle in some cases refers to one use as "common" (*koinē chrēsis*) and others as distinctive according to kinds. So, ingestion is the common function of the mouth, while different animals will employ it for various other purposes. Similarly, tasting is the common function of the tongue in all animals, but some insects also use it as a sting for defense (IV.7). The additional secondary function determines that the organ will differ in different animals despite its common function. Each body part is here to be considered from the point of view of a matrix of functions characteristic of the kind.

The combination of functions illustrates how parts are composed to serve the needs of specific kinds of animals. If an animal needs both to taste and produce speech, it requires a tongue of a certain sort that will offset it from that of an insect say. While the combination of sound-production and tasting are not unique to humans – birds too are vocal and so require broad and flexible tongues (II.17, 660a34–36, cf. *HA* IV.9, 535b1–2) – we might say that the increased need for articulation of sound in a rational animal, which learns by *logos* (*De Sensu* I.1, 437a10–15), positively requires this kind of tongue for a human being.[24] The elephant provides another such case where we need to descend to the lowest species to understand why the trunk is composed just as it is.[25]

We have seen how the combination of functions constrains the composition of a part. But the list of determinants is much longer. Sometimes the animal's size and weight dictate a certain arrangement of the parts. So the size of the horse and of certain oxen requires them to have a bone in their hearts, while the weight of the elephant requires it to have solid hoofs. The environment is a key factor in other cases. So fish are generally saw-toothed as their watery environment requires them to cut up the food fast (III.1, 662a2–15). For the same reason they have no or only a short tongue: the water that enters the mouth gives them little opportunity to taste the food (*PA* IV.11, 690b24–31).

Further examples abound when we turn to parts that support other parts or compensate for the lack of other parts. So animals with a heart have a liver, to aid its function, while animals with a blooded lung have a bladder to deal with the excess moisture that comes from drinking a lot. Other parts are present or composed in specific ways because the animal lacks other parts that an animal of such a kind might be expected to have. So Aristotle says that flies and bee-like insects have long forelegs to remove obstacles in front of them, thereby compensating for their lack of clear vision (IV.6, 683a26–31). Another example is the bladder of the tortoise: a hard-scaled animal would not normally possess a bladder, but because the tortoise has a large lung and so takes in much liquid, which it cannot secrete through the skin, it requires a bladder (III.8, 671a15–26). Conversely, some animals lack a certain part because they have another part that already does the job for them. So many-toed animals have no horns because nature has provided them with other means of protection, such as claws or teeth (III.2, 66b30–34).[26]

These are different examples of how animals need specific parts to serve the totality of their functions. Their explanations are final-formal in the sense that they do not mention material parts, but in that those parts are

[24] As Lennox 2001a: 149 notes, the comparison between animal parts and artifacts "may suggest a view congenial to modern 'functionalist' theories. Eye material will vary from one class of organism to another (cf. *PA* II.10, 657a25–658a10), but such variations are related to differences in the organisms' lives which require differences in function, structure, and material." On the point, see also Johansen 1997: 283–287.

[25] The elephant's trunk is probably the most celebrated case of one part with many functions, cf. *PA* II.16, 658b32–659b35; and Gotthelf 1997b.

[26] On the principle "nature does nothing in vain," see Lennox 1997.

viewed from the point of their enabling or constraining the functions of the animal's other parts. However, there are also explanations that explicitly refer to a material cause as a complement the final-formal, as in the case of the human hair:

> **[T8]** With respect to the head, mankind is the most hairy of animals, *from necessity,* on account of the moistness of the brain and on account of the sutures (for where there is much moisture and heat there must be much growth), and *for the sake of* protection, so that it may provide covering, warding off the extremes of both cold and heat. And since the human brain is the most moist, it is also most in need of this protection; for what is moist boils and freezes most easily, while what is in the opposite state is less easily affected (II.14, 658b2–10).[27]

Here the final-formal causal and material cause line up: the hair is both hypothetically necessary for protection and necessary because of the moistness of the human brain. Sometimes, however, the matter works in the absence of a final-formal explanation. Why do four-footed, egg-laying animals not have upper and lower eyelids? Because scaly skin is too hard to shape into eyelids (*PA* II.13, 657b11–14). Of course, there is a background final causal explanation for the scaly skin: it serves the animal's protection. But relative to the *explanandum*, the absence of eyelids, the explanation counts as material. Similarly, hard-skinned animals with dry flesh have no epiglottis, because this matter cannot be shaped into this part. Instead, their larynx opens and shuts to prevent the food slipping down the wrong way causing the animal to choke (III.3, 664a35–665a5).

Sometimes the material explanation is not determined, as in the manner of scaly skin or dry flesh, by the fact that the matter has already been coopted for another function, but simply by there not being enough of it to go around. So, the bear does not have much hair on its tail because all the hairy material has already been used on covering the rest of its body. "For nature everywhere gives to another part what it takes from elsewhere" (II.14, 658a36–37).[28] Also, the material explanation here requires us to consider facts about the species as a whole. We would expect the final causal explanations to be holistic since, as we saw, the various ends and uses of the body parts connect both with each other and ultimately with

[27] Similarly for the viscera: "So, then: what each of the viscera is *for the sake of* has been stated, and they have come to be *of necessity* at the internal limits of the blood vessels. That is, it is necessary that a bodily fluid diffuse, and that this one, from which the body of the viscera come to be when it becomes constituted and solidified, be bloody" (III.10, 673a22–b1).

[28] J. Lennox draws my attention to another such example: the lack of upper teeth in animals with horns, which in turn necessitates additional stomachs. The case is analyzed by Gotthelf 1987b.

defining features of the animal. We saw this point amply illustrated in the case of the flesh and the various structures that followed upon flesh: an animal was defined by touch, which required flesh and from there a range of supportive structures that worked out differently according to the animal's other needs given its environment, way of life, etc. However, it is noteworthy that also the material causal explanations we have looked at are holistic, drawing in each case on other facts, down to the availability of matter given its distribution to other parts of the body. The final and the material causal explanations we have looked at are all holistic focusing on what is necessary, possible, and best for the animal in question.

The downward pressure of holism toward explaining the parts at the greater level of complexity and so also of specificity does not stop at the ultimate species. For Aristotle draws in also differences in sex and age within the lowest species. So, the female deer does not possess horns, while the male does (III.1, 661b32–662a2). There is a final cause here in terms of usefulness: the male is stronger and has greater use for the horns. Children, like the lower animals, are described as 'dwarf-like' (*nanōdē*) because of the disproportionate size of their upper body (IV.10, 686b6–25). While these observations about differences within the same species in a way go beyond the brief of the *PA*, accounting for the parts common to and different to different kinds of animals, there is a sense in which they lie in extension of the brief. For as in the case of children, the feature is one that matches a difference between kinds of living beings and have a shared cause. Female deer are brought in to illustrate a general point about animals that nature only provides them with tools that they can use, while children are mentioned in the context of discussing animal kinds that are 'dwarf-like.' One might say that intraspecies differences are adduced to confirm general causal points about the species rather than for their own sake.[29]

7.3 Holism and Indivisible Species

Aristotle recommended in **[T4]** that

> Whenever kinds are spoken of by people in a clearly defined manner and have both a single common nature and forms in them not too distant – we should speak in common according to kinds, like bird and fish and any

[29] As S. Connell suggests to me, this could help explain, to some extent at least, the collection of such sub-specific differences in the *HA*, which D. Balme took to mean Aristotle was not an essentialist, cf. Balme 1987c: 298.

> other there may be there may be that, though it is unnamed, embraces, like a kind, the forms within it. But whenever they are *not* such as this, we should speak one by one, e.g. about mankind and any other such kind. (644b1–7)

How does this recommendation fare in view of the holistic accounts we have considered? One might say that the individual references to indivisible species that we have come across do not by themselves contradict Aristotle's claim in **[T3]**; they just qualify as those cases where we cannot speak of the parts of a kind in virtue of the features that all animals of a more common kind possess. But this would not do justice to the holistic pattern of the evidence. For the pattern does not suggest so much that the individual species provide exceptions to be separately accounted for but rather that the interconnectedness of the causes of animal parts is such that an adequate explanation of a part *generally* tends towards the level of the indivisible species.

It could be objected that the interplay of many causal elements is not a problem for generality if the contribution of each of these causes can be understood generally. So, to use an analogy, one might say that one complicated structure built out of Lego bricks can be accounted for in terms of the same bricks as another complicated structure: the bricks themselves are the same. Similarly, if each of the differentiae can be understood independently across the species, then any combination of these differentiae, even if unique to that species, can be understood as a combination of those modules.

However, as we have seen, causes do not work in the manner of Lego bricks or Democritean atoms. Parts of animal bodies are not modules that are the same across different structures. Animal bodies are not compounds of identical modules. The differentiae are not the same across the animals they differentiate. Aristotle indicated that 'blooded' would differ according to the kind of animal involved. Earlier I suggested that indexing 'blooded' to a particular species need not be an obstacle to talking about it also in more general terms, as long as we saw blooded in each species as a specification of a more general attribute. However, while this remains true, the upshot of Section 7.2 is that it is the interplay of the explanatory factors that determines the parts of an animal. We can only work out the parts of an animal by considering how all the differentiae work together.

This point goes back to Aristotle's anti-Platonic approach to division: we cannot arrive at an explanation of an animal's constitution by beginning with a simple differentia (*PA* I.3 644a6–11). We need to keep all of a number of differentiae in play to understand how particular parts are worked out. To give one final example, we see how the conjunction of three differentiae determines the constitution of the lung:

> **[T9]** The lung differs in many ways in animals. Some have a blooded and large one, others a smaller and spongy one. The live-bearing animals, on account of the warmth of their nature, have a large one with much blood. The egg-layers have a dry and small one, though it is able to open wide when being expanded, as for instance with the four-footed, egg-laying land-dwellers, e.g. the lizards, tortoises, and all such kinds, and in addition to these the animals that have a winged nature and are called birds. For the lung of all of these animals is spongy and foam-like; and in fact when foam is stirred a large amount becomes small, and the lung of these animals is small and membranous. (III.6, 668b24–33)

Four-footed, egg-laying, land-dwelling are here the differentiae that pick out animals, like the lizard, which have small lungs capable of expanding.[30] However, this combination of differentiae is not necessary for having this kind of lung since also birds have it. Being oviparous is the common differentia of the lizard and the bird here. Being oviparous in turn correlates with being less warm than the live-bearing animals, which require a larger blood-filled lung. Yet being oviparous is clearly not on its own sufficient to pick out animals with this structure, as there are other oviparous animals (insects, for example) that lack a lung, let alone this kind of lung. It seems necessary therefore to involve a range of differentiae to circumscribe the relevant kinds even though there is one differentia, which in this context is particularly explanatory of the part.[31]

To understand then how a specific part is worked out we need in principle to survey all the differentiae of the animal kind. The challenge before us is how one designs a body for an animal that is both, say, oviparous, warm-blooded, water-dwelling, four-footed, etc.[32] And even

[30] Yet a further consequence of having this sort of lung for four-footed egg-layers is that their spleen is "small, firm, and kidney-like" (III.7, 670b13–14).

[31] The correlation between being live-bearing and being warm allows us here to highlight the explanatory factor amongst a group of differentiae. But this is not always so. There is, for example, a correlation between the shape of the spleen and the form of an animal's feet, but no indication of why that should be so. "The horn-bearing and cloven-hoofed animals, such as goat, sheep, and each of the others, have a round spleen, unless, on account of the animal's size, it grows more quickly lengthwise, as happens with the spleen of the ox. The many-toed animals, however, such as pig, human being, and dog, all have a long spleen, while those with solid hoofs, e.g. horse, mule, and ass, have a spleen intermediate between these and a mixture of the two; compared with the one group they have a wide one, but compared with the other a narrow one" (*PA* III.12, 673b32–674a4).

[32] The differentiae here may be variously ordered according to their importance to the part to be explained. As I have been at pains to emphasize in this chapter, the differentiae are not necessarily part of the essence of the animal though they may follow in various ways from the essence.

with a good solution on our drawing board, this is not always nature's way. For sometimes the character or the quantity of matter will dictate another solution. To say that nature works for the best is here to admit that she will find the best solution *all things considered.*[33] But all things are many and variable and the greater the complexity the more likely it is that the explanation will refer to the indivisible species.

A first reading of [T4] might have suggested that descending to the indivisible species would be the exception. However, we have seen that a complex of differentiae and material constraints determine the composition of the parts. From this point of view, it is unsurprising if Aristotle's accounts of the parts refer to indivisible species since his explanatory holism tends toward accounting for animal parts at the level of lower kinds. However, it is important not to overstate this conclusion for at least two reasons.

First, that greater complexity of explanatory factors exerts a downwards pressure *toward* explaining the part at the level of the indivisible species does not mean that the explanation generally descends *all the way down* to this level. If the aim is to explain why a part is as it is in certain animals, there can be good reason to pick the "commensurate universal" as referred to in *APo* I.4 at a higher level of generality than the indivisible species. So several insect species have a sting for the same reason, just as it is true generally of hard-skinned animals with dry flesh that they have no epiglottis. Here the explanations are rightly at a higher level of generality than the indivisible species.

Second, as we are considering parts of animals that are not too distant and whose parts differ only by degrees, differences in degree between parts that are quite small may not be tracked by Aristotle's accounts. Think, for example, of the finer differences between the feet or beaks of various bird species, which Aristotle passes by in silence.[34] There is little doubt that Aristotle could have given accounts of these differences given his wide arsenal of explanatory terms, yet the differences of degree may be too slight to merit attention since our focus is not on the features of particular species as such. So long or short beak, straight or crooked, is of interest (662b5–16), but the smaller variations of short or crooked beaks that

We should not expect therefore that the order of the differentiae in the definition of the animal kind is followed in the explanation of a particular part of that animal's body.

[33] As J. Lennox points out to me, the statement of the principle that "nature does nothing in vain" in *IA* 2 (704b15–16) explicitly adds "among what is possible."

[34] Beautifully illustrated in Lennox [1987b] 2001b: 163–164.

characterize particular species fall under the radar.[35] Again, we should bear in mind that Aristotle focuses on explaining the differences between the parts that obtain between animals, rather than the differences between species as such. The descent to the indivisible species, when it occurs, is motivated by the need to explain significant variation in the parts.

To sum up. Aristotle in the *PA* is interested in explaining the body parts of animals. This requires direct or indirect reference to essential features of animals, to which those parts contribute, directly or directly. Some, like blood or flesh, can be directly explained by reference to the form and function of general animal kinds, nutrition or touch, respectively. Other parts are explained in relation to these parts, bones or sinews, for example, which in turn may serve to explain further features within the body. Differences between the parts in different kinds of animal can sometimes be explained at a high level of generality. Here differentia invoked may be one or few, like the difference between oviparous or viviparous, blooded or non-blooded. However, sometimes the explanandum calls for a wider range of explanatory factors spanning function, life form, environment, matter, etc. Greater complexity of explanatory factors tends to be correlated with lower levels of generality in the explanandum, with the greatest complexity reserved for indivisible species. This is unsurprising since it is at this level that nature ultimately has to reconcile all the different constraints on the realization of form and function.

[35] *GA* V.1 suggests, Lennox [1987b] 2001b argues, the explanation that such differences make no value difference to the life of the animal.

CHAPTER 8

Analogical Unity and Biological Kinds in Aristotle's Parts of Animals

*Christopher Frey**

Everything whatsoever is like everything whatsoever, in a way.
—Plato, *Protagoras*, 331D

Introduction

Aristotle's *Parts of Animals* is replete with analogies. A bird's feathers are analogous to a tortoise's horned plates, a whale's lungs are analogous to a fish's gills, our hands are analogous to a lion's front paws . . . one finds dozens of them.[1] But the ubiquity of these comparisons belies just how curious it is to find them in a supposedly scientific treatise like *PA* at all. Analogies typically occur in metaphorical speech as when Pericles pronounces that the city, when it lost its young men to war, was like the year losing its spring. Aristotle praises metaphors that employ analogical comparisons as especially vivid and rhetorically effective, and he recognizes that they can facilitate understanding.[2] But he insists, quite rightly, that their purpose is not to express anything essential about their terms. Metaphorical analogies are not to be found among the definitions and demonstrations a serious scientific investigation aims to offer and Aristotle chastises others for relying on them in their explanations of natural phenomena.[3] So why do analogies occur so frequently in *PA*? What role do they play?

* I'd like to thank Sophia Connell, Klaus Corcilius, Jessica Gelber, Devin Henry, Sean Kelsey, Jim Lennox, Mariska Leunissen, and Michail Peramatzis for discussions and comments that have informed and improved this chapter. I use Lennox's translation of *Parts of Animals* and the Revised Oxford Translation for other texts with occasional modifications.

[1] *PA* IV.11, 691a14–17; I.5, 645b1–10; and IV.10, 687b27–688a2.

[2] Cf. *Rh.* III.4, III.10, 1411a1–3, and *Poet.* I.21, 1457b8–24.

[3] Against Empedocles, for example, Aristotle says, "It is equally absurd to suppose that anything has been explained by calling the sea 'the sweat of the earth', like Empedocles. Metaphors are poetical and so that expression of his may satisfy the requirements of a poem, but as to knowledge of nature it is unsatisfactory" (*Meteor.* II.3, 357a24–29). As G. E. R. Lloyd rightly notes, "The very fact that proportions may be no more than 'metaphorical' warns us that the mere discovery of a similarity in a

The value of the analogies in *PA* isn't merely poetic or didactic. They are, I will argue, indispensable to the scientific investigation Aristotle undertakes in that work. For many analogies in *PA* express relations strong enough to ground a unique variety of unity. This will be my focus. What is *analogical unity*? What sort of relationship does an analogy have to capture for there to be such a unity? What role does it play in the scientific study of animals and their parts?

I begin by contrasting analogical unity with two different but related varieties of unity: *formal unity* and *generic unity* (Section 8.2). Aristotle associates each of these varieties of unity with a distinctive way for organisms and their parts to be the same or different. I then examine the numerous analogies in *PA* to discern which of the proportional relationships they express yield analogical unities (Section 8.3). The most promising interpretations of these passages, I contend, risk analogical unity's collapse into one of the other varieties of unity Aristotle accepts. I attempt to preserve analogical unity's unique explanatory role by turning to Aristotle's use of analogy in his *Metaphysics* (§IV). I argue that Aristotle employs the same concept of analogy in both *PA* and the *Metaphysics*. The broader perspective this consonance affords will allow us to see how the analogical unities in *PA* are genuine, irreducible, and explanatorily efficacious. But before I begin, it will be fruitful to say more about analogy's significance to Aristotle's project in *Parts of Animals* (Section 8.1).

8.1 Analogy and the Scientific Study of Animals

According to Aristotle, every science (*epistēmē*) is concerned with a determinate and unique subject genus (*genos*).[4] A science's subject genus is what it is about. For example, arithmetic's subject genus is number and geometry's is spatial magnitude. A subject genus is often a well-defined kind. But if it is not, it must still possess some sort of unity; there must be something that ties the individuals a subject genus comprises together in a way that makes it reasonable to study them through a single form of scientific inquiry.

There are several ways to secure the required level of unity for a subject genus when a univocal definition is unavailable. For example, there is no well-defined account, at any level of generality, that is common to the ten

relationship does not tell us anything fundamental about the members of each pair of relata" (Lloyd 1996: 158).

[4] Cf. *APo.* I.7, 75b3–14; I.28, 87a38–39; *Metaph.* IV.1, 1003a24–25; IV.2, 1004a4–6; and VI.1, 1025b18–21.

categorially distinct ways *to be* (*to on*) is said (*Metaph.* IV.2, 1003a33). But for a science of being *qua* being to be so much as possible, its subject genus must be united somehow. Aristotle supplies this unity by invoking what has come to be called focal meaning. One meaning of *to be*, namely, to be in the sense of substance, is prior in account to each of its other meanings. This priority yields enough unity for *being* to be investigated scientifically despite the irreducible multiplicity of ways *to be* is said.[5]

Analogy is another means to unite what would otherwise be distinct according to the classifications that definitions determine. We frequently encounter things that don't belong to any shared kind that nevertheless seem to be closely connected. It is often appropriate in such cases to introduce one predicate to refer to them all even though there is no single, univocal account that they equally satisfy. And Aristotle's biological explanations often involve classifications that aren't determined by univocal accounts but are, in this way, analogous.

In *Posterior Analytics*, Aristotle makes room for scientific explanations that employ analogical unities. He says,

> you cannot get one thing which pounce [i.e. a hard, porous structure in certain kinds of cuttlefish] and [fish-]spine and bone should be called; but there will be things that follow from them too, as though there were some single nature of this sort (ὥσπερ μιᾶς τινος φύσεως τῆς τοιαύτης οὔσης). (*APo.* II.14, 98a20–23)

As pounce is to cuttlefish so fish-spine and bone are to fish and blooded animals, respectively. The unity this analogy affords is an explanatory unity; it reflects a common cause and allows the analogous features to either explain or be explained by other features the analogs share.[6] So there will be contexts in which one can employ analogically united classes in a way that is explanatorily indistinguishable from the way one would employ a well-defined kind or species.[7] And Aristotle encourages us to adopt a method that allows us to use what is "common by analogy" (τὸ κοινὸν κατ'ἀναλογίαν) in precisely this way (*PA* I.5, 645b1–14, 27).[8] So, when

5 Priority in account is an asymmetric dependence relation: *x* is prior in account to *y* iff (i) the account of *x* does not contain the account of *y* and (ii) the account of *y* contains the account of *x*. Cf. *Metaph.* VII.1, 1028a31–32 and IV.2, 1003b12–18. I briefly discuss another way to supply the requisite unity for a subject genus in n. 50.

6 The commonality of this cause will likely be analogical as well, for "things which are the same by analogy will have their middle terms the same by analogy too" (*APo.* II.17, 99a15).

7 Cf. Gotthelf 1997a: 208–209; 1997b: 194–195; Henry 2014: 165–166.

8 Cf. *PA* I.1, 639a15–b5 and I.4, 644a25–b15 where Aristotle's focus is on what is "common according to kind" (κοινῇ κατὰ γένος).

animals belong to a kind or species, they possess "a single common nature," but when animals are merely analogically united, we can nevertheless proceed "as though there were some single nature of this sort."

Analogical unity transcends the gaps that separate discrete kinds. What definition keeps apart as many, analogy brings together as one. We can employ these analogical unities in our explanations of animals' parts. But, more importantly, I will argue (in Section 8.4) that the class of animals itself, that is, the subject genus of the scientific study to which *PA* belongs, is also an analogical unity. If correct, a properly delineated scientific inquiry into animals and their parts could not exist if it were not for the variety of unity analogy affords.

8.2 Three Varieties of Unity

Like so many of the philosophically rich concepts he discusses, Aristotle insists that *unity* and *one* are said in many ways.

> Again, some things are one in number, others in form, others in kind, others by analogy; in number those whose matter is one, in form those whose account is one, in kind those to which the same figure of predication applies, by analogy those which are related as a third thing is to a fourth.[9]

This passage mentions three ways for a (numerical) multiplicity to be one which I will discuss in turn: (i) in species/form (εἶδος) [*formal unity*], (ii) in genus/kind (γένος) [*generic unity*], and (iii) by analogy (ἀναλογία) [*analogical unity*] (*Metaph.* V.6, 1016b31–35).[10]

Formal Unity

According to one usage, genus/kind and species/form are relative designations that occur at different levels of generality. When a class of individuals share a significant number of attributes, one can often use the more determinate realizations of those attributes as differentiae (διαφορά) to divide it into smaller subclasses. We then call what we divide a genus and the division's results species. For example, galliformes (i.e. terrestrial poor fliers) are both a species (of the genus bird) and a genus (of the species

[9] Cf. V.9, 1018a7ff., X.1, 1052a15ff.; and *Top.* I.7, 103a6–16. We can set aside the first variety of unity, being one in number, since our concern is what unites a numerical multiplicity.

[10] "For what is common some have according to analogy, some according to kind, others according to form" (*PA* I.5, 645b27–28; Cf. *HA* I.1, 486b18–21).

partridge and quail) since it is the intermediate result of two such divisions.[11]

But there is also a nonrelative usage of species/form that applies only to those species that aren't subject to further determination. In *PA*, Aristotle calls these "narrowest" specific classifications the indivisible or atomic forms (τὸ τῷ εἴδει ἄτομον), the undifferentiated forms (τὸ εἶδος ἀδιάφορα), and the final or ultimate forms (τὰ ἔσχατα εἴδη).[12] There is a parallel, nonrelative usage of genus/kind. Antipodal to the atomic forms are the greatest kinds (μέγιστα γένη), namely, those kinds that aren't the result of division by differentia.[13]

Aristotle employs the nonrelative use when he says that things one in form are "those whose account (λόγος) is one." This means that formally united individuals have a common, determinate essence. For if one properly and completely defines something, then one knows what it is (τί ἐστι); the definition will be an account that captures a thing's essence – the "what-it-is-to-be" (τὸ τί ἦ εἶναι) for something. The definition of something's final form is the closest possible expression of its substance (οὐσία). For "it is the last forms that are substantial beings and these, e.g. Socrates and Coriscus, are undifferentiated in respect of form" (*PA* I.4, 644a23–25).[14]

So, Socrates and Coriscus are one in form – they are both human beings – and the account of *human being* is indivisible.[15] That is, if one were to propose further divisions of the form *human being* into those humans who are pale and those who are tanned or into those who are at least six feet tall and those who are shorter, this would yield, at best, accidental classifications. When one arrives at something's form by the method of division, one reaches determination's limit. Any division that transgresses this limit fails to further reveal what a thing is; it ceases to capture something's essence and reflects only how it may be accidentally qualified.[16]

[11] Cf. *HA* IX.7–36, IX.49B, 633a30-b2; Balme 1962; Pellegrin 1986; and Lennox 2001a.

[12] *PA* I.4, 644a24–25, 29.

[13] Cf. *PA* IV.8, 683b26–30; *HA* I.6, 490b7–491a3; and II.15, 505b25–30.

[14] Cf. *Cat.* 5, 2b7–8, *PA* I.4644a29–31; *Metaph.* VII.4, 1030a3–18; and *APo.* II.3, 90b30.

[15] For "the final differences must be equal in number to all the animals that are indivisible in form" (*PA* I.5, 643a18–20) and "when we have the whole, such and such a form in this flesh and in these bones, this is Callias or Socrates; and they are different in virtue of their matter (for that is different), but the same in form; for their form is indivisible (ἄτομον γὰρ τὸ εἶδος)" (*Metaph.* VII.8, 1034a5–7). Cf. V.6, 1016a33–b1, X.1, 1052a32-b1; and Devin Henry's contribution (Chapter 6) to this volume.

[16] Cf. Lennox 2001a: 174. Aristotle uses the language of formal difference to mark an accidental division only once: he says that humans can differ formally if they live in hot or cold countries

Animals' parts can also be one in form. There is a maximally determinate account that serves to define what each part is. And, as with animals, several parts are one in form when they are "without any difference according to their general account (κατὰ τὸν καθόλου λόγον μηδεμίαν ἔχει διαφοράν)" (*PA* I.5, 645b25–26).[17] Though Coriscus' eyes and Socrates' eyes can differ accidentally (say, in color or shape), what it is to be a human eye is the same for each.

Aristotle connects the formal unity of animals and their parts as follows.

> **[Unity 1 – Formal]** Distinct individual animals are one in form just in case most of their respective parts are one in form.[18]

For, "sometimes the parts are identical in form as, for instance, one man's nose or eye resembles another man's nose or eye, flesh flesh, and bone bone; and in like manner with a horse, and with all other animals which we reckon to be of one and the same form; for as the whole is to the whole, so each to each are the parts severally" (*HA* I.1, 486a16–21).

Generic Unity

Generic unity is present "whenever kinds are spoken of by people in a clearly defined manner and have both a single common nature (μίαν φύσιν κοινὴν) and forms in them not too distant (μὴ πολὺ διεστῶτα)" (*PA* I.4, 644b1–5). This quote contains (i) a claim about unity – kinds have univocal definitions that express a single, general essence or nature common to the kind's more determinate forms[19] – and (ii) a claim about difference – a given kind's forms are "not too distant." The former points to a common feature of generic unity and formal unity: both kinds and forms have univocal accounts. For example, not only the kind *bird* but also what it is to be a bird are predicated of both herons and swans. Being a bird is not the most determinate expression of what it is to be a swan. That would be given by the account of the swan's form. Nevertheless, a kind's

(*Long.* 1, 465a3–10; Pellegrin 1987: 317). Following Balme 1962: 95–96, I treat this sole example as an infelicitous usage.

17 To call this account general is not to imply that it is an account of a genus; it is a general (i.e. not individual) account of a specific form, not an account of a (general) kind.

18 This formulation is causally neutral and expresses only a correspondence. I will keep the question of which, if either, side of the correspondence is explanatorily prior to the other in abeyance until Section 8.3.

19 Cf. *Metaph.* X.8, 1057b37–1058a6. In Section 8.4, I will discuss an alternative interpretation that takes this common nature to be material rather than formal (Henry 2014).

account is a legitimate, though comparatively general and determinable, expression of a sparrow's essence.[20]

Animals' parts can also be one in kind. And, as before, there is a correspondence between the generic unity of animals and the generic unity of their parts.

> **[Unity 2 – Generic]** Animals are one in kind and formally distinct just in case (i) most of their respective parts are generically united and (ii) a sufficient number of their respective parts have forms that differ in account by degree.

A heron's legs and a swan's legs share a generic account; they are both *bird legs*. But how are we to understand the "not too distant" differences in degree between formally distinct animals and their parts?

Aristotle says that "animals that differ by degree and the more and the less have been brought together under one kind" (*PA* I.4, 644a16–18). Animals differ by degree when a sufficient number of their parts have accounts that differ by degree.[21] And parts have accounts that differ by degree when these accounts include differences in "bodily affections" (σωματικοῖς πάθεσιν) that are expressible by the relations *more* and *less* (*PA* I.4, 644b7–15). Aristotle mentions more/less relations between bodily affections that occupy positions within the continua of change that the contrarieties large/small, long/short, soft/hard, smooth/rough, broad/narrow, dense/rare, and dry/wet fix.[22] But there are myriad perceivable and measurable differences in parts' quantity, quality, shape, arrangement, and number that can be included in the account of its form.

So, whereas differences in degree between two formally identical parts are accidental, differences in degree between two formally distinct parts that are one in kind are essential. It is an accident that my legs are longer than my neighbor's. But herons are essentially long-legged and swans are essentially short-legged.

[20] See Gotthelf 1985 for evidence that Aristotle calls not only the accounts of forms but also the accounts of kinds at various levels of generality, definitions.

[21] "In other cases the parts are identical [in form], save only for a difference in the way of excess or defect, as is the case in such animals as are of one and the same kind .... Among them, most of the parts as a rule exhibit differences through contrariety of properties such as color and shape, in that some are more and some in a less degree the subject of the same property; and also in the way of multitude or fewness, magnitude or smallness, in short in the way of excess or defect .... As a general rule, most parts and those that go to make up the bulk of the body are either identical [in form] with one another, or differ from one another in the way of contrariety and of excess and defect (καθ' ὑπεροχὴν καὶ ἔλλειψιν). For the more and the less (μᾶλλον καὶ ἧττον) may be represented as excess and defect" (*HA* I.1, 486a21–b17).

[22] Cf. *PA* I.4, 644a16–21, b13–15; IV.12, 692b3–9; and *Metaph.* VIII.2, 1042b22–25.

Analogical Unity

To understand generic unity, we had to answer three questions.

[G1] What unites many forms into a kind?
A common, general, determinable, essence-revealing, univocal account.
[G2] What distinguishes distinct forms within a kind?
The accounts of a kind's distinct forms differ by degree.
[G3] What is the logical relation between a kind and the forms it unites?
It is the relation between a determinable and its determinations as captured in the method of division by differentia.

We must now answer parallel questions about analogical unity.

[A1] What unites many kinds into an analogical unity?
[A2] What distinguishes distinct kinds within an analogical unity?
[A3] What is the logical relation between an analogical unity and the kinds it unites?

Let us begin with the second question. The accounts of animal forms that are one in kind are "not too distant" (μὴ πολὺ διεστῶτα) insofar as they differ only by degree. In contrast, the accounts of animal kinds that are one by analogy are "too far distant" (πέχει πλέον) insofar as they differ in a way that, in some sense, goes beyond what relations of more or less can capture. Aristotle says,

> For those animals that differ by degree and the more and the less have been brought together under one kind, while those that are analogous have been kept apart. I mean, for example, that bird differs from bird by the more or by degree (for one has long feathers, another short feathers), while fish differs from bird by analogy (for what is feather in the one is scale in the other). (*PA* I.4, 644a16–22)

What does Aristotle mean when he says that analogous parts, say, a bird's wing and a fish's fin, do not differ by degree?

In the *Metaphysics*, Aristotle describes what it is to be incommensurable by degree as follows.

> Since things which differ may differ from one another more or less, there is also a greatest difference, and this I call contrariety. That contrariety is the greatest difference is made clear by induction. For things which differ in kind have no way to one another, but are too far distant and are not comparable (τὰ μὲν γὰρ γένει διαφέροντα οὐκ ἔχει ὁδὸν εἰς ἄλληλα, ἀλλ' ἀπέχει πλέον καὶ ἀσύμβλητα). (*Metaph.* X.4, 1055a3–7)

But Aristotle can't mean that the degree to which the bodily affections of analogous parts differ is, *per impossibile*, so great that it extends beyond the contrarieties that fix the limits of intelligible more/less comparisons. Nor can the incomparability at issue be like that which obtains between affections that don't stand in any more/less relations, say, a color and a sound. For the bodily affections of a bird's wings and a fish's fins occupy positions within the same qualitative and quantitative continua of change. So, one *can* make more/less comparisons between analogous parts' bodily affections: a wing can be softer and rougher and larger than a fin.[23] Consequently, there is a possible series of changes to a wing's affections that would result in a body that is materially indistinguishable from a fin.

However, at some point during this series, the material changes will traverse a discrete boundary between kinds. It is the accounts of the parts' kinds that fix these boundaries. The range of material affections that a wing can possess and remain the part it essentially is will be far narrower than the totality of variations its material affections can intelligibly undergo. Wings and fins have "no way to one another," because the degree to which a wing would have to change to traverse this path would involve the violation of the most general expression of a wing's essence.[24]

Let us turn now to analogical unity itself. As with the other varieties of unity, there is a correspondence between analogous animals and analogous parts.

> **[Unity 3 – Analogical]** Animals are one by analogy and generically distinct just in case most of their respective parts (i) are analogically united and (ii) differ in account *simpliciter* (i.e. not by degree).[25]

[23] For example, the flesh in different kinds of animal is one by analogy (*HA* I.4, 489a23–26; *PA* III.5, 668a25–27). Nevertheless, the "flesh of an insect's body is neither shell-like nor is it flesh-like in the way of the internal substance of shell-covered animals; but it is something intermediate in quality" and the insect's "whole body is hard, the hardness, however, being of such a character as to be more flesh-like than bone, and more earthy and bone-like than flesh" (*HA* IV.7, 532a31–b1; and *PA* II.8, 654a27–30. Cf. *HA* IV.1, 523b15–18).

[24] So, the analogical relations that obtain between parts depend upon an explanatorily prior determination of the kinds to which parts belong. This is connected to, but strictly different from, the question of whether the analogical relations that obtain between parts depend upon an explanatorily prior determination of the kinds to which *animals* belong (Charles 2000: 318; Pellegrin 1987: 228–229), or vice versa (Lennox 2005: 96–97; Henry 2014: 161–164). But the kinds/forms of an animal's parts depend upon, and are determined by, the kinds/forms of the animals to which they belong. So, I ultimately accept a version of view that animals' kinds/forms are explanatorily prior to the generic and analogical unities of animal parts.

[25] Cf. *HA* II.1, 486b20–22. Some interpreters reject the idea that animals can be analogically united: "Whereas whole animals can be specifically and generically identical, analogical identity applies only to the parts of animals. For analogy depends on two objects being compared as parts of two different

The univocal account that defines a form is predicated equally, in the same sense, to each of the many numerically distinct individuals it unites. And the univocal account that defines a kind is predicated equally, in the same sense, to each of the formally distinct species it unites. Both formal unity and generic unity are varieties of synonymy. In contrast, those "things that are one by analogy," Aristotle insists, "are not all one in kind."[26] There is no univocal account, at any level of generality, that defines an analogical unity and can be predicated equally, in the same sense, to each of the generically distinct kinds an analogy unites. Analogical unity is a variety of homonymy.[27]

To illustrate, consider three analogous parts: eyelids, nictating membranes, and hard outer eyes (*PA* II.13, 657a25–658a10). We can introduce a neologism, say, *eye coverings*, that applies to all and only these three kinds of part. But whatever this new word means, it doesn't capture anything essential about the kinds to which it is predicated. The univocal accounts that define these three analogous kinds do not differ in degree. They are not more/less variations upon, and more determinate realizations of, a common, determinable account. The (general) essences of analogous kinds are entirely distinct.

We could also introduce a new name for the widest class of animals that possess these analogous parts, say, *animals-with-eye-coverings*. But again, this common predication will apply homonymously across the boundaries of several already established animal kinds.[28] For even if a common predicate "applies in a like manner to the whole range of the homonym, it does not define any one of the objects described by the term."[29]

So analogous kinds lack a common, univocal account. However, they aren't homonyms "that only chance to have the same name" in the way that the side of a river and a financial institution are each *banks*.[30] Unlike chance homonyms, it is no accident that analogs receive the same name.

systems, in this case whole animals" (Wilson 1997: 339). But the proper method of explanation of an animal's features begins with the identification of the widest class of animals that possess it. If generically distinct parts can be one by analogy, there is no reason not to treat the widest class of animals to which the analogous parts belong in a parallel fashion. Cf. Gotthelf 1997a, 1997b: 193–194 n.18.

[26] *Metaph.* V.6, 1017a2.

[27] "When things have only a name in common and the definition of being which corresponds to the name is different, they are called *homonymous* . . .. When things have the name in common and the definition of being which corresponds to the name is the same, they are called *synonymous*" (*Cat.* 1,1a1–12).

[28] The *animals-with-eye-coverings* would include human beings, birds, live-bearing four-footed animals, and egg-bearing four-footed animals.

[29] *Top.* VI.10, 148a25–27.

[30] *EN* I.6, 1096b26.

For there is indeed something that analogous kinds have in common. Analogies express proportional relationships that capture similarities between their terms.[31] Analogies can't reveal anything essential about the terms they relate, but they can reveal that certain sorts of similarity obtain between them and these similarities can ground a genuine and unique way for a multiplicity of essentially discrete kinds to be one. But analogical unities don't result from just any old similarity. In Section 8.3, I will explore the analogies in *PA* to determine what sorts of proportional relationships are capable of making distinct kinds one by analogy.

8.3 Biological Analogies

In what proportional relationships does analogical unity consist? If we look to the analogies Aristotle discusses in *PA,* one relationship stands out. Aristotle says,

> By analogously I mean that while some have a lung, others have, not a lung but instead something different which is to them what a lung is to those that have one; and some have blood, while others have its analog, possessing the same capacity (*tēn autēn echon dunamin*) that blood has for the blooded. (*PA* I.5, 645b6–10)

The kinds of part this passage mentions are analogous because they possess similar capacities or perform similar functions within their respective animals. The function of both lungs and gills is to cool the body; the primary role of both blood and its (unnamed) counterpart is the provision of final matter from which tissues come to be. Though far from exhaustive, (Table 8.1) comprises many of the parts Aristotle claims are analogous in his biological works and each is plausibly such by virtue of a common capacity/function.

But similarity in capacity is not the only proportional relationship one finds in Aristotle's analogies. First, though it is plausible to claim that feathers, scales, hair, and horned plates each perform a similar protective function, Aristotle doesn't cite this capacity when he says these parts are analogous. Instead, he says they correspond in position (ὅμοιον χώρᾳ); they each cover the outside of an animal's body.[32] And when Aristotle discusses the mytis in soft-shelled animals, he says that "its location makes

[31] Cf. *Metaph.* V.6, 1016b34 and *EN* V.3, 1131a29–b17.

[32] Cf. *PA* IV.11, 691a14–17, IV.12, 692b10–13; *GA* V.3, 782a16–20; *HA* II.13, 505a22–27; and *HA* III.10, 517b14–15. Leunissen argues that they each play a protective role regardless and are thereby similar in capacity at Leunissen 2014: 173.

Table 8.1 *Analogous parts*

Analogous parts	Common capacity	Passages
lungs :: gills	cool the body	*PA* I.5, 645b1–10
sinew :: sinew's counterpart	hold parts together	*GA* II.3, 737b1–4
bone :: pounce :: fish-spine	support flesh	*A Po.* II.14, 98a20–3; *HA* III.8, 516b33–517a5; *PA* II.8, 653b30–6, II.9, 654b27–29
heart :: heart's counterpart (e.g. octopus' mytis)	concoct nutriment, seat of sensation, first principle of movement	*PA* IV.5, 678b2, 681b29–30; *MA* 10, 703a15
blood vessels :: blood vessels' counterpart	contain concocted nutriment	*PA* III.5, 663a3–6
nail :: hoof :: claw :: horn	defense	*HA* I.1, 486b20, III.9, 517a6–10; *PA* II.9, 655b2–4; *GA* V.6, 786a34–b4
brain :: brain's counterpart	cool the heart (or its counterpart), cause sleep	*Somn.* 3, 457b27–30; *PA* II.7, 652b23–25, 653a10–12
flesh :: flesh's counterpart	organ/medium of touch	*HA* I.4, 489a23–26; *PA* II.7, 653b20–30, III.5, 668a25–27
hands :: front paws of polydactylous animals	prehension	*HA* I.1, 486b20, II.1, 497b18–20, *PA* IV.10, 687b27–688a4
hair :: feathers :: scales :: horned plates	protection	*HA* II.13, 505a22–27, III.10, 517b14–15; *PA* IV.11, 691a14–17, IV.12, 692b10–13; *GA* V.3, 782a16–20

it clear that this part is the analog of the heart [in blooded animals] (for the location is the same)" (*PA* IV.5, 681b29–30). Second, some analogies emphasize similarities between parts' developmental order. For example, the heart and the analogous organ in bloodless animals are each the first organ that comes to be in their respective animals' embryological development.[33]

Nevertheless, interpreters typically recognize the comparative prominence Aristotle affords functions/capacities and privilege similarity in capacity over other relationships. Mariska Leunissen defends a strong version of this priority claim: similarity in location and developmental order are

[33] Cf. *GA* II.2, 735a23–6, II.6, 742b33–743a1; and *PA* III.4, 666a18–21.

both reducible to similarity in capacity. For both a part's bodily location and the temporal position it occupies in embryonic development are largely determined by the capacity the part realizes in the animal to which it belongs. She says,

> And because scales and feathers are for protection, *that* is why they must be located on the outside of the animals to which they belong and cover them entirely. Similarly, the heart and its analogue are not analogues because they both come into being first during embryogenesis, but rather because they both provide the first principle of growth and motion of the animal, and this requires them – of conditional necessity – to come into being first. (Leunissen 2014: 173–174)

If this is correct, similarity in capacity will always be what grounds analogical unities.

Though plausible, this interpretation, which I will call the *Capacity Account of Analogy*, faces a difficult challenge that Aristotle himself raises.

> One might be puzzled why people have not named one kind that embraces both the water-dwelling and flying animals, comprehending both at once by one higher name. For there are some affections common both to these and to all the other animals. (*PA* I.4, 644a12–15)

Though water-dwellers and flyers share many features, Aristotle insists that any account that encompasses both classes will be ineliminably homonymous. For what it is to swim differs in account from what it is to fly, and neither is a determination of a common, more general account. About this, Aristotle is clear.

> There are probably other attributes which turn out to have the same predicate, but to differ by a difference in form, e.g. the locomotion of animals; it is apparent that locomotion is not one in form, because flying, swimming, walking, and crawling differ. (*PA* I.1, 639a30–b3)

Still, we can correctly describe both swimming and flying as analogous ways of progressing.

The similarity this analogical predication captures doesn't concern just any common attribute; it is a similarity in capacity or function. And Aristotle frequently claims that what an organ is – what it is both in kind and in form – is determined by its function. For "what a thing is is always determined by its function (ἅπαντα δ' ἐστὶν ὡρισμένα τῷ ἔργῳ): a thing really is itself when it can perform its function; an eye, for instance, when it can see" (*Meteor.* IV.12, 390a10–12).

So, if an organ's function determines what it is, why couldn't we use the capacity common to analogous parts to define a (very general) kind that comprises them all? Why aren't fins, wings, and legs determinations of the kind *limb*? Why wouldn't this supposed generic unity among limbs allow us to introduce the kind *progressors* that comprises all animals capable of self-locomotion and excludes sessile animals? For that matter, why isn't *animal*, without any qualification, a kind? Or even more generally, why isn't there a kind that comprises both animals and plants since the roots of plants are similar in capacity to (and thereby analogous to) the mouths of animals?[34] Why isn't *living organism* a kind? If similarity in capacity or function is the only analogical relationship that matters, analogical unities are in danger of collapsing into functional kinds and analogous homonyms are in danger of being synonyms all along.

One way proponents of the *Capacity Account of Analogy* might avoid this outcome is by distinguishing capacity (δύναμις) and function (ἔργον) more sharply than I have to this point. Leunissen does just this: she claims that capacities are "significantly broader" than functions and argues that a single capacity (e.g. locomotion) can be realized by parts with essentially different functions (e.g. flying, swimming, and walking).[35] Only common functions can ground generic unities. Similarities in capacity are at best marks of analogical unity. So, flying, swimming, and walking are functionally (and so generically or formally) distinct activities. Wings, fins, and feet – the parts required for flyers, swimmers, and walkers to perform their respective essential functions – will also be generically distinct. But this does not preclude these parts from being similar in capacity; each part is equally capable of enabling animal progression. So, both these parts and the animals that possess them differ in kind, but they can still be one by analogy.

It is unclear, however, why capacities, even when conceived broadly or abstractly, can't determine kinds just as readily as functions. For without saying more, the natural understanding of the logical relationship between a capacity and a function is that between a determinable and its determination. What is walking if not a more determinate way of locomoting or progressing? Why isn't being something that progresses belong to the essences of animals who do so? I will return to these rhetorical questions

[34] Cf. *HA* I.2 488b29–32; *Juv.* 1, 468a9–12; *DA* II.4, 415b28–416a6; and *PA* IV.10, 686b31–687a1.

[35] Leunissen 2014: 172. Lennox sometimes describes this "broadness" very simply as "being predominately greater than" or differing "more extensively" than differences by degree, Lennox [1987b] 2001b: 161, 170. Other times he describes it in terms of comparative abstractness: "In distinguishing extensive kinds from one another, this term [sc. 'analogy'] usually refers to a relationship between structures which at a very abstract level perform a similar function for their possessors, but do so by different means, and are not structural variations on a common theme, i.e., are not open to more/less comparison" (Lennox 2001b: 179 n. 8).

shortly, but it will be instructive to first discuss Aristotle's own response to the puzzle he raises.

Aristotle immediately dismisses his own puzzle. He insists that water-dwellers and flyers are "correctly defined" as belonging to different kinds that are not determinations of a broader genus. He says,

> For those animals that differ by degree and the more and the less have been brought together under one kind, while those that are analogous have been kept apart. I mean, for example, that bird differs from bird by the more or by degree (for one has long feathers, another short feathers), while fish differs from bird by analogy (for what is feather in the one is scale in the other). (*PA* I.4, 644a16–22)

What Aristotle says may be entirely correct, but as an explanation, it is lacking. That is, to claim, without further elucidation, that the difference between wings and fins cannot be a difference in degree is to beg the question that currently concerns us. This is because, as I argued in Section 8.2, it is the accounts of parts' kinds that set the boundaries within which they can differ by degree. So, the question whether two parts are incommensurable or differ by degree just is the question whether the parts are one by analogy or one in kind.

But perhaps I am wrong that analogical relations between parts, and distinctions between accounts that differ by degree and those that are incomparable by degree, depend upon explanatorily prior determinations of the generic and formal unities of parts. If we could know what analogical relations obtain between parts prior to, and independently of, knowing parts' kinds, then what Aristotle says after he raises his puzzle would be appropriately responsive.

This suggests an alternative to the *Capacity Account* that merits consideration, which I will call the *Hylomorphic Account of Analogy*. The *Hylomorphic Account of Analogy* maintains that an animal's parts are not defined solely by their functions but also by their material characteristics. For example, according to Allan Gotthelf, an animal's parts,

> are not defined (only) functionally: Aristotle's distinction between a part and its analog is precisely one between two parts with (essentially) the same function but which differ in material constitution and structure, and do so by more than 'the more and the less'. The material constitution and structure of such parts thus enters into their nature for Aristotle, and this seems ignored by functionalist interpretations. (Gotthelf 1985: 238–239)

So, if we could discern parts' material constitutions, we could then determine whether two parts are one in kind (i.e. have a common underlying matter) or whether they are one by analogy (i.e. have similar capacities/functions but distinct material substrata).[36] For example, Devin Henry defends a version of the *Hylomorphic Account* and claims that "even in cases where two animals appear to have similar attributes or where their parts execute similar functions, we can be sure that they constitute distinct natural kinds if it is discovered that their attributes are not [more-and-less] modifications of a common material substratum" (Henry 2014: 164).[37]

Can we determine, without appealing to their forms or functions, whether two parts do or do not have a common underlying material substratum? One immediate concern is that analogous parts often share sundry material similarities. A part's form/function conditionally necessitates much of its material character.[38] It is unsurprising, then, that analogous parts with similar capacities/functions will be materially similar.

Some of these material similarities are relatively weak. For example, feathers, hair, and scales all arise from the moist material present under the skin, so many of their "affections occur in a corresponding manner."[39] But some analogous parts' material similarities are far more extensive. For example, the tongue and its analog are both "fleshy," bone and cartilage are materially earthen and differ primarily by degree of hardness, and nails and hooves are almost entirely earthen and manifest only slight differences in their perceptible, bodily features.[40]

So, not only do the material substrata of some analogically united parts differ by degree, they can differ to a lesser degree than the material substrata of generically united parts. If this is correct, we can't use organs'

[36] For "that to which the differentia and quality belongs is the subject (τὸ ὑποκείμενον), which we call matter. Things, then, are said to differ in kind whose proximate subject (τὸ πρῶτον ὑποκείμενον) is different and cannot be resolved into one another or both into the same thing" (*Metaph.* V.28, 1024b6–11, Cf. X.3, 1054b23–31, X.8, 1057b37–1058a6, and 1058a21–4).

[37] See also Lennox 2001b: 169; 2005b: 96–97.

[38] For example, at *PA* III.3, 664a36-b3, Aristotle argues: (1) the larynx is for the sake of vocalizing, (2) in order for something to vocalize, it must be smooth and hard, (3) cartilage is smooth and hard, so (4) the larynx is made of cartilage. Cf. *PA* I.1, 639b24–30, 642a1–13, I.5, 645b15–20, Lennox 1996b: 196–198; Cooper 1985: 133; and Frey 2007.

[39] "Hairs differ in respect of hardness and softness, length and shortness, straightness and curliness, quantity and scantiness, and in addition to these qualities, in their colors . . .. Some such affections occur in a corresponding manner also in animals which have not hair but something analogous to it, as the feathers of birds and scales in the class of fish" (*GA* V.3, 782a1–19).

[40] On the tongue and its analog, see *PA* IV.5, 678b10–11. On bone :: cartilage, see *PA* II.9, 655a23–34 and II.8, 653b35. On horn :: nail, see *HA* I.1, 486b20, III.15, 519b23–26; *GA* II.6, 743a14–745b20, V.6, 786a34–786b4; and *PA* II.9, 655b2–12. A strong case for the near material identity of nails and hooves occurs at Leunissen 2014: 180.

functionally independent material similarities and differences to determine the kinds and analogical unities of animals or their parts.[41]

If neither the *Capacity Account*'s appeal to function nor the *Hylomorphic Account*'s appeal to matter are entirely satisfactory, how are we to proceed? How are we to understand the distinctive variety of unity analogy affords and distinguish it from the more familiar varieties of unity that belong to kinds and forms?

8.4 Metaphysical Analogies

Though the method of division by differentia proceeds from kinds to forms, it is an organism's final form that is both causally and substantially prior to its kind(s). Individual organisms are primary substances. To be, for an organism, is to live.[42] And the formal, final, and efficient cause of the specific vital activity in which an organism's being consists is the organism's form, namely, its soul *qua* nature.[43] So, "what is indivisible in form is a substantial being" most of all.[44]

When we classify organisms, organs, and vital activities at the level of kinds, we move from what is determinate – a materially realized actuality – to what is determinable – a general potentiality. For being in actuality is prior in both account and substance to being in potentiality.[45] A kind's account, then, is an expression of the potentiality that remains when we

[41] To be fair, Henry accepts that analogous parts can be materially similar and, in rare cases, so similar that they are nearly materially indistinguishable. In response, he emphasizes that when we ask whether parts have a common material substratum, the sense of underlying matter at issue is the part's *proximate* matter. This allows him to say that "when a new substance comes into being the subject from which the change proceeds does not endure but is 'wholly transformed' (ὅλον μεταβάλλῃ)" and so "while nails and hooves might be formed out of the same raw materials, those materials will be completely transformed into different proximate matters in the process of generation: nail-matter in the one case; hoof-matter in the other" and, if this is right, "then the two analogues are not modifications of the same material subjects, even if they are generated from the same raw material" (Henry 2014: 168). But now it would seem that either (i) we cannot determine what a part's "transformed" proximate material substratum is without appealing to the part's form/function, or (ii) we cannot discern (perceptually) whether two parts do or do not have a common underlying (proximate) material substratum.

[42] *DA* II.4, 415b12–14.

[43] For "nature as substantial being is both nature as mover and nature as end. And it is the soul – either all of it or some part of it – that is such in the animal's case" (*PA* I.1, 641a26–29; cf. *DA* II.4, 415b9–12).

[44] *PA* I.4, 644a29–30. Cf. *Metaph.* VII.8, 1034a3–4.

[45] Cf. *Metaph.* IX. Moreover, though a kind can be even if any one of the forms that fall under it is not, a kind cannot be if all the forms that fall under it are not (Cf. *Cat.* 5, 2a35–b6, 13, 15a4–7; and *Top.* VI.4, 141b28–29).

remove the actual, determinate activity of living from the account of the form that is this essential activity's cause and end.

We cannot tell the same story about analogical unity. We cannot begin with an organism's actual parts, its actual activities, or the essence of the actual individual organism itself and arrive at a determinable potentiality beyond what is present at the level of kinds. An organism's greatest kind is where the path of increasing potentiality terminates. So, what an analogical unity captures is not a potentiality, it is not determinable, and it is not related to an organism's essence in the way that kinds are.

How, then, are we to understand what it is to be one by analogy? It is helpful to look to a different context in which Aristotle employs analogies. In the *Metaphysics*, Aristotle asks whether the first principles of things are one or many (*Metaph.* III.4, 999b24–1001a3). He ultimately answers this question in a typically Aristotelian fashion: "The causes and the principles of different things are in a sense different, but in a sense, if one speaks universally and analogically (ἂν καθόλου λέγῃ τις καὶ κατ' ἀναλογίαν), they are the same for all" (X.4, 1070a31–33).

Aristotle provides two illustrations.[46] First, he claims that white, black, and surface are to color as light, darkness, and air are to the day and the night. This analogy compares generically distinct items. Nevertheless, it successfully brings an important commonality into relief. The first three items on each side of the analogy are instances of the three fundamental principles of change within the domains the fourth items designate: *form* (white/light), *privation* (black/darkness), *matter/substratum* (surface/air), and *domain of change* (color/day and night).

Second, Aristotle claims that health, disease, the body, and the medical art are to medicine as the form of a house, the kind of disorder appropriate to houses, bricks, and the house building art are to house building. Again, the items this analogy compares differ in kind. But the first four members of each series are instances of the fundamental principles of substantial generation for the productive activities the fifth items designate: *form* (health/the form of a house), *privation/disorder* (disease/the kind of disorder appropriate to houses), *matter* (the body/bricks), *efficient cause* (the medical art/the house building art), and *productive activity* (medicine/house building).

Instantiating a principle is nothing like belonging to a kind. Neither white nor light are determinate ways to realize the determinable feature *formal-principle-of-change*. There is no account associated with change's

[46] *Metaph.* X.4, 1070b10–21 and 1070b22–35.

privative principle that expresses darkness' essence in any way. To call air a substratum is not a classificatory act.

I'd like to suggest that the concept of analogy Aristotle employs in these passages from the *Metaphysics* is the very same concept he employs in many of the analogies we find in *PA*. If correct, this consonance will allow us to better understand what it is for animals and their parts to be one by analogy.

We already noted that Aristotle can consistently assert that swimming, walking, and flying are formally distinct ways of progressing and also reject the claim that swimmers, walkers, and flyers are more determinate forms of the kind *progressor*. This is not just because a univocal definition of *progressor* would be so far removed from the most determinate ways that formally distinct animals move themselves that it would fail to reveal much of anything essential about them. Rather, it is because these formally distinct ways of progressing aren't determinations of a more general determinable activity. If we were to predicate *progressor* of swimmers, walkers, and flyers, we would be doing so homonymously. Swimming, walking, and flying are one by analogy. And this analogical unity consists in each being an instance of a principle of the vital activities of the animals of which they are respectfully predicated.

In general, we cannot provide univocal definitions of what is one by analogy. To understand an analogical unity, we must instead consider the homonymous instances in turn and come to appreciate their subtle similarities and differences. For the "meaning [sc. of what is one by analogy] can be seen in the particular cases by induction, and we must not seek a definition of everything but be content to grasp the analogy."[47] We must not seek a definition, because there is nothing that could be conveyed in a univocal account that would capture that by virtue of which a multiplicity is one by analogy. We can describe the similarities common to analogous terms. But what an analogical unity is can only be conveyed thoroughly by the analogy itself.

This applies not only to the analogous animals and analogous parts Aristotle discusses but also to the predicate *animal* (ζῷον) itself. For Aristotle makes a tripartite distinction between animals' vital activities: "those common to all, those according to kind, and those according to form." He immediately makes it clear that the first group concerns the predicate *animal*: "I call 'common' those that belong to all the animals." Finally, he associates each of these three commonalities with the variety of

[47] *Metaph.* IX.6 1048a35–36. Cf. *APr.* II.24, 69a16–19.

unity they realize: "For what is common some have according to analogy, some according to kind, others according to form" (*PA* I.5, 645b20–28). Aristotle mentions numerous attributes, parts, capacities, and activities that all animals possess, and he repeatedly claims that while they may be one in form (or in kind) if one restricts one's attention to forms (or kinds) of animal, they are, for the class of all animals, one by analogy.[48] Consequently, no attribute common to all animals could be included in a univocal account that would define a purported kind *animal*.[49] These common attributes, and *animal* itself, are predicated homonymously.[50]

[48] Aristotle claims that what is common to all animals is one by analogy at *HA* I.2, 488b30–32 and I.3, 489a17–19. A partial list of what is common to all animals: perception, the sense of touch, memory, appetite, desire in general, pleasure, pain, being most excited by the desire for and the pleasure derived from copulation, a natural origin of heat, the organ of touch, mouths, the part by which residue departs, and stomachs (cf. *DA* II.3, 414a29-b14; *Sens.* 1, 436a7–16; *HA* I.2, 488b29–489a8, I.3, 489a17–19, VI.18, 571b8–10; *PA II*.3, 650a2–14, II.8, 653b21–24, II.10, 655b28–30, and III.1, 662a19).

[49] The greatest kinds to which Aristotle appeals – birds, fish, cetacea, insects, the hard-shelled, the soft-shelled, and the soft-bodied (or, alternatively, the blooded and the non-blooded) – are narrower than the class of all animals. Cf. *HA* I.6, 790b7–14, II.15, 505b30; *PA* IV.8, 683b26; and Stoyles 2013.

[50] There are two passages that *prima facie* conflict with this interpretation.

i. This is clear even from the account (κατὰ τὸν λόγον); for we define animal (ζῷον ὁριζόμεθα) by the possession of perception, and primary in this account is the primary mode of perception [viz. touch]. (*PA* II.8, 653b22–24)

Though this passage does speak of definition, it mentions only an account (λόγος) and not, as is typically the case when questions of essence are at issue, substance (οὐσίᾳ, λόγος τῆς οὐσίας, etc.; Gotthelf 1985: 221–237 contains a nice overview of these expressions in the biological works). But there is no problem with an analogical unity having an account. Even accidental compounds, e.g., Socrates-seated, have accounts that include their accidental attributes (cf. *PA* II.3, 649b21–27 and *Metaph.* VII.5, 1031a1–14). Moreover, there is neither a problem employing *animal* in explanations in a way that parallels the use of kinds, nor a problem with *animal* being a genus in the relative sense, i.e., a class that permits division by differentiation. Regardless, there is nothing here that indicates this so-called definition isn't disjunctive or otherwise expressive of an analogical unity's homonymy.

ii. We must be careful not to ignore the question whether there is a single account of soul (πότερον εἷς ὁ λόγος αὐτῆς ἐστι), as there is one account of animal, or whether we must not give a separate account for each, as we do for horse, dog, man, god (in the latter case the universal, animal – and so too every other common predicate – is either nothing or posterior). (*DA* I.1, 402b5–8)

There is nothing inconsistent with *animal* being one by analogy in *PA* and being treated as one in kind with a univocal definition in other contexts. I will note, though, that regardless of whether Aristotle maintains that there is a sense of *animal* on which it is one in kind, a parallel problem about unity presents itself. Aristotle insists that *life* is said in many ways (*DA* II.2, 413a20–26). There is no univocal definition that applies to these homonymous ways to live (Cf. *Top.* VI.10, 148a25–30; *Juv.* 467b21–23; and *DA* II.3, 414b25–27). So, if there is going to be a subject genus for a science of life (and of life's principle, soul), the many ways to live must realize some other unity. In this case, Aristotle appeals neither to analogical unity nor to focal connection. Instead, he argues that life's homonyms are *hierarchically united* (*DA* II.3, 414b28–415a13). Each of life's homonymous meanings occupies a position within a properly grounded hierarchical series and the principle of this series effects the requisite unity. I argue for this interpretation in much more detail in Frey 2015, 2021.

In Section 8.1, I discussed how every science requires an appropriately unified subject genus. The study of animals and their parts is no exception. Regarding parts, Aristotle says that "each of the parts of the body is for the sake of something, and what they are for the sake of is a certain action." And this is true for the animals that comprise these parts considered as wholes as well. For "it is apparent that the entire body too has been constituted for the sake of a certain complete action" (*PA* I.5, 645b15–18). Each individual organism of which one predicates *animal* has a final form, that is, a soul, that is the principle of a single specific way of living – a "certain complete action." These principles of life aren't one in kind. But they are, like the other cases of common principles we have discussed, one by analogy. It is this analogical unity that makes their scientific study possible.

In sum, the unity analogy affords, unlike the empty classifications of chance homonyms, is explanatory in a way that makes it indispensable to the investigation Aristotle undertakes in *PA*. Without analogy, we are blind to many genuine explanations of biological phenomena. Without analogy, the biologist's task becomes unnecessarily and unedifyingly repetitious. And without analogy, we are unable to unite animals and their parts in a way that would secure the legitimacy of their scientific study.

CHAPTER 9

Uniform Parts in Aristotle's Biology

*Stasinos Stavrianeas**

Introduction

On one influential reading of Aristotle's biological works, final causes dominate explanations of animal parts and the processes that produce them, in the sense that, where present, final causes take priority over other causes. Teleological explanations are usually contrasted to material explanations as exemplified in the zoologies of earlier thinkers. Their theories are found wanting in that they attempt to explain the constitution of animals from the bottom up, that is, from some elementary material up to animate bodies and their parts, a strategy insufficient for accounting for the great diversity of parts among animals. This may give the impression that for Aristotle animal forms are responsible for the transformation of elemental matter through various stages into animate tissues and organs. However, the work of several scholars in recent decades has shown that material processes play a much more influential role in the constitution of such bodies.[1] In both his biology and chemistry, Aristotle investigates processes or laws that govern material interactions that result in the constitution of higher-level bodies. These bodies serve as the material available to

* This chapter was presented at the conference: Critical Philosophical Approaches to Aristotle's *Parts of Animals*, at Cambridge (2019). I benefited from comments by the participants, who I would like to thank warmly, as well as the organizers Sophia Connell and Myrto Hatzimichali, for stimulating discussions at the conference. I am gratefully indebted to Sophia Connell who has offered extended comments on previous drafts, and I also would like to thank Stavros Koulouments and Spyridon Rangos for comments and discussion. The remaining shortcomings are, of course, mine. During the authoring of the chapter, I have benefited from my participation in the research project: Aristotelists – The collaborative corpus + of Aristotle. A project cofunded by the European Regional Development Fund and the Operational Programme Competitiveness, Entrepreneurship and Innovation (code number T2EΔK-00093).

1 See Lennox's seminal paper on material and formal natures (1996b). The thesis that material facts can be explanatorily basic is presented, explored, and defended by Connell (See Section 10.2).

organisms and out of which their parts are formed. The constitution of such bodies depends on three distinct levels of material constitution. From the top-down: (1) instrumental animal parts, that perform the complex functions of an organism, necessitate the presence of (2) uniform parts, each one of which possesses a specific characteristic necessary for corresponding aspects of the complex function of the instrumental parts (*PA* II.1, 646b10–26), such uniform parts depend for their constitution on a variety of inanimate uniform mixtures capable of undergoing specific *concoctive* processes inside the organism, transforming them into uniform animate parts; (3) at the bottom level, the constitution of inanimate uniform mixtures is a result of a chemical process (*mixis*) in which a small set of primary elemental potentials (hot, cold, dry, and moist) produce the rich variety of further material dispositions. The question I tackle in this chapter concerns the relation between the bottom level, the constitution of inorganic mixtures, and the second level, the constitution of animate uniform parts. The relation between these two levels and the contrast between the two types of processes, *mixis* and concoction, producing the corresponding bodies has not received sufficient attention in the literature. I draw in both Aristotle's biology and his chemistry so as to examine the transformation from inanimate to animate materials.[2]

In Section 9.1, I discuss the place of uniform parts in the three levels of material constitution, their teleological relation to what they constitute (nonuniform animate parts), and what goes into their constitution (elemental mixtures). In Section 9.2, I turn to the related account of the formation of animal parts in *GC* II.7 that reveals part of the motivation for introducing the level of inanimate mixtures as a necessary part of the three levels of material constitution. Section 9.3 briefly discusses the case of blood and how its characteristic heat inheres in it, as evidence that uniform animate parts are not mixtures. Section 9.4 argues from the evidence of the *PA* and *Meteorology* that uniform inanimate materials are in a different category to animate ones and should be ranked at the first level of material constitution. Finally, Section 9.5 concludes that concoction, a distinctively teleological process, is what produces animate uniform parts out of inanimate ones.

[2] As Lennox comments, the discussion of uniform parts in *PA* II is "massively indebted" to *Meteorology* IV (2014: 274). The *GC*, on the other hand, covers much of the same material, even if only briefly, and sheds some light on the questions we will tackle. I will not take a stance as to the relative chronologies of these works.

9.1 Structure and Teleology in the Material Constitution of Animate Bodies

The most general division of animal parts in Aristotelian biology is between homeomerous (uniform) and anhomeomerous (nonuniform) parts. This distinction is sometimes drawn in terms of a structural difference: uniform parts (such as blood, flesh, bone, etc.) are divisible into parts that have the same name and definition as the whole to which they belong, while nonuniform parts (such as leg, hand, or face) cannot so be divided; they are divisible in dissimilar portions that do not bear the name of the whole.[3]

This structural difference, however, is only the result of a more profound difference in the functional architecture of natural bodies. A living body capable of leading its characteristic way of life needs a number of complex organs (uniform parts) to perform these activities. Such organs must be formed out of materials suitable for the various aspects of the activities they will undertake. Material properties are contributed by different materials (uniform parts) and these in turn must ultimately be formed out of the elements, the building blocks of natural bodies.[4] The distinction between nonuniform and uniform parts helps to analyze and explain the complex structure of an animal's body step-by-step and also to ground the construction of living bodies in simple materials out of which all natural entities are formed. So natural bodies are constituted by hierarchically layered matter, in three levels, each one of which serves certain function(s) necessary for the higher ones. The distinction is discussed twice in the *PA*. First, in *PA* I.1:

> Air and water are matter for bodies; that is, it is from such things that all the ancients constitute the nature of bodies. But if human beings, animals, and their parts exist by nature, one should speak about flesh, bone, blood and all the uniform parts. Likewise too, about the non-uniform parts such as face, hand, and foot, one should say in virtue of what each of them is such as it is, and in respect of what sort of potential. For it is not enough to say from

[3] See *GC* I.10, 328a8, *GC* I.1, 314a19–20. See also *HA* I.1, 486a4–7 where uniform parts are said to be incomposite (*asuntheta*) because they are divided into parts that bear the same name and, presumably, definition as the whole (cf. *PA* II.1, 647a1–3 where uniform parts are contrasted with nonuniform ones as simple against composite parts). But there are exceptions in the case of uniform parts: e.g. the parts of a blood-vessel are only in a sense blood-vessels (*PA* II.2, 647b17–19).

[4] See *PA* II.1 646b10–27, discussed below for the functional dependence of complex organs on simpler materials, and the comment on the hierarchy of functions entailing an equivalent hierarchy between the parts responsible for such functions (*PA* I.1, 645b28–33). This hierarchy is reflected in the division between nonuniform and uniform parts.

> what things they are constituted, e.g. from fire or earth. It is just as if we were speaking about a bed or any other such thing; we would attempt to define its form rather than its matter .... For the nature in respect of shape is more important than the material nature. (*PA* I.1, 640b16–29)[5]

Aristotle mentions here the three layers of material constitution: (1) the elements (like air and water), (2) the uniform animal parts (flesh, bone, blood), and (3) the nonuniform parts (face, hand, and foot).[6] The predecessors mistakenly tried to analyze or reduce all natural bodies (animals, plants, and their parts) to the elemental material level without offering an account of the intermediate levels in the hierarchy (such as the uniform and nonuniform parts). Living bodies and their parts have a particular shape and visible character (*idean*, 640b28), which needs to be accounted for by the natural philosopher, since they too exist by nature. But the mistake of the predecessors is not simply one that concerns incompleteness, that is, that they omit to account for some natural entities (e.g. parts of living bodies). In failing to account for the higher levels of material organization, they fail to account fully for the basic level as well; they fail to explain how elemental matter is gradually formed into uniform stuffs that in turn are useful for the parts an organism needs. Aristotle's theory of mixtures, by contrast, can fill such a gap since it explains how the four elementary contrarieties produce further and more complex dispositions when the bodies in which they inhere get mixed. In this passage, we have only a hint concerning this second mistake: the ancients fail to explain "in virtue of what each of them is such as it is, and in respect of what sort of potential." The reference to the "potential" points to the activities that a part can sustain inside an organism (most typically a function), explaining in the case of nonuniform parts the necessity of their shape and structure but also to some extent the material constitution necessary for carrying out their function. In the case of a uniform part, such as blood, the potential will typically point to a disposition the part possesses due to its material constitution making it suitable for, in this instance, nourishing the body.[7]

[5] Translations from the *PA* are from Lennox 2001a. Translations from other works come from the Barnes' edition of Aristotle's Complete Works (1984), unless otherwise indicated.

[6] I omit the further and highest level of animal bodies as wholes, since in *PA* II.1 Aristotle does not acknowledge that level (by contrast to *GA* I.1, 715a10–11). Plausibly, he does not think of animal bodies as constituting a different level of material structure compared to the level of nonuniform parts. Or maybe in a work such as the *PA* dealing with animal *parts*, the highest level is the level of nonuniform parts as Lennox suggests (2001a:180).

[7] Lennox labels such powers "differential dispositions": "A given uniform body has the dispositional differential it has in virtue of the way that heating and cooling have acted upon the elemental

If so, the expression "in virtue of *what* each of them is as it is" may point not just to considerations that pertain to final causes but also to considerations that relate to the material processes and constituents that result in the presence of corresponding potentials in such bodies. Thus, on the one hand, the higher levels influence the lower ones, while, on the other, the lower-level materials with their relatively simple powers can interact so as to constitute higher-level materials possessing further powers. So, an investigation devoted uniquely to the lowest level will fail to grasp fully the operations even at this elementary level of producing a richer inventory of material powers. These thoughts are reinforced by what Aristotle says at *PA* II.1, where the three levels of material constitution are officially introduced with an important correction: it is the four elemental potentials and not the four traditional elements of the physiologists that we should treat as the basic material:

> Since there are three compositions, one might put first composition from what some people call the elements, e.g. earth, air, water, and fire. And yet, perhaps it is better to speak of composition from the potentials, and not from all of them, but as stated previously in other works. That is, moist, dry, hot and cold are matter of the composite bodies, while the other differences, e.g. heaviness and lightness, density and rarity, roughness and smoothness, and the other bodily affections of this sort, follow these. Second is the composition of the nature of the uniform parts within animals – e.g. bones flesh and the other things of this sort – out of the primary things. Third and last in the series is the composition of the nature of the non-uniform parts – e.g. face, hand and such parts. (*PA* II.1, 646a17–24)[8]

The correction is justified with a back-reference to *Meteorology* IV where the constitution of uniform bodies, animate and inanimate alike, is accounted for by the operations of the two active potentials (hot and cold) on the two passive ones (dry and moist).[9] However, we are not told explicitly what these contrarieties produce (first level of composition). Moreover, it is unclear what the "primary things" are from which the uniform bodies are composed (second level). One plausible suggestion is that the elements themselves are both the products of the first level

ingredients from which it is constituted .... Such compounds acquire and lose these differential dispositions of necessity whenever acted upon by the appropriate sorts of heating and cooling" (2014: 273). *Meteor.* IV, he argues, is an investigation of the *thermodynamic laws* that explain how dispositional contraries (e.g. hard and soft) absent at the level of elements come to be present at the level of uniform animate or inanimate bodies.

[8] Cf. *GA* I.1, 715a8–11; *HA* I.1, 486a4–7.

[9] *Meteor.* IV.1, 378b10–14.

composition and what goes into the constitution of uniform bodies at the second level.[10]

However, some details of the text cast doubt on such a reading. First, although Aristotle says that bodies at the first level are composed out of the potentials, he does not deny that in some sense they are composed out of the elements. Maybe he means to attribute this role to the elements, at least in their pure, unmixed form.[11] Second, Aristotle says that other affections follow from these basic ones, and although some of these "derivative" affections characterize the elements as well (e.g. heaviness or lightness), there are others that belong only to mixtures of the elements,[12] such as certain inorganic uniform stuffs. If so, the products at this first level of composition must include, on top of impure versions of the elements, all uniform inanimate stuffs. Finally, it is reasonable to think that the "primary things" that go into the composition of the uniform animate parts, at the second level, include whatever goes into the constitution and maintenance of the organism, for example, food and drink, which is processed to feed the parts.[13] We could therefore read the class of things composed out of the elemental potentials, that is, the "primary things" constituting the uniform parts, more generously, as a class that includes all inorganic stuffs.

The next level, that of the material constitution of animate uniform parts, is particularly crucial because it stands at the border between the

[10] Lennox 2001a: 180. See also Lennox 2014: 274, suggesting that the products at the first level include the uniform inorganic bodies. This is the line I pursue here with the difference that I take "the primaries" as referring not only to the contrarieties but also, derivatively, to the pure forms of the elements.

[11] The distinction between pure and impure versions of the elements is briefly discussed in *GC* II.3, 330b22–30. Aristotle distinguishes between mixed versions of fire or air as we encounter them in our daily experience, from their simple and unmixed versions, i.e., some idealized or abstract version of them (Joachim 1922: 217) or the elemental potentials or contrarieties. Some commentators propose that at 330a30–34 the term "elements" refers to the four contrarieties (Williams 1982: 160; Frede 2004: 303) while others reject this reading (Crowley 2013: 163–166). At any rate, it seems safe to say that the term refers primarily to the pure versions while only secondarily to the mixed ones and to assume that elemental bodies are never pure but always the product of some composition, i.e., a *mixis* of different elemental portions (Cf. for a detailed discussion of the *GC* passage see Katz: 2022: 71–76; and also Frey 2007: 181).

[12] As Lennox 2014 argues, the differential potentials or dispositions emerge only at the level of mixtures (274 n. 4).

[13] Aristotle says that all nourishment that goes into the constitution of organs contains water and earth, or that blood is a composite of watery and earthy material (blood is made of water and earth and air, *Meteor.* IV.10 389a19ff. See also the comment at 389b15 concerning all uniform parts: when they lose their warmth, they become water and earth). But this does not mean that animals feed immediately from the elements. Nor that the elements immediately compose blood but rather that uniform parts are ultimately composed from them. *PA* II.1, however, is concerned with what immediately composes what is inside living bodies.

world of inanimate bodies governed by some, to use Lennox's term, *thermodynamic* laws (that appeal to material and efficient causes) and the world of animate bodies whose generation is accounted for by final causes. At the bottom level, the activity of elemental potentials produces both elemental transformation and inanimate uniform matter by the process of mixing (*mixis*), a thermodynamic process that need not be determined by final causes. On the other hand, as we will see below, the uniform parts are for the sake of the nonuniform ones, and their design follows a teleological plan for the animal's body. So it is at this level that the operation of final causes makes a crucial difference to the constitution of living bodies. These teleological relations are what Aristotle aims to clarify with the introduction of three levels of composition.

Indeed, Aristotle embarks on a long argument, for the rest of *PA* II.1, to show that these three levels of composition are related teleologically: the "primaries" are necessary for the sake of the uniform parts, while the uniform parts are for the sake of the nonuniform ones (*PA* II.1, 646b5–6, 646b11–12). His argument appeals to the principle that the order of temporal priority in generation is directly opposed to the order of teleological priority (*PA* II.1, 646a24–26).[14] The principle is supported by appealing, first, to the craft-analogue: the finished house is temporally posterior to the bricks and stones, but the latter, while teleologically posterior, exists for the sake of the house (*PA* II.1, 646b2–5). The case is similar between uniform and nonuniform animate parts; and this picture is extended to the relation between the products at the first level of composition (what the elemental potentials constitute) and the animate uniform parts.[15] This can be read in two ways: either all compositions from the bottom up are (ultimately) for the sake of the nonuniform animal parts, and for the bodies of organisms as wholes, or there is some (implicit) qualification of the teleological claim. The first reading, however, is

[14] The relation between temporal and natural priority is connected with one of the initial *aporiai* expressed in *PA* I.1: "We need also to determine about these causes [cf. the final and the efficient one], which is naturally first and which second. Now it is apparent that first is the one we call for the sake of which; for this is an account, and the account is an origin" (639b13–15). Again, here it is the account (*logos*) that determines the order of natural priority that is opposed to the order of temporal priority. The opposition between temporal priority and teleological priority is repeated throughout *PA* I.1, 640a10–33, 640b11–16, 641a7–13, 641b30–37, 642a13–22.

[15] "Thus the matter of the elements is necessary for the sake of the uniform parts, since these are later in generation than the elements, and later than the uniform parts are the non-uniform parts; for these have already attained their end and limit, having achieved a constitution of the third sort, as often happens when generations are completed" (*PA* II.1, 646b6–10).

objectionable.[16] First, it is not the case that everything composed out of the elemental potentials goes into the constitution of some animal part or other. Second, even if we restrict the scope to what goes on inside animal bodies, what serves as matter for their uniform parts is not conditionally necessitated by them in the way that uniform parts are conditionally necessitated by the nonuniform ones.[17] Rejecting the first reading, we need an alternative that will allow for the lower level matter to be teleologically related to higher level bodies without existing for the sake of them. Such an alternative under the name of "secondary teleology" has been proposed and defended by several scholars: an organism is able to use already constituted material with the appropriate characteristic for some function without thereby being responsible for its production.[18] Aristotle's chemistry of uniform bodies shows how a variety of material properties that can be of use for an organism, or even necessary for certain functions, are produced by material interaction at the level of uniform bodies.

Now, the main argument by means of which Aristotle supports his claim concerning the teleological relation between uniform parts and nonuniform ones (*PA* II.1, 646b10–26)[19] turns on the idea that

[16] According to the first option, the form reaches all the way down to the four elements determining every feature of an animal. As Connell argues in this volume that this claim "faces the problem that Aristotle's biological works are full of explanations of parts which refer to materials" (p. 225). And the problem may be wider, for the thesis seems to entail that elements (pure elements or their contrarieties) in general, and not only their portions that constitute uniform bodies, exist for the sake of the latter.

[17] As Lennox puts it: "The necessitation goes from entity [animal] to function [of a uniform part] to differential potential to some uniform part" (2014: 278). But the matter of the uniform parts may be produced by means of some chemical procedure determined by the materials that go into the mixture and not existing for the sake of something, or so I will argue below, even though it may be used by the organism for a purpose.

[18] See Leunissen (2010a: 18–20), who distinguishes between cases where the form of the animal is responsible for the coming to be and the presence of a part in an animal's body (primary teleology) and cases where the form uses already constituted material and is responsible only for the presence but not for the coming to be of that part (secondary teleology). In the latter cases, "the operation of the formal nature is secondary to the operation of material necessity that produced the material." My suggestion is that this happens universally for the inanimate uniform material that goes into the constitution of the uniform animal parts with the difference that in the case of some parts such as, most crucially, blood and its by-products, the form is responsible for their final processing (see Section 9.3).

[19] "Thus animals have been constituted from both of these parts, but the uniform parts are for the sake of the non-uniform for of the latter there are functions and actions, e.g. of the eye, the nostril and the entire face, of finger, hand and the entire arm. And since the action and movements present in animals as a whole and in their non-uniform parts as a complex, it is necessary for their components to have distinct potentials; for softness is useful for some things, hardness for others . . .. Now in the uniform parts such potentials are distributed part by parts (one is soft, another hard . . .) while in the non-uniform parts they are distributed to many and in combination with one another; for different potentials are useful to the hand for squeezing and for grasping."

nonuniform parts, such as the hand, have a complex function since they need to possess hardness, in order to be able to press, but also softness in order to be able to touch, and flexibility in order to be able to grasp. The performance of such a complex set of functions necessitates the composition out of several uniform materials each one of which will contribute the corresponding differential disposition (flesh possesses the potential for touch, bones the potential for pressing). And yet the presence of these materials is not sufficient for its constitution, since the materials need to be present at the right time and under the right amount of heat and be arranged in the proper way so as to form a hand.[20] Thus, the constitution of animal bodies depends on the availability and transformation of simple uniform materials endowed with the appropriate potentials necessary for constituting animate uniform parts that will compose nonuniform ones and ultimately a functioning living body. The potentials of uniform parts mentioned in *PA* II.1 (softness/hardness, moisture/dryness elasticity, flexibility, brittleness) are such material dispositions whose constitution depends on thermodynamic laws (i.e. on efficient and material causal explanations). Once present in the organism, however, they can be managed for specific ends and their functionality is explained in the context of the nonuniform part they compose. This observation is what drives the analogical claim that the element potentials are for the sake of the uniform parts as the uniform parts are for the sake of the nonuniform ones. However, it is only at the latter level that the relevance of specific material dispositions for the life of the organism becomes clear. And the same holds for the element potentials. If so, then the presence of the element potentials and mixtures thereof can be said to be for the sake of the organism in the sense that the organism can make good use of these materials by molding them, locating them and arranging them so that they can benefit the organism.

One conclusion we can draw from the division of levels of material constitution in *PA* II.1 is that Aristotle resists the idea that animal parts are formed directly from the elements because he wants to explore the non-teleological thermodynamic processes that generate a much richer variety of contrarieties than the four elementary ones. After all the material dispositional contrarieties, we find in animals are also found in nonliving

[20] I owe this point to Sophia Connell. On this, Lennox notes: "In order for parts like flesh or sinew to function biologically for the sake of the animal, they must differentiate at precisely the right place and time, and in precise and complicated relationships to each other, if they, or the organs they constitute, are to serve the needs of the whole organism" (2014: 273).

matter, and so the chemical procedures accounting for their production must be universal. The particular problem, how animate parts are constituted from inanimate matter, occupies Aristotle in the *GC* to which we shall now turn. His solution to the puzzle that results from the previous philosophers' failure to explain this phenomenon satisfactorily will help us understand the composition of uniform animate parts.

9.2 The Composition of Animal Parts in the *GC*

The question concerning the formation of animal parts from lower-level material bodies is addressed once more by Aristotle in *GC* II.7:

> Let us return to the elements of which bodies are composed. The theories that there is something common to all the elements, and that they are reciprocally transformed, are so related that those who accept either are bound to accept the other as well. Those, on the other hand, who do not make their coming-to-be reciprocal – who refuse to suppose that any one of the 'elements' comes-to-be out of any other taken singly, except in the sense in which bricks come-to-be out of a wall – are faced with an absurdity. How, on their theory, are flesh and bones or any of the other compounds to result from the elements? Indeed, the point we have raised constitutes a problem even for those who generate the 'elements' out of one another. In what manner does anything other than, and beside, the 'elements' come-to-be out of them? Let me illustrate my meaning. Water can come-to-be out of Fire and Fire out of Water; for their substratum is something common to them both. But flesh too, presumably, and marrow come-to-be out of them. How, then, do such things come to-be? (*GC* II.7, 334a16–26)

The puzzle here concerns the formation of animal parts or higher-level composites out of lower-level material, and particularly the four elements. Aristotle considers two alternative theories that are equally weak in explaining this phenomenon. The first neither allows for the transmutation of the elements nor posits something common between them. The second allows for transmutation by positing something common between the elements, namely their matter. The way the two theories fail is instructive and leads to Aristotle's positive proposal. According to the first, the bodies formed by mixing elemental stuffs are mixtures where the elemental portions survive intact. But if this was the case, then when the mixture dissolves what should come out from each portion of it would be the corresponding elemental stuff having survived intact in that portion. But, in fact, any portion of the mixture may dissolve into any one of the elements constituting it (e.g. any part of flesh dissolves into water or earth). The second view posits something common between the elements, to avoid the above result and in order to explain transmutation. And yet such a theory, complains Aristotle, cannot explain how elemental portions come together to generate flesh or bone etc. For instance, if water and earth produce flesh and flesh is neither water nor earth, flesh must be identified with their matter. But if flesh consisted of the matter of water and earth, it could not possess any of the characteristics explained by the fact that flesh is made of water and earth (*GC* II.7, 334b5–7). Aristotle's solution to the puzzle is the following:

> Perhaps we may suggest the following solution. (A) There are differences of degree in hot and cold. Although, therefore, when either is fully real without qualification, the other will exist potentially; yet, when neither exists in the full completeness of its being, but both by mixing destroy one another's excesses so that there exist instead a hot which (for a hot) is cold and a cold which (for a cold) is hot; then what results from these two contraries will be neither their matter, nor either of them existing in its full reality without qualification. (B) There will result instead an intermediate: and this intermediate, according as it is potentially more hot than cold or vice versa, will possess a power-of-heating that is double or triple its power-of-cooling, or otherwise related thereto in some similar ratio. (C) Thus all the other bodies will result from the contraries, or rather from the elements, in so far as these have been mixed: while the elements will result from them (ἐξ ἐκείνων), in so far as these exist potentially in a special sense-not as matter exists potentially, but in the sense explained above. And when a thing comes-to-be in this manner, the process is a mixing; whereas what comes-to-be in the other manner is matter. (*GC* II.7, 334b8–20)

The Aristotelian solution described in sections (A) and (B) is a version of the theory of mixtures already presented in *GC* I.10. Roughly, in mixtures the contrary qualities of the components are balanced out, the result being an entity in which the constituents survive potentially, that is, they can be retrieved when the mixture is dissolved. However, in section (B) Aristotle seems to refer to a different sense of power or potentiality, for he says that the intermediate, that is, the mixture that is *potentially* hotter, possesses a greater power of heating.[21] This cannot be the sense in which an elemental body – say water, which is actually moist and cold – is potentially dry or

[21] For this sense of power or potentiality (*dunamis*), see Anagnostopoulos 2022: 170. This sense of *dunamis* seems to be at play in *PA* II.3 649b17–18 where Aristotle claims that bodies extracted from a mixture which are made of water are actually as well as *potentially* fluid, meaning that they are actually fluid and possess the corresponding (passive) *power* of fluidity.

hot. For water is potentially hot not because it has a power to heat more or less but because it can become hot if it loses its coldness. If so, then elemental mixtures acquire powers of heating and cooling that are a function of the corresponding ratios between the powers of their constituents.

The next section, (C) above, seems to confirm this reading. For, we are told, first, that other things (presumably uniform bodies, which are the target of *GC* II.7) are formed *from* the mixed elements, that is, when a ratio of the primary contrarieties is reached that corresponds to the characteristic properties of a given uniform body. Secondly, we are told that the elements come to be from contraries that exist somehow potentially in the mixture but not in the way that matter exists potentially. This reservation points to a different sense of potentiality, namely the relatively compromised power of a contrariety when it is found inside a mixture.[22] Hence, when elements come together and their contrarieties are compromised in some given ratio, what gets produced is a mixture that has the power (the second sense of potentiality) to heat or cool, while when such a mixture is decomposed into what was surviving in it potentially (in the first sense of potentiality) what is produced is the matter (of that mixture) or, simply, elemental matter.

To return to the initial puzzle concerning the production of uniform parts, Aristotle wishes to explain their formation without claiming that they come to be out of the elements surviving in them and without saying that they come to be from elemental matter. His solution employs the two senses of potentiality we distinguished above. On the one hand, the contraries, when mixed and having reached a balance along the continuum defined by each pair of contrary properties, have a power to exercise their active or passive power to some degree. From this perspective, the elements are not matter for the mixture, but they contribute their formal properties (the contraries) altered to some extent. On the other hand, this balanced mixture when dissolved gives rise to portions of elemental matter that will possess the corresponding essential contraries in full actuality.

Through this account of mixtures, Aristotle can establish that there is a natural process explaining the constitution of the rich variety of the formal properties that uniform bodies have. They have these properties as a result of the interaction of the formal properties (contrarieties) of the elements. The elements are not just the material out of which uniform bodies are formed but they determine the corresponding natures of these bodies.

[22] For a different reading, see Anagnostopoulos 2022: 174ff.

As we saw in the *PA* II.1, the moist, the dry, the hot, and the cold are said to be the matter of what they compose (i.e. the uniform mixtures at the first level of constitution) but they also explain the other differences and qualities of these bodies, that is, their essential differences (646a17–21). However, *PA* II.1 seems to be in tension with *GC* II.7; according to the former, bodies formed at this first level of constitution subsequently form the uniform animate parts, whereas the account of mixtures at *GC* II.7 explains their formation directly from the elements. This tension may be mitigated by Aristotle's claim in the *GC* that uniform parts have a double nature; their name applies, on the one hand, to their matter and, on the other, to their form.[23] As argued below, *GC* II.7 refers to uniform animate parts in their material aspect, while *PA* II.1 refers to their formal element. So let us turn again to *PA* II.2–3 and this double nature of animate parts and to the variability of contrary properties required to explain the differences between uniform bodies.

9.3 Uniform Parts: Their Matter and Form

In *PA* II.2–3, Aristotle examines more closely the contrarieties that go into the constitution of the uniform parts. His analysis is not exhaustive, referring back to *Meteorology* for a detailed consideration of the different types of potentials and their corresponding functions (*PA* II.2, 649a33–35). But his discussion, however brief, gives a good sense of the great diversity of contrarieties available to animal bodies (each contrary, e.g., hot or being hotter, is applied in different senses). So, at a first level, when the bodies in which elemental contraries inhere are mixed, they produce not only further differences and affections but also variations of these contrarieties, for example, different ways of being hotter. At a second level, the nature of an organism can use the variations produced by the elemental contraries to produce similar variations within each type of uniform part, for example, blood, serum, marrow, etc. The following passage is indicative:

> The relative differences between things of the same kind are for the sake of the better – that is, the differences both of other parts and of blood from blood. One sort of blood is thinner, another thicker, one purer, another more turbid, and again one colder, another hotter, both in the parts of one animal (for the upper parts of one animal are distinguished from the lower parts by these differences), and between one animal and another ....

[23] *GC* I.5, 321b19–22.

> Thicker and hotter blood is more productive of strength, while thinner and cooler blood is more perceptive and intelligent. (*PA* II.2, 647b29–648a3)

Inside an animal's body, differences in blood are controlled by the animal's natural heat, which is responsible for the concoction of nutriments and the production and distribution of blood. So, while at a first level materials when mixed are able to produce a variety of different affections, at a second level the animal's nature can further act on these materials to form uniform parts with the appropriate constitution. Indeed, *PA* II.2 gives us crucial details about the constitution of uniform parts and, in particular, the distinction between two ways in which the relevant contrary property may be said to belong to them (i.e. depending on whether we refer to it as matter or as matter coupled with the form). Taking heat as an example, Aristotle distinguishes things called hot because they have derivative heat from those that possess their own, intrinsic heat; the former are *near* to being hot incidentally rather than in themselves. His illustration is a person with a fever who happens to be cultured, thus licensing the claim that the cultured person is hotter than the healthy person. He applies this distinction to uniform parts:

> Nor is it possible in some such cases to say without qualification that something is, or is not hot. For what the underlying subject happens to *be* is not hot, but *coupled* it is hot, as if someone were to give a name to hot water or hot iron. In fact it is in this way that blood is hot. Such cases, namely those in which the underlying subject is hot in virtue of being affected, also make it apparent that cold is a certain nature, and not a privation.[24] Perhaps even the nature of fire may turn out to be some such thing; that is, perhaps the underlying subject is smoke or charcoal, the former being always hot (for smoke is a vapour), while charcoal, when extinguished, is cold. (Lennox's trans. with modifications *PA* II.2 649a14–649a24)

Aristotle refers here to whatever it is that underlies fully functional blood, that is, a mixture (e.g. sanguineous fluid) capable of acquiring the heat characteristic of the animal kind and can thereby function as blood. Thus, he distinguishes between the mixture underlying the production of blood (analogous to water that accidentally acquires heat when boiled) and

[24] I adopt the reading of manuscripts PSU followed by Bekker and Peck: τό ψυχρόν φύσις τις ἀλλ' οὔ στέρησις ἐστιν, instead of the reading of EYZ followed by Lennox: τό ψυχρόν οὔ φύσις τις ἀλλά στέρησις ἐστιν (the cold is not some nature, but a privation). The latter reading is in accordance with what is said in *Meteor.* IV.2, 380a8; *GA* II.6, 743a36, IV.4, 784a33. I defend the reading I adopt in Section 9.5.

functional blood (analogous to heated water).[25] And this distinction seems also present in the next chapter:

> It is evident that blood is in a way hot, i.e. in so far as it is what it is for blood to be blood; blood is spoken of just as we would speak of boiling water were we to signify it by a certain term. But the underlying subject, i.e. whatever it is that is blood is not hot; and in itself blood is in one way hot, and in another not. For heat will belong to its account, just as white belongs in the account of white human being; but in so far as blood is hot in virtue of an affection it is not hot in itself. (*PA* II.3, 649b20–27)

The claim that blood is in one way hot and in another not may be less puzzling if we distinguish two ways of referring to blood, that is, as the underlying subject and as the subject coupled with the affection that plays the role of form (*GC* I.5 321b19–22). Indeed, what lacks heat is blood in the sense of a uniform mixture of water-earth that serves as the underlying subject of functional blood (*PA* II.2, 649a18). What happens in the production of blood is that the formal nature bestows to this underlying subject the specific type of heat that corresponds to the vital heat of the kind. Indeed, Aristotle characterizes the derivative heat that inheres in the underlying subject of blood as an affection; in *GC* I.10, he explains that affections inhering in a subject do not get mixed: "Nor can white get mixed with the body, nor in general affections and states (*hexeis*) with things; for we see them persisting unaltered. Nor anything else which is not separable" (*GC* I.10, 327b15–18). The implication is that functional blood cannot be a mixture; it is the coupling of an affection *with* a mixture, like pale human is a coupling of an affection with an individual substance. And the reason must be that this heat is not separable from it, as mixture ingredients normally are (*GC* I.10 327b21–22). The constituents of blood, that is, the constituents of the underlying substratum of blood, are not mere portions of the elements but a mixture of such portions. For if we look at the wider context of the *PA* passage, the mixture that constitutes the underlying subject for blood cannot be identified with the elements but rather with some uniform mixture made out of them.

[25] This distinction is between (a) the material substratum that may or may not be (accidentally) hot and (b) the compound formed out of this substratum and the heat; (b) is now a new entity (so to speak) to which the level of heat is essential, as, for instance, in the case of the person who has a fever the heat of the fever is essential to him/ her *qua* patient but accidental to the person who may or may not fall ill (accidentally).

9.4 Two Levels of Uniform Bodies: Inanimate Mixtures and Animate Parts

After the account of hot and cold (*PA* II.2), *PA* II.3 discusses the various senses of moist and dry. The presentation of the case of blood is prefaced by the following introductory remarks concerning the different ways in which a body can be said to be moist or dry:

> These [the moist and the dry] are spoken in various ways; for example, some things are moist or dry potentially, some actually. For ice and all solidified moist things are called actually and incidentally dry, being potentially and in themselves moist, while earth, ash, and such things, having been *mixed* (*mixthenta*) with moisture, are actually and incidentally moist, yet in themselves and potentially dry; but when these have been decomposed, the ingredients that consist of water are fluid and both actually and potentially moist, while the ingredients that consist of earth are all dry, and it is in this way most of all that the dry is spoken of *properly and without qualification*. (my italics, *PA* II.3 649b10–20)

The two distinctions concerning the different ways in which the moist and the dry may be said to belong to a body (actually/potentially, in itself/incidentally) most probably refer to inanimate uniform stuffs, as opposed to animate ones. For, first, we have an explicit reference to *mixis*, whereas as we saw in the case of blood, the heat coupled with it is not mixed in it. Second, incidentally belonging, that is, the fact that the stuffs possess a contrariety incidentally, is not qualified here, as it was in the case of blood. Finally, decomposition results in the separation of earthy and watery materials, which regain their dryness and moisture respectively. This process, which results in unqualified cases of dryness or moisture or, equivalently, to the elements themselves, is unlike what happens to decomposed blood:

> And so it is with dry and moist [as it is with the hot that affects blood]. Hence, those who are in the nature [sc. of the organism] are hot and moist but when separated solidify and appear cold, such as blood, while others, like bile are hot and thick, but when separated from the nature of their possessors suffer the opposite – they cool and become moist; for while blood becomes more dry, yellow bile becomes more moist. It must be posited that *more and less* participation in the opposites is present in these things. (Lennox's trans. with slight alterations, *PA* II.3 649b28–35)

Two aspects can help us understand the contrast between animate and inanimate uniform bodies. First, animate bodies decompose when separated from a higher compound, an organism, whose nature is responsible for keeping their heat and moisture at a certain level. So, functional animate uniform parts retain their properties only so long as they are part, and under the influence, of the nature of a living being.[26] By means of its characteristic heat, this nature keeps the character of the uniform part intact. By contrast, inanimate uniform bodies depend on the natures of the mixed ingredients to determine their properties. Second, while animate uniform parts when separated from an organism become dry or moist depending on their constitution, they do not become dry or moist in the primary or unqualified way as in the case of ingredients of inanimate mixtures (649b20). Blood or yellow bile that lose the heat they are coupled with when inside an organism become, in the case of blood, more solid while moist, or, in the case of yellow bile, more fluid while thick.[27] So they move toward the opposite contrary but only to some degree, that is, in a qualified way. That is how we could understand the claim that blood and yellow bile participate to the contraries by the more and the less.[28] If so, the two passages from *PA* II.3 suggest that while inanimate mixtures decompose into elemental matter, animate uniform parts decompose into inanimate mixtures. This reading, if correct, encourages the thought that the composition of inanimate and animate uniform parts differ accordingly: at a first level of composition, the contraries compose inanimate uniform parts as well as impure versions of the elements, while at the second level, some inanimate uniform parts are turned into fully functional parts inside organisms.[29] And there is more here, since this is not just a difference of lower and higher levels of material constitution but is due to different operations in nature, namely *mixis* and concoction, respectively. So let us turn briefly to these processes.

With respect to the operation of *mixis* (649b12), when, for instance, earth is mixed with some moisture it is actually and incidentally moist, while remaining in itself and potentially dry.[30] Accordingly, in the discussion of mixtures in the *GC*, when two or more bodies enter into a mixture,

[26] See the reference to the nature of the organism at *PA* II.3, 649b28 (sc. *HA* I.1, 487a3).

[27] Notice also that the result of decomposition (i.e. what is said to become drier) is blood itself, presumably the matter of functional blood, and not some elemental portion, say, of water or earth.

[28] This reference to the more and the less echoes the claim in *GC* II.7, 334b8–17 that the balance of the contrary properties in a mixture is subject to the more and the less, depending on the portions or powers of the elemental bodies that go into the mixture.

[29] The various functions of the hotter (and the colder) analyzed in *PA* II. 2 can be understood, similarly, as the result of mixing the elemental bodies into uniform mixtures. If so, the uniform inanimate stuffs are the products (together with the impure version of the elements at the first level of composition) and serve as matter for second level compositions, i.e., animate uniform parts. See the remark in *GC* II.8 335a11–14 concerning the food of plants.

[30] There is a difference between our passage and the *GC*. In the *GC*, the elements survive potentially in the mixture (see *GC* II.7) while in the *PA*, it is the elementary qualities that survive potentially in

there is an equilibrium between their elemental powers, each of them changing out of its own nature toward the dominant one; yet neither becomes the other; instead, both become an intermediate (*GC* I.10, 328a28–30). For example, both the earthy and the watery materials that enter a mixture become actually and incidentally something intermediate with respect to their contrary properties of dryness and moisture respectively (and something equivalent occurs with all other elemental contrarieties). This suggests the view that at the level of mixtures, all four contraries are on an equal footing and that no contrary is defined as a mere privation of its contrary.[31]

Things are different in cases of concoctions, as in the heating of blood. Functional blood is cold sanguineous fluid (watery/earthy material) coupled with heat. As Aristotle remarks, because the material is blood only when *coupled* with heat, cold is a nature and not a privation. What he presumably means is that the coldness of the sanguineous fluid is a positive characterization of its nature and not simply a lack of the heat it receives when it becomes functional blood. The mixture considered as the material that constitutes blood is naturally cold. Its coldness counts as a privation only if we consider the process from the perspective of the concocted blood (e.g. in cases of a failure to heat it properly). And this suggests that concoction processes (of blood and of any other animate uniform part) belong to a domain where the contraries are not on equal terms. Rather, one contrary, heat, acts as the positive pole whereas its opposite, cold, as its privation. Thus, at the level of uniform animate parts, coldness is not a quality neutrally opposed to heat, as in the case of inanimate mixtures, but rather the privation of a quality that is normally the result of a positive process, namely concoction. Concoction, thus, is not a neutral process whereby all contraries are of equal value and herein lies its difference from *mixing*. This becomes clear if we attend to evidence outside the *PA*. *Meteorology* IV's discussion of concoction licenses the claim that cold is a privation but not the claim that heat may also be a privation as stated in the *PA* (II.1, 649a21–22), for when Aristotle considers the active factors in

the mixture. This could be explained in terms of the emphasis of *PA* II on elemental qualities rather than the elements.

[31] Indeed, failure of *mixis* due to unequal quantities of bodies that share contrary properties results simply in the weaker quantity being absorbed in the greater (*GC* I.10, 328a24–26) By contrast, in the case of concoction failures, a new material body is produced displaying in some of its characteristics the traces of such a failure. And that is why in the case of concoction there is an opposite negative process, inconcoction, whereas this is not the case with *mixis* (see Section 9.4).

natural changes, he specifies cold as one of the passive factors together with dry and moist, leaving heat as the only formative quality:[32]

> It must be understood that cold is in a sense matter. For as dry and moist are matter (being passive), and find their principal embodiments in earth and water which have cold as a defining characteristic, it is clear that all bodies that are made of either element alone tend to be cold unless they have an external source of heat like boiling water or water strained through ashes, which contains the heat from the ash .... Things composed of more than one element (*koina*) contain heat, having *most* of them been formed by concoction by heat, though some are the products of decay like the waste products of the body. So as long as blood, semen, marrow, rennet and the like keep their nature they are warm, but once they perish and lose their nature they lose their warmth, for all that is left is their material factors, earth and water.[33] (*Meteor.* IV.11, 389a29–b12, Lee's trans. with slight alterations, my italics)

Aristotle's examples are all of uniform living parts.[34] And his reference to warmth as a necessary condition for their nature suggests that he has in mind stuffs similar to what serves as the underlying subject of blood. If so, heat here is a formative principle transforming the lower level mixtures of earth and water to higher level composites, animate parts capable of serving specific functions. Conversely, when these higher level composites are deprived of the heat coupled with them, they turn into a privative state (considered from the perspective of the state they were in before) and result in earth-water mixtures.[35] Heat, then, is considered as a formative element for higher level animate uniform bodies that serve teleological functions, and they are formed out of lower level inanimate uniform materials, characterized by cold.

9.5 From the Thermodynamic Process of *mixis* to the Teleology of *pepsis*

The proposal then is that the first level of constitution in the threefold distinction referred to in *PA* II.1 concerns inanimate uniform substances

[32] See also *Meteor.* IV.2, 380a8–9; *Met.* Λ.4 1070b12; *GA* II.6, 743a36; IV.4, 784a33–34; and the comments in Lennox 2001a: 194–195.

[33] Note that once heat leaves blood what is left is not cooled blood but cool earth and cool water.

[34] With the exception of rennet. But its constitution is very much like that of semen in that it contains vital heat, and its operation has a setting effect for the milk analogous to the one semen has (*GA* II.4, 739b21–23).

[35] Note also that at *PA* II.3, 650a2 Aristotle argues from the formative power of heat (transforming the wet and the dry) to the conclusion that animals must possess such a power internally, so that they will be capable of transforming the wet and the dry of their nourishment. In this explanation, it seems that the chemical analysis of *Meteorology* and the principles or theorems established there have a role to play in grounding zoological explanations.

that are the result of mixing, while at the second level these mixtures compose animate uniform bodies as a result of a process that is teleologically determined, given the limitations of the materials available. Indeed, concoction is presented as a positive process of formation (*Meteor.* IV.11), and I wish to conclude by attending to two features of such processes that corroborate the above contrast between mixture and concoction. The first concerns the distinctive domain of concoction processes:

> Concoction is maturity produced from the opposite passive characteristics by a thing's own natural heat, these passive characteristics being the matter proper to the particular thing. For when a thing has been concocted it has become fully mature .... In some cases [A] the end of the process is a thing's nature, in the sense of its form or essence. In others [B] the end of concoction is the realization of some latent form, as when moisture takes on a certain quality and quantity when cooked or boiled or rotted or otherwise heated; for then it is useful for something and we say it has been concocted. Examples are must, the pus that gathers in boils,[36] and tears when they become rheum. (*Meteor.* IV.2, 379b19–32, Lee's trans.)

[A] and [B] present two kinds of end in concoction, aimed at by the active *dunameis* that operate as efficient causes. The discussion is concerned with the effect heating and cooling have on already constituted bodies. And, though it is not entirely clear how this claim restricts the scope of the discussion, it seems that these processes are not operating on bodies that are getting mixed, for both [A] and [B] differ from the results of mixing. Mixtures, unlike cases under [A], do not constitute a mature state of their ingredients, while unlike cases under [B], mixing is the mutual destruction of the excesses between two (or more) bodies. Thus, concoction concerns a whole different range of processes from mere mixtures.

The second point relates to the *logos* acting as efficient cause in concoction:

> Unqualified and natural generation is a change effected by these *dunameis* [i.e. hot and cold] when present in the right proportion [*logos*], in the matter underlying each and every nature, this matter being the passive properties of which we have spoken [i.e. the wet and the dry]. The hot and the cold produce generation by mastering the matter: when they fail to master it the result is half-cooked and undigested. But the most general contrary to unqualified generation is decay. (*Meteor.* IV.1, 378b32–379a3 Lee's trans. with alterations)

First, the reference here to unqualified generation and decay directly opposes concoction to mixing, since mixing, as we saw in the *GC*, is

[36] *De Respiratione* 479b33.

neither mere alteration nor generation of a body out of the destruction of another but a hybrid process.[37] Second, the reference to the proportion of heat and cold here already points to the fact that these are not two independent causal forces operating separately. Hot and cold constitute, what Lennox labels, the agent continuum of primary qualities as opposed to the passive continuum constituted by the dry and the moist (Lennox 2001a: 195). Every natural body will have a characteristic temperature corresponding to a determinate, specific point along this continuum, whose extremes are heat and cold. Any given point on the continuum is the result of a proportionate relation of extreme heat and cold, and this proportion is what the nature of an organism exploits to produce and allocate what is necessary for the constitution of its uniform parts.

The variety of respects in which a natural body can be more or less active seems to be a result of the chemical or thermodynamic interaction and mixing of the elements. These processes are not directed by an external efficient cause or source of heat but are determined by the natures of the constituents. These constituents produce a rich enough inventory of natural bodies with all sorts of secondary passive qualities and corresponding *dunameis*. This seems a very powerful pool of materials with which living natures can form useful parts that allow or assist the functioning of the complex activities of their nonuniform parts. Such a rich variety of differently disposed material bodies is not available at the level of the simple elements. Without this variety, the formative power of an organism would not have at its disposal the necessary resources. So, this level provides Aristotelian science with an inventory of qualities and processes capable of explaining the constitution of the rich variety of animal parts. The alternative would be to claim that animate forms can reach all the way down to the elements and transform them into what is conditionally necessary for the organism. Such a position would require attributing an extraordinary power to organismal forms. And this seems to go against the very spirit of Aristotelian science according to which nature advances by small steps from nonliving to living bodies.[38] Secondly, it would fail to

[37] *GC* II.10, 327b7–9; Frey 2007: 187. For an extensive discussion of mixtures in Aristotle, see Fine 1996.

[38] This point is made by reference to living beings and extended to all natural entities at *PA* IV.5, 681a12–14. In fact, nature passes continuously from soulless things into animals by way of those things that are alive yet not animals, so that by their proximity the one seems to differ very little from the other. See also *HA* VII(VIII), 588b4–7. This suggests some degree of uniformity in natural processes, even for entities in different domains. Thus, operations whose results are similar, e.g., the production of tactile contrarieties, must be accounted for by similar processes in both animate and inanimate beings.

account for all the biochemical processes present in animate and in inanimate nature. The process of *mixis* and the bodies it constitutes, offers a level of natural explanation that bridges the gap between the elements and organisms by operations that are both veriable, as part of Aristotle's chemistry of inanimate bodies, and autonomous from the scientific study of animate ones.

CHAPTER 10

Blood and Animal Character in Aristotle's Parts of Animals

*Sophia M. Connell**

Introduction

On the Parts of Animals (*PA*) Book II is our main source for any Aristotelian explanation of animal character (*ēthos*).[1] In brief, Aristotle connects character to the material properties of blood, whether it is hot, cold, thick, thin, turbid, or pure (*PA* II.2, 651a16). There have been two main ways to understand this. The first posits that blood consistency is conditionally necessary for character. The soul of the animal (and more specifically its nutritive soul) must work on the materials that enter to create the type of blood required for its character.[2] The second account sees the statements in the *PA* as a "complete" explanation of character in terms of a 'material nature.' This also implies that while having a character is part of the account of the substantial being of many animals, the specific character it ends up with is not.[3]

In this chapter, I support and further develop the second position. Although blood consistency is not directly teleological, character is closely

* I would like to thank James Lennox for his incredibly astute detailed comments on a previous draft, which helped me understand the issues better. I am also indebted to Mariska Leunissen, Stasinos Stravrianeas, and Claire Bubb who read and commented on earlier versions.

1 Character, along with parts, activities, and ways of life, occupies Aristotle's attention in the *Historia Animalium* (*HA*). While the *HA* describes character dispositions (especially in Books VII(VIII) and VIII(IX)), it doesn't provide any explanations for these.

2 Gelber 2015 and assumed in Johansen 2012: 286; Lloyd 1992: 152; Connell 2016: 142–144. The idea that form 'goes all the way down' comes most straightforwardly from concentrating on *De Anima* and the *Metaphysics* without the *PA*. For a recent account along these lines, see Hennig 2020.

3 Leunissen 2017: 9, 20. Within the *PA*, there is a question about what the substance (*ousia*) or essence (*ti esti*) of animal is, whether it is just the form or is rather the composite (matter and form) taken universally. This controversy involves different interpretations of *Metaphysics* E and Z, which are not relevant to the present chapter. However, the question of how we are to take the essence or 'substantial being' of the kind is important for my account. In short, the *PA* appears to treat the material aspects of the parts as included in the substantial being of the kind, even though these are not parts of the form or contained in the form/soul (*PA* I.1, 641a14–17). It is the composite taken universally that interests Aristotle in this work (e.g. *PA* I.5, 645a35). See Gotthelf 1985; Lennox 2008, 2021, chps. 4–5; Henry 2021 and Henry's chapter in this volume (Chapter 6).

intertwined with an animal's way of living. Animals' preferences and feelings, based in their material nature, play a part in structuring their more advanced functioning as the living beings that they are. While the materials that exist in the blood vessels are not put there for this purpose, they are often utilized to bring about such advantageous outcomes, displaying a coordination of material and formal natures.

The chapter will proceed as follows. Section 10.1 will detail what is said in *On the Parts of Animals* about the connections between blood (or its analogue) and character. Section 10.2 will set out the 'material nature' central to determining an animal's character. Section 10.3 will detail character in the *PA* and its connections to an animal's "way of life" (*bios*). Section 10.4 provides a way to understand the benefits of character in teleological terms. The final Section (10.5) details types of blood as a basis for the final cause of character. In short, a complex causal scheme, including the material nature, food and environment, along with the animal's optimal way of living, determines character.

10.1 *Parts of Animals* on Blood

In the *PA*, blood is initially classed as a uniform part (*PA* II.1, 646b6–8).[4] This is then qualified: the moist uniform parts are divided into those that nourish the nonuniform parts (i.e. blood) and others that are residues (e.g. urine) (647b26–27). Blood is the "most complete" form (*teleiotatos*) or "final stage" of nutriment (*eschatē trophē*), matter for the whole body (651a12–15). It is contained in the blood vessels (649b23–24) and heart (*PA* III.4, 666a34–36).[5] As a part of the body, it is subject to the homonymy principle; when not serving its role, it is blood "only in name" (*GA* II.1, 734b24; *PA* I.1, 641a2; *Meteor.* IV.12, 389b29–340a10; *Metaph.* VII.10, 1035b19–25, VII.11, 1036b30). Unlike other parts, however, blood is not attached to or continuous with other parts of the

[4] Lennox 2001a: 200. The view that blood is not a part of the body (Frey 2015) rests on its insensitivity (650b3–5, 656b19–21, 666a11–18) and a passage at *PA* II.1, 656b19–22 excised by Peck and Ogle. Lennox treats it as "odd" to say that blood is no part and is perhaps an awkward way of expressing its uniqueness among parts, as the material for other parts (226). The fact that it is insensitive (*HA* III.19, 520b13–16; *PA* II.3, 650b4–6, II.5, 651b4–7; II.7, 652b4–7?) is no reason to class it as a non-part, since other internal organs are likewise unable to sense. In the *Parts of Animals* and elsewhere, blood is most often treated as a uniform part (I.1, 640b18–20, II.2, 647b10–16; *HA* 511b1–10).

[5] Cf. *HA* III.19, 520b10–14; *Juv.* 3, 469a4–5. All parts of the body come to be from blood (*Juv.* 3, 469a1; *Resp.* 8 474b4–5; *GA* I.23, 725a11–13). Blood is matter for the body (*GA* III.1, 751b2; *PA* III.5, 668a4–6); it also sustains the animal's life (Johansen 2012: 106–111; King 2021).

body (*PA* II.3, 650b6–7), which might make it seem that blood outside the body is like the blood inside it.[6] For Aristotle, blood is hot only accidentally. "If we give a special name to hot water or hot iron: that is the mode in which blood is hot" (*PA* II.2, 649a16–17).[7] As a permanent substratum, blood is not hot (650a2–4). But in terms of what it is as nutriment of this animal, having been acted on by natural heat, it is hot (Lennox 2001a: 197–198). Thus blood, like other parts, is radically different from parts outside living organic systems and does not remain 'blood' once it exits the body.[8]

Because blood is produced by an animal's nutri-generative soul, some scholars think of it as directly connected to the animal's form. Frey posits that "an organism's capacities for nutritive movement constitute a single, unitary capacity for form" (2015: 6). For Johansen, "blood is finely differentiated to suit the particular animal concerned – dog blood differs from cat blood" (2012: 108). For this and other reasons, he holds that blood "contain[s] the form that defines the living being in question" and is "isomorphic" with the animal (109). This position invites certain qualifications. Although blood is specific to an animal kind, as is most apparent in the case of the even more concocted menstrual blood, in the reproductive context (Connell 2021a), it is "isomorphic" only in the sense of having a potential for that specific form.

The idea that blood "contain[s] the form" of the animal attracts some scholars to speculate that nutritive soul makes blood in such a way as to directly sustain the animal's essence. Alternatively, others view blood as the ultimate *material* nature of an animal and see variations that result in character differences as less directly connected to essence.[9] This latter point of view is supported by the fact that the longest and most detailed discussions of blood in the *PA* focus on its materiality. *On the Parts of Animals* II.4 can be broken down into four sections (A–D).[10] Aristotle first sets out that blood has two key material parts, earthen fibres and watery serum.

[6] Aristotle talks of a finger not existing "if severed from the whole" (*Metaph.* VII.10, 1035b24) but blood cannot be severed in that way. As Gill notes, "blood can survive outside the body as the cold material it is in its own right" (2021: 27).

[7] As Lennox points out, this is puzzling, given the idea at *Meteor.* IV.11, 389b8–15 that blood "in its nature" is hot and once it is out of the body, it cools and become mere 'matter.' See Althoff (1992: 40–46), who interprets the parallels from *Meteor.* IV in detail; and Kullmann 2007: 391. For a full discussion of the dependence of the *PA* on the 'chemistry' of *Meteor.* IV and *GC* II, see especially Lennox 2021: ch. 9 and Stavrianeas' chapter in this volume (Chapter 9).

[8] In this, I am sympathetic to Frey's analysis against those who hold that "an organism's organs and tissues are . . . identical in account to an appropriate class of inanimate, natural bodies" (2015: 4). See also Lloyd 1992: 164–165.

[9] E.g. Leunissen 2017, ch.1.

[10] All translations are from Lennox 2001a.

(A) 650b14–26: Some blood has what are called fibres, but some, such as that of deer and gazelles, does not (Τὰς δὲ καλουμένας ἶνας τὸ μὲν ἔχει αἷμα τὸ δ' οὐκ ἔχει). It is for just this reason that this sort of blood does not solidify – the part of the blood that is watery (τὸ μὲν ὑδατῶδες) is colder, which is also why it does not solidify; while the part of the blood that is earthen (τὸ δὲ γεῶδες) solidifies when the moisture evaporates, and the fibres are made of earth (αἱ δ' ἶνες γῆς εἰσιν). Now some of these animals also have a more subtle intelligence (ἔνιά γε καὶ γλαφυρωτέραν ἔχειν τὴν διάνοιαν τῶν τοιούτων), not because of the coldness of their blood, but rather because it is thin and pure; for what is earthen has neither of these properties. For those animals with finer and purer moisture have quicker perception (Εὐκινητοτέραν γὰρ ἔχουσι τὴν αἴσθησιν). Indeed, because of this even some of the bloodless animals have a more intelligent (συνετωτέραν) soul than some of the blooded ones, as has been said before, e.g. the bee, the ant kind, and any others there may be of this sort.

More watery blood is the material basis for intelligence and also timidity.

(B) 650b26–32: The animals that are excessively watery are more timid.[11] This is because fear cools; accordingly, those having such a blend in the heart are predisposed to this affection (προωδοποίηται οὖν τῷ πάθει τὰ τοιαύτην ἔχοντα τὴν ἐν τῇ καρδίᾳ κρᾶσιν), since water is solidified by the cold. This is also why the other bloodless animals are, generally speaking, more timid than the blooded, and when afraid become immobile, discharge residues, and in some cases change their colors.

More earthen blood supports spiritedness.

(C) 650b32–651a11: However, those with excessively fibrous, thick blood are more earthen in nature, and both spirited in character and excitable because of their spirit (γεωδέστερα τὴν φύσιν ἐστὶ καὶ θυμώδη τὸ ἦθος καὶ ἐκστατικὰ διὰ τὸν θυμόν). For spirit is productive of heat, and solids that have been heated give off more heat than liquids; and the fibres are solid and earthen, so that they become like sparks in the blood, and produce a boiling in the spirit (αἱ δ' ἶνες στερεὸν καὶ γεῶδες, ὥστε γίνονται οἷον πυρίαι ἐν τῷ αἵματι καὶ ζέσιν ποιοῦσιν ἐν τοῖς θυμοῖς). This is why bulls and boars are spirited and excitable (ἐκστατικοί); their blood is most fibrous, and indeed the blood of the bull solidifies most quickly of all. But their blood does not solidify when these fibres have been extracted; for just as if one

[11] Note, this idea is found in the Hippocratic corpus, for example *Epid.* 6.2.20 where "watery blood" is found "in those who cower under fear or have troubled sleep" (Littré 5.288.5–6).

were to extract the earth from mud, the water would not solidify, so it is with the blood, since the fibres are made of earth. But if they are not extracted the blood is solidified by cold like moist earth. This is because the moisture evaporates when the heat is squeezed out by the cold, as was said before, and moist earth is solidified not by heat, but by being dried out by the cold. And there is moisture in these bodies on account of the heat that is present in animals.

Having detailed the effects of the predominance of either earthen or watery parts in blood, Aristotle offers an explanation:

> (D) 651a11–19: The nature of the blood is the cause of many features of animals with respect to both character and perception, as is reasonable, since blood is the matter of the entire body; for nourishment is matter, and blood is the last stage of nourishment (Πολλῶν δ᾽ ἐστὶν αἰτία ἡ τοῦ αἵματος φύσις καὶ κατὰ τὸ ἦθος τοῖς ζῴοις καὶ κατὰ τὴν αἴσθησιν, εὐλόγως· ὕλη γάρ ἐστι παντὸς τοῦ σώματος· ἡ γὰρ τροφὴ ὕλη, τὸ δ᾽ αἷμα ἡ ἐσχάτη τροφή). It therefore makes a great difference whether it is hot or cold, thin or thick, turbid or pure. Serum is the watery part of blood, either on account of its not yet having been concocted or its having undergone decay, so that serum exists on the one hand of necessity, and on the other for the sake of blood.

A related mention of blood as the cause of character and perception occurs earlier in *PA* II.2.

> The relative differences between things of the same kind are for the sake of the better, that is, the differences both of other parts and of blood from blood. One sort of blood is thinner, another thicker, one purer, another more turbid, and again one colder, another hotter, both in the parts of one animal (for the upper parts of one animal are distinguished from the lower parts by these differences), and between one animal and another. (647b29–35)
>
> Thicker and hotter blood is more productive of strength, while thinner and cooler blood is more [productive of] perception and intelligence. And the same difference obtains among the attributes analogous to blood. This is why both bees and other such animals are more intelligent (φρονιμώτερα) in their nature than many blooded animals, and why amongst blooded animals those having cold and thin blood are more intelligent than their opposites. But those with hot, thin, and pure blood are best; for such animals are at once in a good state relative to both courage and intelligence. It is for this reason too that the upper parts differ in this way compared with the lower parts, and again the male compared with the female, and the right side of the body with the left. (*PA* II.2, 648a2–13)

The causal story is at first difficult to make sense of.[12] From what Aristotle says directly after the passage from *PA* II.2, variations in parts can be connected to the goals of the animal in question in two ways: as essential *or* as advantageous/disadvantageous.

> And likewise with the other parts, both the parts such as these and the non-uniform parts: they should be assumed to differ in some cases relative to each animal's functions and substantial being, in other cases relative to what is better and worse (τὰ μὲν πρὸς τὸ βέλτιον ἢ).
>
> For example, among those with eyes, some are hard-eyed, some are moist-eyed, and some do not have eyelids, while others do, but either way it is for greater accuracy of sight. (648a13–18)

The passage has been taken to indicate that not only character but also blood consistency is determined by the essence and contained in the form of the kind.[13] However, as Lennox explains, while some differences are among parts that "perform basic functions" (and are thus explained with reference to an animal's substantial being), other differences are explained by how they make life better for an organism (*PA* I.1, 640a36; Lennox 2001a: 189).

Aristotle never says that blood differs to accommodate animals' different characters in the way he says that mouths differ to accommodate functions such as eating, protection, speech, or breathing (*PA* III.1, 662a16–27). Blood consistency, when it related to character, doesn't look to be directly explained by the substantial being of the kind. Instead, the effects of blood on character in the *PA* are explained in terms of its materiality, as being watery (651a29, 651a3) and/or earthen (650b17, 650b30, 651a2).[14] And these are the effects of the qualities of water and earth (650b16, b26, 651a14, 650b17, 19, 651b3–8).[15] These material explanations are connected to what makes a particular type of animal's life go better. The passage from *PA* II.4 connects character capacities to what is good for an animal, but in a general way that covers all living beings and not as

[12] As Lennox remarks of 650b22–23: "The causation here appears to be the reverse of what we found in the earlier passage to which we are referred [i.e.648a4–13] – here the nature of the blood appears to be determining character" (2001a: 202).

[13] Johansen 2012; Gelber 2015; Connell 2018b.

[14] Matter in a compound is not still bronze or earth but becomes "brazen" and "earthen" (*Metaph.* VII, 1933b9–16; IX.7, 1049a19–b3).

[15] Frey opposes a view he terms "the prison model," which has it that materials in animals are the same as inanimate materials on the grounds that nutrition has a transformative effect on matter (2015: 15–17). While I agree with the latter point, this doesn't mean that the *dunameis* in the living body aren't the same as those outside it. Some animals are 'watery,' some 'earthen,' and this affects how they lead their lives. In order to explain this, Aristotle frequently refers to the powers of elemental natures (A–C).

something that is contained in the account of the substantial being of any specific kind.[16] The best thing is to be able to give an account in terms of the essential features, but if that is not possible, one must explain why the parts are better for the animal more generally.[17] The fact that some advantage is brought about by a particular constitution of blood is not referred to 'their essential nature' but rather to their 'material nature,' which, as we will see, is related to substantial being in a different manner.

10.2 Material Natures

Certain traditional interpretations of Aristotle's psychology and metaphysics convey the thought that living natural things are governed by their forms and that their matter does not have its own, independent nature.[18] However, philosophers of Aristotle's biology who have a detailed understanding of his argumentative strategies, particularly in the *PA*, find a significant role for material nature in Aristotle's psychology, metaphysics, and biological practice.[19] This was first clearly articulated in a paper by Lennox entitled "Material and Formal Natures," which explains how the *PA* is "in large measure a study of the interaction of formal and material natures in living things" (182–183). The material natures of animals play "a more central role in Aristotelian science" than is usually given to them. The exact expression 'material nature' is rare in the corpus;[20] the best example being *PA* I.1, 640b28–29 where is it stated that "the nature of the shape is more commanding than the nature according to the matter" (Ἡ γὰρ κατὰ τὴν μορφὴν φύσις κυριωτέρα τῆς ὑλικῆς φύσεως.).[21] In constraining (and aiding) formal nature, material natures shape animals, as is seen in the case of the distribution of nutriment to horns and other

[16] In that context, Aristotle is sometimes attracted by the idea that certain blood types as absolutely best: i.e. the thin and pure blood of male humans (648a9); given the different lifestyles of different animals, this seems problematic. Lennox 2001a: 188. "Such claims are more difficult to defend because the standard of evaluation is not clear. Why should one sort of character and life be better than another?" See Gotthelf 1988 for discussion of evaluations across kinds (127–128). Stavrianeas 2018: 58–60 thinks this sort of account in terms of a "tension" in the animals' formal nature between what is best from the point of view of the wider kind and what is best for the particular type of animal. See also Freeland 1994: 171–172, 175.

[17] "Hence it would be best to say that, since this is what it is to be a human being, on account of this it has these things; for it cannot be without these parts. If one cannot say this, one should say the next best thing, i.e. either that in general it cannot be otherwise, or that at least it is good thus" (*PA* I.1, 640a33–640b1).

[18] E.g. Hennig 2020.

[19] Charles 2021; Lennox 2021.

[20] However, see *Phys.* II.1, 193a28–29; II.2, 194a12–13; II.8, 199a30–31; *GA* IV.4, 770b15–17.

[21] Aristotle sometimes uses the expression 'necessary nature,' which seems to have a similar import. I thank Mariska Leunissen for pointing this out.

body parts in *PA* II.3. Through an analysis of this process, Lennox shows that:

> There are certain features of living things that are sufficiently explained by reference to their material natures; and there are certain material facts about certain kinds of animals that are as *explanatorily primitive* as other facts about their living functions. (Lennox 1996: 219)[22]

This approach is taken up in portions of Allan Gotthelf's paper "The Elephant's Nose' (1997b). In seeking to provide a pseudo-axiomatic structure of explanation for the elephant's trunk (*PA* II.16, 658b26–659a29), Aristotle includes material facts, such as what materials are available to the animal in order to construct such an organ (Gotthelf 2012b: 189). There is then a separate difficulty about how the differences in animals based on material factors are related to their essence. The idea of something's being "explanatorily basic" allows that it can be "essential to the subject kind" (Gotthelf 2021b: 188). Thus, a material fact can be part of the essence. Although this might seem to pull against essence being form and not matter, that idea comes from certain traditional readings of Aristotle's *Metaphysics*, which do not engage with his scientific works (Henry 2021: 9).[23]

The most oft-cited example of an explanation in terms of material nature is the account of why animals have different methods of reproduction (*GA* II.1, 732b28–733b16).[24] The best method, live-bearing, is not possible for animals that are colder and drier.[25] Thus, while all animals must have some way of generating, how they reproduce is determined by where they sit on a scale of material potentials between hot/fluid and cold/dry (Charles 2000: 338). To resist the idea that "material natures" are "explanatorily basic," opponents must make every part of the body hypothetically/conditionally necessary (e.g. *PA* I.1, 642a12–14).[26] This would mean that different forms determine every feature of an animal.[27] But the

[22] One motivation for this view is to be able to include premises about matter in demonstrative science, either as first principles (Henry 2019: 183) or as middle terms (Leunissen 2010: ch. 7).

[23] For the most comprehensive treatment of how to reconcile this aspect of the *Metaphysics* with the biological works on this point, see Lennox 2008.

[24] Henry 2013: 238–239; Leunissen 2017: 22–26; Stravrianeas 2018: 64–65.

[25] "Humans exhibit the most perfect mode of reproduction because they have the best material nature" (Leunissen 2017: 25).

[26] Gelber repudiates "brute facts" (Charles 2000: 334) as "explanatorily primitive" (Lennox [1996b] 2001b: 183), although her own position is nuanced.

[27] Koslicki 1997: 84 writes, "The bodily composition of living things is, on Aristotle's view, thoroughly necessitated (hypothetically) by their respective natures. Neglecting the occasional deformity, every body-part has to be exactly the way it is." This is to disregard that different animal carry out the same living functions differently due to their material natures.

assumption that "most variations, such as ... in size or quality of the instrumental parts and sensory organs, are explained by the functional needs of the particular kind of organism" (Gelber 2015: 51), faces the problem that Aristotle's biological works are full of explanations of parts that refer to materials. The differences between animals are often due to the materials that make up their bodies; this is why some breathers have an epiglottis and others do not (*PA* III.3, 664b20–665a9; Lennox 2001a: 253). Amongst the uniform parts, the same sort of reasoning explains the formation of "hard fat" (sometimes translated "suet").

> Among blooded animals the ones with bodily blood have harder fat. For hard fat is earthen, which is why it solidifies .... That the nature of these animals is full of such an element is apparent from their being horned and having knuckle-bones; for all these parts are dry and earthen in nature. (*PA* II.5, 651a28–33)

Furthermore, conditionally necessary features, if mentioned, are usually so general as to cover most animals in a broad grouping – such as bones having to be rigid and eyes having to be fluid (*PA* II.8, 653b35–37; II.13, 657a33; *DA* III.1, 425a4; *Sens.* 2, 438a13–26; *GA* V.1, 779b23–26).[28]

Many of Aristotle's explanations of animals' features rest on thought experiments about what combinations will optimize their living functions, within the limits of their bodily materiality and material environment. These employ the principle that "nature does nothing in vain" along with the idea that "nature realizes the best of the available possibilities" (*IA* 2, 704b17).[29] For each type of living being, "nature does, among the possibilities, what is best" (*PA* IV.10, 687a15–16; *IA* 12, 711a18). There are a number of ways in which parts and features can be combined, which are aided and limited by materials, those out of which the animals is composed and those that surround it and which it must negotiate in order to live well.[30] The "better" is used of that which "most enables each of the parts

[28] Part of the difficulty is that after *PA* I, Aristotle does not mention conditional necessity; I owe this point to James Lennox. On references to essence in the *PA* and how rare they are, see Gotthelf 1985. It might even seem that some material features unify 'natural kinds' in Aristotle's bio-ontology (Henry 2011: 210–211).

[29] While Lennox's view (1997) is that these are weaker and stronger versions of the same principle, Henry (2013: 231) considers them to be more unified. For a discussion of these ideas in the *IA* see Stravrianeas 2021.

[30] See *Pol.* IV.3, 1290b25ff. for a comparison of this schema with the construction of possible political constitutions. For an account of how these ideas of 'hypothetical design' and the 'optimality principle' work in practice, see especially Leunissen 2010a: ch. 4; Henry 2008, 2013; Johansen (Chapter 7) this volume.

fully to perform its proper function" (*Juv.* 4, 469a28–31); many of these functions are general across almost all animals, such as "locomotion," but they are also more specific, such as "swimming" (*PA* IV.13, 695b17–20; *PA* I.1, 639b1–3). Materials factors alter what is best – the shape and material constitution of the swimming animal explains how many fins they have and where these are located (*PA* IV.13, 695b26–696a9, 20–32). In terms of the material conditions of their surroundings, the best walker will not travel well in water; instead, what is best for it is what "enabl[es] the animal to fulfil the needs of its distinctive mode of life" (Gotthelf 2012: 188).[31] While it is true that "elemental powers in living organisms bodies do not operate independently of, but are rather subordinate to form or soul" (Gelber 2015: 58), in accounting for how these parts work in coordination with other aspects of that sort of life, material factors, both internal and external, are explicitly referenced by Aristotle.

The account of cephalopod ink provides an illustration of an explanation that combines formal and material natures. Aristotle notes that sepia (a type of cephalopod) have a greater supply of ink because they use it more (679a10), which suggests that nature is producing more for this employment. Then a reason for its abundance is given that relates to the way of life (*bios*) of sepia: the fact that they live near the land. Since they have no other means of defense, they must use this extra ink (679a11). The ink sac itself is created by nature for the sake of defense (*PA* IV.5, 678b36–679a6). The material nature, then, is coordinated with the formal one.

> These animals, as they have no blood, are cold and liable to take fright. While in some other animals fear causes a disturbance of the stomach, and in some the discharge of residue from the bladder, in these creatures its effect is to make them discharge their ink; and though this is an effect due to necessity, like the discharge of urine in the others, yet nature makes good use of this residue at the same time for the animal's defense and preservation. (*PA* IV.5, 679a25–31)

Here formal nature fashions a part in anticipation of a necessary material effect of fear. Ultimately, the coordination of these features must fit to the animal's way of life.

In differentiating specific forms of animals, such as human and horse, there is no direct evidence that Aristotle thought of their nature without

[31] "Fish come to be from water . . . . The capacities of the water – wet and cold – are preserved (but not the water), and predominate in the fish: fish are mainly wet and cold, and so can survive in a wet and cold environment" (King 2001: 117).

reference to materials. Indeed, the materials play an important part in determining differences in parts, as he explains in the *GA*:

> Each thing is then for the sake of something and comes to be through the same cause and through the remaining which exist in the *logos* of each of them. Either it [the part] is for the sake of something or is itself the end. Those characteristics which are not like that which exists in generation, we must seek the cause of these in the change and the process of generation already there, so that in the very constitution these differences take hold. For having an eye is of necessity (for such exists in the being of the animal), the sort of eye is necessary also, not that sort of necessity, but another manner of it [i.e. material necessity], that is naturally made to act and be acted on in this way or that. (*GA* V.1, 778b11–19)

Different types of eyes develop due to material factors but this allows for the different ways that resultant species see to be part of what it is to be that particular kind of animal.[32] For example, owls have to be able to see at night because this is how they hunt and live as the sort of animal they are. In a parallel fashion, the basis for character, that is, blood consistency, is due to material interactions. But the resultant characters are like the differing functionality of eyes, which serve the ways of life of the animals in question. While character and the material natures that support them are not specified in the definition of the kind, these features are integrated into the animal's manner of existing in its environment.[33]

10.3 Aspects of Character in the *PA*

So far, we have seen how materials influence and shape the lives of animals for Aristotle. Explaining and discussing what an animal is like (its substantial being) and how it lives and functions often requires reference to the material aspects of it and to materials that affect its objectives. Analysis of the *PA* passages in Section 10.1 reveals two aspects to character and one effect of its bodily basis.[34] The first is a very general ability to experience feeling (*pathos*).

[32] See Gotthelf and Leunissen 2010: 331–337; Connell 2016: 333–337.

[33] Lennox's note on *PA* 648a1–10 reads: "Blood types ... are correlated with differences in character and intelligence (noesis). It would be wrong to see this as accounting for the character differences by reference to differences in blood – ultimately, the blood/blood-analogue differences will need to be explained by references to the role of differences in character and intelligence in the animal's life" (2001a: 187).

[34] Character also includes intelligent capacities (*HA* VII(VIII).1, 588a16–588b3; VIII(IX).1, 608a12–21). My main focus in this chapter is on 'affective' responses. The role of blood in perception and intelligence is complicated and cannot be covered here. See van der Eijk 1997; Gregoric 2007: ch. 7; Connell 2021b.

Some animals will only have the most basic of feelings, that is, pain and pleasure.[35] All others ought to have some version of fear, anger, and affection in addition.[36] Different animals will feel these emotions in the presence of different stimuli, which will depend on the way they picture or represent objects of perception. A mouse will respond to perceiving a cat with a feeling of fear, a dog with a feeling of aggression.[37] The body of the animal must be able to assist in registering this feeling. This may be mainly a perceptive matter: so long as the animal is able to properly perceive, it will crudely classify the object as something to be afraid of or aggressive toward.[38]

Whatever it is that initially happens in the body when an animal is angry, whether boiling of blood (*DA* I.1, 403b1) or cooling in the upper parts (*Juv.* 479b23), this is just one aspect of character. The second is whether the body facilitates a continuing effect (B, C, 650b26–651a1). The term Aristotle uses for this is *proōdopoētai* (650b28), often translated "preparing the way." The same term is employed in discussing certain temperaments in the *Problemata*: "The announcement of something alarming, if it occurs at a time when the temperament is rather cold, makes a man cowardly; for it has already prepared a way for the entrance of fear (*proōdopepoiēke gar tō phobō*)" (*Prob.* 30, 954b10–13). In the *Rhetoric*, a similar idea is expressed:

> [The elderly] are cowardly, and are always anticipating danger; unlike that of the young, who are warm-blooded, their temperament is chilly; old age has prepared the way (*proōdopepoiēke*) for cowardice; fear is, in fact, a form of chill. (*Rh.* II.13, 1389b29–31)

Everyone feels fear, but in a body that is cold, that feeling is exaggerated and sustained, which leads to the character disposition of cowardliness.[39]

The passages from *PA* II.4 (B, C) also mention involuntary bodily effects of feelings. For those that have liquid blood, fear leads to physical immobility[40] and sometimes also to voiding fluids[41] or change of color (IV.5, 679a13;

[35] This class of non-locomotive sentient living beings (*DA* II.2, 413b2–4) must be small, mostly consisting of sea anemones and other attached sea creatures (*HA* VII(VIII).1, 588b4–27; Güreman 2015).

[36] I take these to be the three main axes of animal emotion (see *HA* VII–VIII; Connell 2024).

[37] Lennox 2001a: 203; Corcilius and Gregoric 2013: 63; Connell 2021b: 231.

[38] For different accounts of this sort of proto-conceptualisation, see Gregoric and Grgic 2006; Gasser-Wingate 2019, 2021.

[39] For other similar uses, see *Somn. Div.* 463a26, a29; *Mem.* 453a26–27. This terminology is also used by Aristotle to explain why some animals are prone to certain deformities – their material nature "prepares the way" for such accidental causes (Connell 2018a).

[40] Lennox 2001a: 55: "as we say, fear leads them to freeze."

[41] The pseudo-Aristotelian *Problemata* Book 27 is devoted to account of the bodily causes and effects of "fear and courage". The author associates fear with the transfer of heat from the upper to the lower parts of the body and with paleness and trembling. Other bodily consequences of fear include

IV.11, 692a21–25; *HA* II.11, 503b2–8; VIII(IX).37 622a3–14). Another automatic response occurs in those with thick blood who are "excitable" – a term that indicates impetuosity (*EN* VII.8, 1151a1, 20; *Poet.* 15, 1455a34; *Phys.* IV.13, 222b16, 21). Presumably, these animals attack automatically, being mobilized rather than frozen by their feelings.

We can note that Aristotle keeps character dispositions general. Indeed, his concern seems to be not with individual differences between forms of kinds but with broader grouping, that is, which animals are fearful, which bold, which skilful, etc. (Leunissen 2017: 12). Materials admit of or support vital activities to different degrees (*GA* II.3, 736b30–34). The best or noblest animals are in general those that are hottest and most fluid. They are better able to perceive, and some are capable of theoretical understanding. In terms of reproduction, they gestate internally and complete their offspring.[42] As for locomotion, these better animals have four points of motion and a forward trajectory (*IA* 7–8). If we include character, such animals are the most courageous, a spiritedness tempered by intelligence.[43] The worst or most ignoble animals are cold and dry. They are less alive in the sense that they are less able to access perception, a sort of knowing. They cannot generate another like in kind. As for character, it is more prominent in more advanced animals.[44] Those animals that merely respond to the environment automatically, for example, by voiding fluid, do not really have a character since character implies voluntary responses. At this extreme, lack of character is due to how cold and dry an animal is.[45]

As Leunissen explains, this indicates that "there is simply no evidence in the corpus to suggest that Aristotle thinks that animals have the particular character profiles they have for the sake of something or that the particular character profile is part of the definition of the substantial being of

thirst, rapid and sharp heartbeat, trembling of voice, hand, lower lip, shivering, silence and speechlessness, as well as loosening of the bowels, desire to urinate, and involuntary emission of semen. See Mayhew's introduction (2011).

42 Henry 2013: 238: "In a perfect world, then, all animals would be live-bearers."

43 This is one way to read the comments in *HA* VII(IX).1, 608b4–8 that humans have "the most perfected nature"; being hot and fluid, their character traits are more pronounced (Leunissen 2017: 20).

44 On degrees of vitality in Aristotle, see *Protrepticus* VII, 43.27–44.26; XI, 58.7–59.6; *GA* 1.23, 731a33–b5.

45 In general, animals that are 'worse' are made of materials that make them less able to control their bodies and environment through heat (see Connell 2016: 368–370). Those animals which are cold "cannot be as powerful, strong or sensitive, since heat is required for so many soul functions, then there will be types of animal that do not best meet the highest goals that all living beings share" (370). Animals that are cold and dry are less perfected (*Resp.* 13, 477a15–25; Leunissen 2017b: 25).

animals" (2017: 9 n. 11). Character being a 'differentia' also "does not mean that it is necessarily an essential characteristic of a whole animal kind; that is, that it is intrinsically or primarily connected to the form or the formal nature of that animal" (20). Leunissen cites Balme who noted that "some of the differentiae are variable accidents, such as the geographical variations in sheep, goats, cattle, and changes in birds' coloration and note" (1987: 19). While a whole class of differentia being variable accidents is possible, we cannot rule out that Aristotle provided "no formal final explanations" (Leunissen 2017: 20) of character because he never got around to writing an explanatory work on this topic.[46] One can make a case for character being integrated with the substantial being of an animal kind. Leunissen's position is compatible with this insofar as on her view essence does not "include" character but does "accommodate" it.[47] Thus, we might note that the sustaining of spirit is important for animals that defend their territory. And a continuity of feeling is crucial to many animals' ways of life, for example, in care of young, where affectionate feelings must be sustained.[48] In such cases, there is a connection to the 'way of life' (*bios*), which organizes this particular animal's attributes, activities, and parts.[49] Character must be subordinated to the way of life of the particular animal since character is a certain way of responding to the environment and of acting that is conducive to their well-being.[50] It is in this sense that character is related to essence and can have a final cause.

10.4 Types of Teleology

The body is like the saw (*PA* I.5, 645b17–20): it is there for the actions that characterize an animal's life. As a part of that body, blood's primary role is nourishing; but it has the secondary role of underlying affective states and

[46] And this might explain why some of his students spend a good deal of time on character, including Theophrastus and Eudemus (see Hatzimichali 2021).

[47] Mariska Leunissen, personal correspondence, April 2021.

[48] All emotional responses appear also to require some sort of memory/retention of images (Connell 2024).

[49] "Each one of these ways of life will be correlated with an indeterminate number of activities performed by many part; and those activities will be performed in different ways depending on the animal's character, for example, whether it is spirited or sluggish, timid or aggressive" (Lennox 2010c: 336). Lennox argues that "way of life" (*bios*) takes "precedence over its particular activities, character traits and parts" in effect organizing or coordinating them (318).

[50] Lennox 2010c: 342: "Animals can perform activities related to feeding, mating, or care of offspring with greater or lesser intelligence, skill, resourcefulness, courage, gentleness, and so on, given their particular way of life." Lennox's reading of character here does not ground seeing blood consistency as the only explanation but provides a more formal account, which makes it a matter of coordinating functions under the heading of way of life.

dispositions that allow animals to live in the way that they do. Even though all parts are for the sake of the overall animal's life, not all are directly put in place by nature to serve those ends. The talk of nature "using" materials or the results of materially necessary interactions is common.[51] Kullmann (1985) was the first to term this 'secondary' teleology; Mariska Leunissen elaborates.[52] On her picture, primary teleology is directed toward features that are fundamental or 'vital,' such as the heart. Secondary teleology uses materially necessary interactions, or 'co-opts' these, to serve the functions of the animal. These parts or aspects are not part of the substantial being of the kind but are parts which instead make life better (2010a: 19–20, 89–99). The close alignment of two types of teleology and two types of part has been challenged.[53] Secondary teleology doesn't always produce subsidiary or luxury parts but is sometimes closer to the particular life of the animal (Henry 2019: 191–192). However, there is more agreement than disagreement here in terms of the complex connections between nonessential features and an animal's essential nature.

For Leunissen, many features of particular kinds of animal are not in the definition of their substantial being (2010: 126); they are also not conditionally necessitated. While an animal cannot be without certain parts, such as a heart, other parts make life go better; these are not simply necessary features for formal essences. The complexity of Aristotle's use of the "nature does nothing in vain" principle shows how each nature has a range of possibilities that are determined in part by the animal's material nature. This doesn't mean that all differentiation between types of animal is due to matter unconnected to the formal essence of kinds.[54] As Gotthelf explains:

> Notions of 'better' and 'well' enter when there are options in the survival of the animal. 'Better' selects the more efficient options from among real possibilities for an animal's particular being. (2012: 55)

[51] References in Henry 2019: 179–182.

[52] Possibly inspired by Lennox's commentary on *PA* 663b20–24 "The discussion of horn production that follows strongly suggest that 'according to the account' means the nature specified in the formal account of the animal under discussion. It is this formal nature that makes use of the necessary nature for a specific end. When Aristotle discusses instances where the formal nature of an animal 'uses' or 'makes use of' materials for various ends, it may be that he has in mind a form of indirect teleology" (2001a: 249). See also Henry 2019: 192 n. 44.

[53] The strictness of this division and its alignment with two different kinds of parts is criticized by Henry (2019: appendix to ch. 8). While he cites cases in which nature uses necessary features of materials, such as hot and cold, this does not undermine Leunissen's distinction between vital parts and parts like horns, which are not strictly necessary.

[54] "Aristotle believes that, hypothetically speaking, nature could have designed the animals that have subsidiary parts in such a way that they would have been able to function without them" (Leunissen 2010a: 216).

The causal connections between horns, character, and other features derive from some complex combination of formal and material factors.

In setting out connections between essence and character, it is useful to employ another idea of different sorts of teleological explanation in Aristotle. Some explanations use the genitive and so are "for the sake of which," while others use the dative and so are "that for which" (415b20–21). One might refer to these as for the sake of 'the end' versus for the sake of 'the beneficiary.'[55] On this reading, parts that make life better for animals can be vital, for example, allowing that the presence of horns is for the sake of the horn-bearing kind. Henry compares such cases to a housebuilder adding window boxes with leftover materials (2019: 189). But a better comparison might be the addition of insulation around the windows, if the house is in a cold region. Afterall, horns are closely connected to how horned animals live their lives in the circumstances that they do (*PA* I.1, 640a33–640b1). Using this distinction, we can see character as an instance of nature using the consistency of blood to benefit the kind in their particular circumstances.

10.5 Accidents of Blood Consistency and the Final Cause of Character

In producing blood, nutritive soul is primarily concerned to sustain the texture of the uniform parts of animals.[56] The resultant consistency of blood is a by-product. The story of how the animals' nature 'uses' this situation to its advantage is complicated by what is actually going on in the blood vessels at any given time. The digestive process separates earthy from liquid: the earthy ingredients descend to the lower parts, the more liquid ones rise to the upper regions.[57] This process is daily and leads to sleep when the separation results in exhalations that impair the senses and particularly the central sense organ, the heart (*Somn.* 1, 454b4–14; 2, 455a26–455b2; 3, 457b20–458a10). The last concoction of nutriment,

[55] Kullmann 1985; see review of the debate concerning two senses of 'for the sake of which' in Johnson 2005: 64–80; 2012: 205. Gelber 2019 argues for a different reading of the distinction.

[56] Gelber 2015: 53: "Aristotle's view is that the presence of a particular krasis in an organism's blood is explained teleologically, by reference to the dispositional properties that the kind's uniform parts must have." These requirements would seem to have nothing to do with the role of blood in character, however. See Leunissen 2017: 21 n. 42.

[57] *Somn.* 3, 458a14–18; *HA* III.19, 521a4–6. Here, as elsewhere, Aristotle's model seems more medical than zoological. Only humans and certain birds spend most of their time upright, which would allow for a significant differentiation between the upper parts and the lower parts and for the vertical travel of fluids. See Bubb 2024.

which turns it into blood, occurs in the heart (*Juv.* 3, 469a5; *Somn.* 3, 456b5–6; *Resp.* 20, 480a6–7).[58] There is evidence, though, that for Aristotle this proto-blood passes through the heart several times before it becomes blood proper.[59] The fluid in the blood vessels will also at various times include residues, both useful and useless (and potentially pathological). The idea of useful residues is most fully developed in the *GA* where both male semen and menstrual blood are more concocted versions of blood. But what males and females actually discharge is a combination of pure seed and useless and/or pathological materials (*HA* IX(VII).1, 581b29–33, *GA* I.18, 726a10–15; *GA* II.4, 739a6–10) (Connell 2021a: 65–69). Thus, what exists in the blood vessels includes various useless residues too.

It appears, then, that Aristotle uses 'blood' in a broader and a narrower manner. In the broader sense, 'blood' is whatever is standing in the blood vessels and heart and will move toward the extremities (*PA* II.10, 656b19–22; Bubb 2020). This includes not only the fully concocted 'blood,' but also unconcocted blood, exhalations and many different residues. 'Blood' in a narrower sense is the fully concocted nutriment for the parts. It is only this sort of 'blood' that serves the ends of the kind as nutriment and is homogenous throughout (*PA* II.2, 647b10–21). The function of 'blood' proper is to be able to become other uniform parts with all their various different elemental and dispositional qualities (e.g. soft, hard, flexible, etc.). The three scales mentioned at 647b29–648a1 – thin to thick, clear to muddy, cold to warm – are a different set of factors. And it is these material potentials, in all the fluids in the blood vessels (i.e. 'blood' in a broader sense), which underlie the character dispositions Aristotle goes on to detail in *PA* II.4.[60]

'Blood' in this broader sense changes with diet, environment, health, and age; if these fluids affect perception and character (D), they do so in a way that is less connected to the essence of the kind. The effects on

[58] Blood is formed even before the rest of the embryo is differentiated (*HA* 521a9–11; *GA* II.4, 740b3; *Resp.* 20, 480a7–8).

[59] Bubb 2020: 147: "There seems to be a constant, low-level movement of blood, not-yet-blood and other materials into and out of the heart and up and down the vascular system."

[60] The discussion of thick and thin blood mirrors that in the Hippocratic *De Victu* as pointed out by Peck (Aristotle 1937: 136). It is possible that the main idea of cold blood and intelligence comes from a Hippocratic source and is ill-adapted to an Aristotelian framework. There is certainly scope to be dissatisfied with Aristotle's account of what underlies intelligence. That the thinnest blood is best is hard to square with the idea of a median blend or balance between intelligence and courage (*PA*) or with the view that the best state of blood is "neither very thick nor very thin" (*HA* III.19, 520b18–23).

perception and cognition are more a case of avoiding interference rather than aiding these processes; any earthy substance in blood and its movement impairs perception.[61] Aristotle does not tell us how the motions from external sense objects are transmitted into the body; however, he makes clear that blood contains perceptual images (*phantasmata*), required for memory, understanding, and character. These side-effects, so to speak, of nutritive processes can be shown to have final causes and be related to the being of the animal in question.

One way of thinking about higher cognitive capacities and their dependence on *phantasmata* is that the animal's nature makes use of an accidental by-product of sentient capacities.[62] This is likened to the role of hearing in understanding. The function of the sentient capacity is to discern sense objects, in the case of hearing, sounds. That hearing can also lead to understanding is accidental, as Aristotle explains in *De Sensu.*

> Accidentally hearing makes the greatest contribution to thought. For speech (*logos*) by being heard is responsible for learning, not as such [sc. as heard] but accidentally. For speech is composed of words [or 'names,' *onomata*], and each of the words is a symbol (*sumbolon*). (*Sens.* 1, 437a5–12)

In a similar way, *phantasmata* "do not have a final cause" of their own. Instead, they facilitate psychological activities "to be explained in extension of the perceptual faculty" (Johansen 2012: 207). On this reading of Aristotelian teleology, "the beneficiary may find further uses for the senses other than the ones that define them as such" (207).[63]

One could see a parallel situation for blood consistency. Blood in the narrow sense as processed through the heart to become final nourishment is for the sake of maintaining the parts of the animal's body through nutrition; this is its primary teleological end. Furthermore, those processes that mean that the blood vessels also contain mixtures of unconcocted nutriment, exhalations, and residues are also for the sake of nutrition insofar as they may be used to benefit the animal. An animal's perceptual capabilities use these internal resources in order to negotiate its environment for the best. A colder animal's nature will direct their feelings toward protection, a warmer animal toward defense – both of which aim to

[61] Leunissen 2017: 18; Connell 2021b.

[62] Johansen 2012, 2015. Bubb 2019 focuses on the nutritive processes intertwined with *phantasmata*, partly because she sees them as dependent on blood ("There have been few serious attempts to discuss the physiology of phantasmata," 276 n. 13). Beyond noting that they are dependent on the body broadly speaking, Johansen doesn't think that the material details are all that important (2012: ch. 10).

[63] See also Johansen 2015: 133–135.

preserve life. In those animals suited to the sustaining of affectionate feelings, these material conditions in the blood vessels are used to promote the lives of others of the kind, especially the nurture of the young. Character provides a good example of the integration of nutritive and sentient soul in animals. Blood becomes flesh (*PA* III.5, 668a28; GC I.5.322a23–27)[64] and flesh is connected closely to pleasure and pain, which are arguably wrapped up with emotional and proto-emotional responses.[65]

Another example of this structure of explanation is sleep. Nutritive processes bring about sleep; sleep is not aimed for but is a "side effect" (Bubb 2019: 308). Aristotle's explanation of it is almost entirely physiological and connected to digestion,[66] but sleep also has a final cause; it is for the sake of preserving life because animals cannot be continually active (*Somn.* 2, 455b22). Thus, *phantasmata*, sleep, and character all have final causes, but the material situations that produce and maintain them are not directly for any of these ends.[67]

Blood consistency is not created by an animal's nature directly in order to facilitate its character. However, the materiality of blood that underlies character dispositions in animals is connected to the ends of their kind and, in conjunction with other features, constitutes an important part of their way of life. Like the accidental effects of sensory soul, certain of the material powers of blood are used by an animal's nature for features that have final causes: memory, understanding, desire, sleep, character, and intelligence. This use maps onto both 'secondary' teleology and the type of teleology that more generally presents the animal as a beneficiary. Perhaps the broader blood that contains impurities is indicative of the limitations of formal nature to turn materials into what is useful at any given time. This could also explain how more powerful formal natures have purer and better blood since they are more able to convert and control the harmful elements that disrupt blood consistency.[68]

64 Nourishment is potentially flesh (*GC* 322a23–27).

65 See especially Dow 2015. Degrees of vitality correspond to how involved animals are in the bringing up of their young, which requires sustained emotional responses, such as affection and attachment (*HA* VII(VIII).1, 588b21–589a5), secured so often in a tactile fashion.

66 "Every incapacitation of the perceptive part is not sleep, but this experience is the one that comes about from the exhalation of food" (*Somn.* 3, 456b18–19).

67 See also Segev 2012 for speculations about a final cause of dreams in Aristotle.

68 Pointed out to me by Stasinos Stavrianeas.

CHAPTER 11

Failure in a Beautiful World
Aristotle on Deformed, Useless, and Weird Parts of Animals

*Mariska Leunissen**

Introduction

It takes a philosopher, so Aristotle claims, to see the true beauty of animals and their designs. While admittedly many of their features – such as blood, intestines, or even their entire appearance[1] – may be disagreeable to our perception, those who investigate their causes will see their functionality and hence the teleological design of nature, which in their turn inspire awe, admiration, and pleasure.

In this chapter, I turn to a small group of features discussed in Aristotle's *Parts of Animals* that challenge Aristotle's professed functional beauty of the sublunary world and that instead appear to reveal shortcomings in the 'craftsmanship' of nature responsible for their production. In my analysis of such deformed, useless, harmful, or otherwise weird parts, I will differentiate between three types of teleological failure that can occur in the production processes of animals (at either the species or at the individual level), one of which has not yet received much scholarly attention and seems to involve 'genuine' mistakes or oversights attributed to the formal natures that are crafting these animals. I will suggest that Aristotle's attributions of such failures to nature signals a methodological weakness in the explanatory project of the *Parts of Animals* as a whole,[2] namely that it overextends Aristotle's theory of natural teleology to features that could have been explained more satisfactorily by reference to chance or necessity alone.

In Section 11.2, I will discuss Aristotle's famous exhortation to the study of biology in *PA* I.5 by way of background to the place beauty and failure

* I would like to thank Sophia Connell, Emily Kress, and Jim Lennox for their extensive and helpful feedback on this chapter.

1 Especially birds are singled out for their ugly plumage: see, e.g., the "terrible color" of the golden oriole (*HA* IX.15, 616b11–12: χρόαν ... μοχθηράν), the "badly colored" acanthis (*HA* IX.17, 616b31–32: κακόχροοι), and the "disgusting color" of the heron (*HA* VIII(IX).18, 616b35: τὴν ... χρόαν ... φαύλην).

2 This critique was first raised by Theophrastus in his *Metaphysics* IX; cf. Lennox 2001c: 259–279.

hold in Aristotle's teleological view of the animal world. Next, in Sections 11.3–5, I will analyze Aristotle's treatment of deformed, useless, harmful, and weird parts and argue that some of these accounts reveal methodological problems in Aristotle's application of his theory of natural teleology.

11.1 Beauty, Function, and Failure in Aristotle's Natural World

Aristotle's claim that the true beauty of the animal world lies in its functionality and that it requires the trained eye and the causal understanding of the natural philosopher to see it, is expressed most explicitly in his famous exhortation to the study of biology in *PA* I.5, 644b22–645a36.[3]

In the first part of his speech, Aristotle counters the contemporary prejudice against the study of animals as being ignoble and unpleasant compared to the study of the heavens by claiming that knowledge of sublunary creatures brings its own kind of value and that it has its own attractiveness: because we live among plants and animals, we can learn a lot *more* about them than about the heavenly bodies that are at such a far remove from us.[4] In this way, while the heavenly bodies may constitute the most divine objects of study and while their knowledge provides the highest kind of pleasure, our knowledge of sublunary living beings takes the first prize in terms of its breadth and thoroughness (644b22–645a4).

The second part of the speech (645a4–36) is the most relevant for our purposes, as there Aristotle acknowledges the superficial ugliness of animals and the disgust we may experience while examining and dissecting them but also argues that "even in the study of animals disagreeable to perception (μὴ κεχαρισμένοις … πρὸς τὴν αἴσθησιν), the nature that crafted (ἡ δημιουργήσασα φύσις) them likewise provides extraordinary pleasures to those who are able to know their causes and are by nature philosophers" (645a7–10). In what follows, Aristotle links this pleasure to a special kind of aesthetic experience available for natural philosophers who understand the causes of animals by providing two a fortiori arguments:

> Argument 1 (645a10–17): Surely it would be unreasonable and absurd for us to enjoy studying their representations[5] – on the grounds that we are at the same time studying the art that made them, such as painting or sculpture – while not prizing even more the study of things constituted

[3] On the rhetorical brilliance and protreptic function of this passage, see Lennox 2001a: 172–174 and Johnson (Chapter 2) in this volume.

[4] Our distance to the heavenly bodies and the limits of our knowledge about them is a common trope in Aristotle: see *Cael.* II.3, 286a3–7 and *Cael.* II.12, 292a14–18.

[5] See *Poet.* 4, 1448b10–16 on us indeed enjoying the study of such animal representations.

> by nature, at least when we can observe their causes. For this reason, we should not childishly be disgusted (μὴ δυσχεραίνειν παιδικῶς) at the examination of the less honorable animals. For in all natural things, there is something wonderful (τι θαυμαστόν).[6]
>
> Argument 2 (645a26–30): If someone were to consider the study of other animals as worthless (ἄτιμον), it would be necessary to think in the same way also about themselves: for it is not possible to look at the things from which humankind has been put together without a lot of disgust (πολλῆς δυσχερείας ἰδεῖν), such as blood, flesh, bones, blood vessels and such parts.

At the core of each of these arguments lies of course Aristotle's theory of natural teleology and especially his notion of a "crafting nature" that is responsible for the good and functionality in the sublunary realm. This "crafting nature" is what Aristotle elsewhere refers to as the animal's "formal nature,"[7] which is to be identified with the animal's soul as its formal, efficient, and final cause.[8] These formal natures are present as internal principles of motion and rest in all living beings; they guide their production and organize them in the best way possible, while preserving and realizing the animal's substantial being or form. For Aristotle, formal natures thus resemble – or indeed even exceed – human craftsmen in their teleological actions, as they produce the original objects that the human craftsman then imitates or represents in paint or stone. Accordingly, if we admire representations of animals, we should admire nature – the "original" and most skillful artist – even more. For in nature, Aristotle claims – while borrowing Heraclitus' famous words – "there are gods too" (645a21: εἶναι γὰρ καὶ ἐνταῦθα θεούς)[9] and we should therefore approach the study of animals without disgust for their visible ugliness (645a22: μὴ δυσωπούμενον) and with the recognition that there is something natural and beautiful in all of them (645a22–23: ἐν ἅπασιν ὄντος τινὸς φυσικοῦ καὶ καλοῦ).[10] Aristotle justifies this claim by an appeal to his theory of natural teleology: that for the sake of which is present most of all in nature and this "end for the sake of which" animals and their parts are constituted takes the place of beauty (645a25–26: τὴν τοῦ καλοῦ χώραν εἴληφεν). For when we are discussing materials or instruments in the arts, we understand that our interest

[6] All translations are mine.

[7] Lennox [1996b] 2001b: 182–194. "Formal natures" are to be contrasted with "material natures," i.e., the elemental constitution and the kinds of food animals process.

[8] See, e.g., *PA* I.1, 641a23–28; *DA* II.1, 412a19–21; and *GA* IV.4, 770b17.

[9] Cf. *Div. Somn.* 2, 463b14–15: "nature is daemonic, but not divine."

[10] See Balme [1972] 1992: 123: "The visual arts aim at beautiful representations of animals; real animals may not have such beauty, but instead they are composed for the sake of ends, which is even more wonderful."

in them is not about them as matter or as instrument but as contributions to the complete shape or artifact: we should approach the study of nature in the same way and hence consider the parts and instruments of animals in light of their contributions to the composite and to their complete substantial being. In short, those who are able to 'see' how parts contribute functionally to the animal and its substantial being will no longer be put off and disgusted by any visual unpleasantness involved in their study.

The analogy – or even the straightforward identification of nature with a kind of craftsman that is responsible for the teleology in the natural world[11] in this second part of Aristotle's exhortation to the study of animals – is a common one throughout his biological works and reinforces the idea that the animal world with all its functional features as the "original work of art" on which our human art depends is as beautiful and good as it can be. The philosopher of nature, who understands the causal relations of the parts to the whole and grasps the final cause, is able to experience this type of functional beauty and its associated type of pleasure.

Aristotle's explanatory practice in *PA* II–IV makes clear that the aforementioned teleological workings of formal nature do not, of course, produce an animal world that is functional and therefore aesthetically pleasing to philosophers *in every way*. First and most importantly, despite his claims about the pervasiveness of natural teleology and there being something amazing and beautiful and good in *all* natural things, Aristotle is no Panglossian: not *everything* that comes to be by nature always or for the most part comes to be or is present for the sake of something. For instance, as Aristotle points out, the coming to be of bile happens due to the operation of material necessity alone and it is not for the sake of something.[12] Bile itself, then, as well as animals such as toads and tortoises that have extremely bilious livers, are genuinely foul (and inedible)[13] and hold no beauty in that respect even for a natural philosopher.[14]

[11] For relevant examples in *PA*, see *PA* II.8, 654a24–26; *PA* II.9, 654b27–32; *PA* II.14, 658a21–23, and possibly *PA* II.10, 656b27–28: "as to the position of the sense-organs, all the organs of sense have been ordered beautifully (καλῶς) by nature."

[12] *PA* IV.2, 677a15–18: "Sometimes nature makes use even of residues for some benefit, yet one should not on this account search for what something is for in every case; on the contrary, when certain things are such as they are, many other such things happen from necessity because of these." Cf. *A. Po.* II.12, 95a14–21 on ice and eclipses and *GA* V.1, 778a29–b6 on features such as eye-color as not being for the sake of something.

[13] *PA* III.12, 673b28–31 and 673b24. The same holds for other useless residues, such as excrement: cf. Aristotle's claims about the disgusting smells of bee poop in *HA* IX.40, 626a26–27.

[14] The presence of ink in certain soft-bodied animals provides the only – complicated – example of a part that is both ugly and functional, in that Aristotle argues that this ink is also a materially necessitated residue that is genuinely disgusting but that in this case nature has co-opted the part's

Secondly, despite the pervasiveness of natural teleology, formal natures sometimes fail to realize their ends. Just as in the case of the crafts, so too in the case of nature mistakes are possible, as Aristotle explains in *Ph* II.8, 199a33–b7:

> Mistakes (ἁμαρτία) happen also among things [that come to be] in accordance with art (for the writer writes something incorrectly and the doctor pours out medicine incorrectly), such that it is clear that they are also possible among things [that come to be] in accordance with nature. If, then, there are some things that come to be in accordance with art in which that for the sake of something [is achieved] correctly, while in those in which mistakes happen it was attempted for the sake of something but it did not come to pass (ἐν δὲ τοῖς ἁμαρτανομένοις ἕνεκα μέν τινος ἐπιχειρεῖται ἀλλ' ἀποτυγχάνεται), in the same way would the situation be in the natural products, and monsters are mistakes with regard to that for the sake of something (καὶ τὰ τέρατα ἁμαρτήματα ἐκείνου τοῦ ἕνεκά του). And in the original compositions, then, the ox-progeny, if they were not able to reach a certain limit and end, must have come to be due to some corrupted principle (διαφθειρομένης ... ἀρχῆς τινός), just as happens now when the seed [is corrupted].

According to this analogy, mistakes in art and nature happen in the same way: they occur when things are undertaken for the sake of something, but when that goal or end does not come to pass. As an example of such a natural mistake, Aristotle mentions the birth of monstrosities, and, as the cause of this failure of the teleological process reaching its end, he identifies the "corruption of the principle." Elsewhere, Aristotle explains the production of monsters – as well as any kind of smaller or larger deviation from the species form that affects the production processes of individual animals, such as a goat born with two heads[15] – in terms of the teleological process of reproduction failing to reach its most natural end (that is, the production of male offspring resembling the father) due to the motions of the male principle relapsing and not mastering the female menses, either due to weaknesses of the male principle or due to imperfections of the matter provided by the female or even a combination of both.[16] Teleological mistakes or failures of this type then are failures of formal natures to gain control over the material natures and to realize their ends,

nastiness *as part of its functionality* and uses it as a means of protection, as the discharge of the ink darkens the water and provides a kind of protective shield: see *PA* IV.5, 678b36–679a7, 679a30–31, and 681b25–28.

[15] *GA* IV.3, 769b25–27.

[16] On monstrosities, see *GA* IV.3, 769b3–b30 and IV.4, with detailed analyses of these texts by Connell 2016: 340–352; 2018a: 207–223.

where these failures are due to the inference of material-efficient causal factors (e.g. the female menses being too moist or copious, or the winds blowing from the south during conception and interfering with the heat required for concoction) – that is, they happen "by accident." In fact, Aristotle believes that the majority of teleological failures – such as most controversially the birth of female animals,[17] but also the coming to be of certain deformities that affect entire species, such as the "incomplete" eyes in moles[18] – are due to similar kinds of imperfections of the matter or "miscalibrations" of the efficient cause, causing formal natures to fail to fully realize their ends.[19] In all of these cases, whether they happen at the individual or species level, "something greater interferes" with the teleological process of reproduction, and the results are monstrous, "bad" (φαῦλα),[20] or imperfect (ἀτελῆ)[21] offspring.

In addition to this well-known type of teleological failure ('type 1'),[22] I would like to introduce two other types of teleological failures that result in features that are either less-than-optimal or "bad" in other ways *despite formal natures having realized their end.* That is, in the cases that I have in mind, it is not the material nature but the actions of the formal nature that somehow produce the bad outcome – either because this outcome was *the best possible, and nature could not have produced something better* ('type 2')[23] or because *nature was mistaken to produce this outcome, and in fact goes on to remediate or fix the 'original' mistake* ('type 3'). Notably, in explaining these cases, Aristotle often makes explicit use of the analogy between nature and craft and presses the image of formal nature as an internal designer of animals.[24]

[17] On the birth of female animals as a kind of teleological failure and on their characterization as "disabled," "deformed," or as a "monster," see *GA* II.3, 737a25–28; *GA* IV.6, 775a15–16; *GA* IV.3, 767b5–13. Cf. *GA* II.6, 743a29–32; *GA* IV.3, 769b3–13; *GA* IV.4, 770b16–17; and *HA* VIII.2, 589b29–590a11. The coming to be and existence of female animals is best explained as a case of 'secondary teleology'; see Leunissen 2017: 139–176.

[18] *HA* I.9, 491b27: ἀτελές; cf. *DA* III.1, 425a9–11. This 'disability' is produced during generation: see *HA* I.9, 491b27–34 and *HA* IV.8, 533a11–12.

[19] Cf. *GA* IV.2, 766b29–767a1 and *Pol* VII.16, 1335a36–b2.

[20] *HA* VII.2, 582b20–22.

[21] *Pol* VII.16, 1335a11–17 and 1335b29–31.

[22] Johnson 2005: 198–199; Stavrianeas 2018: 51–59.

[23] My account of these first two types builds heavily on Stavrianeas 2018, who argues that there are two types of teleological failure in Aristotle's biology: one according to which nature aims at producing a good result but does not achieve it due to defective means, and one where nature produces the best possible result, but where this result is objectively speaking not a good one. On his account, the interference of *something else* keeps nature from realizing the (optimal) end in both cases.

[24] I discuss this image of nature as a craftsman in Leunissen 2012.

The second type of teleological failure, then, consists of formal natures realizing the end and then in fact doing what is best – given the possibilities – for the substantial being of the animal in question, but where the result does not seem to be very good, at least not relative to how this part or feature is realized in other (relevantly similar) animals. In this case, the natural range of possibilities available to nature, so to speak, when realizing a certain part or feature of an animal is itself limited and/or suboptimal (due to factors outside of the 'control' of the formal nature), such that nature 'choosing' the best outcome for that animal still leads to a less-than-optimal result comparatively speaking.[25] Failures of this type include several of the deformities found among the animals that Aristotle characterizes as "dualizing regarding their nature" (ἐπαμφοτερίζει τὴν φύσιν), which includes dolphins, whales, seals, bats, apes, and the Libyan ostrich. These animals partake in a way in *two natures* or two species forms but in another way in *neither*. Because of this, *some of them* end up with features that are suboptimal relative to their realization in single-natured animals. This second type of teleological failure, then, is perhaps only a hypothetical one: even dualizers are the *best they could possibly be in this world* and we only think of some of their features as 'bad' because we can imagine a world in which nature would not have been restricted in its options and/or would have been able to produce something better.[26]

The third and final type of teleological failure that I would like to introduce involves mistakes made by formal natures that resemble those made by the writer who writes something incorrectly or by the doctor who pours out the wrong dose of medicine. In these cases, Aristotle argues that the formal nature of an animal made a *genuine* mistake in realizing the teleological process and in some cases is depicted as even going on to remedy[27] or fix the earlier mistake, for instance, by adding or modifying

[25] Stavrianeas 2018: 52, 57 n.14, and 59–63.

[26] Note that my account of deformities builds on Gotthelf's ([1985] 2012: 230–232) important insight that deformities for Aristotle are identified *in relation to* a relevant contrast class (where this contrast class provides a heuristic role in providing empirical evidence of what the normative expectations are for parts performing their natural function), though this contrast class need not be restricted to wider kinds: female animals are deformed relative to male animals, seals are deformed relative to four-footed land-dwellers, but these are not wider kinds. I also agree with Witt's (2012) equally important claim that Aristotle's identification of animals as deformed entails a normative judgment about their (in)ability to perform their natural function(s). However, despite the similarities in terminology, I believe that the differences in the causal histories of the different types of deformity and the type of teleological failure at stake in these are important and that they require a pluralistic account.

[27] *PA* III.3, 665a8: ἰάτρευκεν ἡ φύσις; on nature's need to provide 'medical treatment' in this case, see Lennox 2001: 253.

another part of the animal. These cases are admittedly rare, but – or so I will argue below – these failures also put the most strain on Aristotle's depiction of the natural world as one of functional beauty.

With these distinctions in place, let me now turn to the various types of 'bad' parts that Aristotle discusses in his *Parts of Animals*.

11.2 Deformed Parts

One group of features mentioned in Aristotle's *Parts of Animals* that challenges his teleological worldview as presented in his exhortation to the study of biology is that of the so-called deformed (πεπηρωμένον) parts that characterize entire kinds. These deformed parts appear to be developmentally stunted and/or morphologically deviant from their homologues as present in other animals and seem to perform their natural function less well than their "regular" counterparts.[28] Sometimes Aristotle even characterizes an entire species as deformed based on its possession of one or more of such deformities. While in the case of the mole discussed above the deformity seems to be due entirely to the development of their eyes being stunted (i.e. due to a type 1 failure of teleology), Aristotle's *Parts of Animals* also includes examples of deformities that involve no such stunting of the process of teleology and that instead constitute more complex failures of teleology. Let me discuss two examples: the deformed feet of seals and the claws of lobsters.

Aristotle discusses the seal twice in his *Parts of Animals*. The first occurrence is in *PA* II.12, where he characterizes the seal as "a deformed four-footed animal" – thus locating the deformity of its species in its feet – and identifies this as the reason for their lack of outer ears (657a22–24: διὰ τὸ πεπηρωμένον εἶναι τετράπουν). Immediately preceding this account, Aristotle explained the lack of outer ears in birds and in four-footed egg-layers with hard scales as being due to them lacking the kind of matter from which outer ears can be formed. This latter lack exhibits not a failure but a limit of natural teleology: nature did not produce outer ears in these animals – the presence of which in the context seems to be presumed as generally being for the better, as they improve hearing – simply because it did not have the required soft, flexible materials. The case of the seal is more complicated though, first, because it is unclear from this passage what the causal relation is between the seal's deformity as a four-footed animal and its lack of outer ears. In the *History of Animals* Aristotle

[28] Cf. Henry 2007: 254; Witt 2012: 86.

characterizes the deformity of seals more specifically in terms of the *placement* and *shape* of their feet (their back feet resemble a fish tail, making it impossible for the seal to walk and causing it to slip),[29] but this in itself does not explain their lack of ears. Secondly, in his *Generation of Animals*, Aristotle explicitly explains this absence of ears in seals as a case of nature not doing anything in vain (*GA* V.2, 781b22–28): because seals spend most of their life in the water, outer ears would be of no use for them and would indeed be harmful as they would fill with water. On this account, the absence of ears is not due to the absence of certain materials or due to some deformity, but it is rather an intrinsic part of the teleological design of seals. The life of seals in water features more prominently in Aristotle's second discussion of the seal in *PA* IV.13 where he does not refer to them as deformed but instead uses the fact that seals "tend towards two natures" to explain why they possess some parts, but lack others (697b1–7):

> And seals and bats on account of them tending in the one case both toward water-dwellers and land-dwellers, and in the other towards flyers and land-dwellers, on account of this they partake of both and of neither. For seals while being water-dwellers, have feet and while being land-dwellers have fins (for their rear feet are entirely fish-like, and their teeth are razor-toothed and sharp).

My suggestion, then, is that it is this *dualizing* nature that explains the deformity of the seal qua land-dweller and that ultimately explains their lack of outer ears. The explanation – if we can use Aristotle's depiction of nature as a craftsman – runs as follows: seals are by nature land-dwellers, but during their early embryological formation, nutriment that belongs to the sea is mixed in, and this changes their natural preference for food that is found in water.[30] This material change in the early fetus then introduces a 'second nature' to the being of the seal, making it a land-dweller and a water-dweller. For the formal nature of the seal, this means that during the remainder of the production of the fetus, this nature of the seal needs to 'negotiate' the demands of the two natures, making the seal tend to its land-dweller nature in some respects and to its water-dweller nature in others, and to produce the necessary design adjustments to enable the seal

[29] See *HA* I.1, 487b23–24: "The seal has mutilated (κεκολοβωμένοι) feet"; *HA* II.1, 498a31–498b4; and *HA* VI.12, 567a6–7. Cf. *IA* 19, 714b12–13: "Seals are four-footed animals but are so in a bad way (κακῶς)."

[30] This materialistic thought-experiment of how dualizers come to be is suggested in *HA* VIII.2, 589b29–590a11; cf. Witt 2012: 103.

to function as well as possible in both environments. Part of this negotiation involves making decisions about which analogous parts to produce and which to 'remove' from the design (e.g. nature decides to give seals feet, not fins, and lungs, not gills), but it also involves adapting those parts to the animal's specific way of life (e.g. shaping the back feet to resemble fishtails and the front feet like flippers, which disables walking but enables seals to swim).[31] The lack of outer ears, then, can be explained by reference to seals being *deformed* four-footers, not because their deformity *causes* the lack of ears but because it puts that lack in the context of other less-than-optimal adjustments nature had to make to the seal's body in order to accommodate its aquatic way of life. Relative to other four-footed land-dwellers, the seal is deformed, for its fishtail-shaped back legs and front flippers make it unable to walk while its lack of outer ears makes it less able to hear while out of water. However, seals are *also* the full and proper realizations of their double natures and have been produced by their formal natures 'in the best way, given the possibilities.' Accordingly, the 'bad' features of the seal are not due to a stunting of its development but are rather produced 'on purpose' by the teleological actions of the formal nature of seals. At the same time, it is not clear that nature, hypothetically speaking, 'could have produced a *better* design,' given the limitations of the natural possibilities and the complexities of the functional needs of dualizing animals. The failure at stake, then, seems to be of the second type introduced above.

The second deformed part I would like to discuss is that of the claws of the lobster in *PA* IV.8. Lobsters, so Aristotle claims, have claws but do so "in a disorganized manner, on account of being deformed" (684a35: πεπήρωνται). In all other animals with claws – such as crayfish and crabs – the *right claw* is bigger and stronger as a result of the teleological actions of their formal natures: all animals naturally use parts located on the right side of their body more and nature always assigns parts for protection and defense – exclusively or more – to what is best able to use those parts.[32] Hence, these animals use their claws not for locomotion (for that, they have lots of feet), but for grasping (683b32–33). Lobsters possess claws "because they are in the kind that has claws" (which I take to mean that they share a certain material nature that is responsible for the

[31] These 'negotiations' by the formal natures of dualizing animals need not always result in deformed features, which is why not all dualizers have deformed parts.

[32] *PA* IV.8, 684a26–31; cf. *IA* 4, 705a29–b5 and 706a21–25; *PA* III.3, 665a23–26; and *PA* II.14, 658a18–24.

production of excess earthy materials, which formal natures subsequently use largely for the production of defensive parts),[33] but these claws exhibit a random variation according to which sometimes the left and sometimes the right claw turns out to be the bigger and stronger. This morphological anomaly affects their functionality: lobsters do not use claws for "what they are for naturally" (683b32), namely, for grasping, but for locomotion. For this reason, the claws in lobsters are constitutive of their deformity and, at least comparatively speaking, a failure in their design. Perhaps the random distribution of extra materials in lobsters is due to material necessity alone (though elsewhere Aristotle suggests that formal natures can redirect flows of extra residues; see my discussion of horns below). However, the implication that the formal natures of lobsters co-opt their irregularly designed claws for a function that is *not their natural one* and one that is explicitly denied to other claw-bearers *and* that does not even allow lobsters to locomote well,[34] suggests that the account provided here is problematic even by Aristotle's own standards[35] and that the teleological failure at stake is perhaps of the third, most severe type. I will return to this suggestion below.

11.3 Useless and Harmful Parts

The biggest challenge to Aristotle's characterization of the natural world as a functional and beautiful product of the teleological actions of nature is the existence of useless and harmful parts. Given the teleological principle according to which "nature always or for the most part does nothing in vain" – invoked by Aristotle to explain the absences of parts as being due to them having been 'taken away' by nature on account of their uselessness or harmfulness[36] – the observed presence of such 'vain' parts suggests a mistake (or at least a failure of omission) in the teleological production-process and one that is due to nature itself. Such parts thus exemplify the third type of teleological mistake I introduced above. Interestingly, in his treatment of these parts, Aristotle sidesteps the question of why nature

[33] Cf. *PA* III.2, 664a5–6: "The cause of this is that both are of the same [material] nature, namely that of horn-bearers."

[34] Cf. *IA* 19, 714b8–19: all hard-shelled animals are mutilated (ἀνάπηρον) because they move in a way that resembles animals that have lost a limb and that is contrary to nature.

[35] Cf. Lennox 2001a: 310.

[36] Cf. *PA* IV.11, 691a27–b4 about the sideway jaw motion in fish, birds, and egg-laying four-footed animals: "It is not useful at all, *which is why it is taken away* (ἀφῄρηται) *from all such animals*; for nature produces nothing superfluous." See also *PA* IV.10, 689b21–25; *Cael* II.8, 290a29–35; *HA* II.17, 508a8–11; and my discussion in Leunissen 2012: 31–33.

'allowed' their presence and instead highlights the teleological actions of nature that compensate for or remedy this mistake by adding or modifying other features of the animals in question to allow them to perform their essential and vital functions. In the sections below, I will discuss three series of parts that represent increasingly worse mistakes implicitly or explicitly attributed to nature.

Useless Feet in Wquid, Cuttlefish, and River Crocodiles

Aristotle discusses several species of animals with useless feet. For instance, in *PA* IV.9, 685a22–685b3, Aristotle says that octopuses, cuttlefish, and squid all have eight feet, but that in octopuses, nature takes material away from the body – thereby reducing the size of their trunk – and uses that material to add length to the feet, while in the cuttlefish and squid, nature takes material away from the feet in order to increase the size of their body. As a result, "in octopuses the feet are not only useful for swimming, but also for walking, while in the others they are useless (ἄχρηστοι); for their feet are small while they have a large trunk" (685a28–30). In both groups of animals, the formal natures create feet and trunks with the material available, but they distribute those materials differently, resulting in some being unable to walk with their feet (though, presumably, they all use their 'feet' for swimming: see 685a14–15). The river crocodile similarly has feet that are useless (*PA* IV.11, 691b7: ἀχρήστους), also because they are 'in every respect small' while their trunk is large.

The presence of such useless feet is teleologically problematic because Aristotle typically appeals to the uselessness of a feature counterfactually as part of the teleological explanation of their *absence*: parts are absent because their presence would have been in vain, and nature does not produce parts that are in vain. In fact, nature even removes parts – such as feet in snakes – from the animal design if needed.[37] In the examples above, nature in effect caused the uselessness of the feet by taking away materials from them and adding size to the trunk and by not removing the feet, now rendered useless for walking, from the animal.[38] Aristotle glosses over these problems and instead focuses on the ways in which the formal natures of these animals *compensate* for the presence of useless feet by adding or

[37] See *IA* 8, 708a9–20, referenced in *PA* IV.11, 690b14–18; cf. *PA* IV.13, 696a10–15.

[38] In this case, perhaps, though the feet are useless for the performance of the natural function of feet, namely walking, Aristotle may have believed that nature (rightly) preserved them for their use in swimming. I thank Jim Lennox for this point.

functionally adjusting other features. For instance, in the case of cuttlefish and squids, nature compensates for the uselessness of their feet for holding on to rocks and grasping food (*PA* IV.9, 685a31: ἀχρήστους πρὸς ...) by providing a proboscis – a part not needed by the octopus, which has feet for performing those functions. In river crocodiles, "nature – instead of feet – has made for it a mouth that is useful for these uses" (*PA* IV.11, 691b8–9: πρὸς οὖν ταύτας τὰς χρείας ἀντὶ ποδῶν τὸ στόμα ἡ φύσις χρήσιμον αὐτῷ ἐποίησεν), for they are able to move their upper jaw, which is something other animals cannot do (in them, nature in fact *removed* the sideways motion of the jaws on account of its uselessness: 691a27–b4). The ultimate design of these animals is thus functional and beautiful thanks to the goal-directed actions of nature that compensate for its failure in producing feet useless for walking and grasping.

Useless and Harmful Horns in Antelopes, Gazelles, Bison, and Male Deer

Even more problematic for Aristotle's teleology is the presence of horns in antelopes, gazelles, bison, and male deer. For these parts are not just useless, but in some cases also harmful, which suggests again that nature made a mistake in producing them and/or in not removing them.

Aristotle argues that horns exist by nature on the heads of certain live-bearing animals and that they are present for protection and strength (*PA* III.2, 662b23–30). However, while horns are thus characterized as being present for the sake of something, their coming to be is not conditionally necessitated; rather, Aristotle claims that the bodies of large four-footed live-bearing animals produce a lot of earthy residues and that nature as a "good housekeeper"[39] tends to make use of what has come to be of material necessity for the sake of something good. That is, given the defensive potential[40] of these excess earthy materials, nature can use them to produce protective parts such as teeth and tusks in some animals and horns in others (663b22–35):

> We must say what the character of the necessary nature is, and, how nature according to the account has made use of things present of necessity for the sake of something .... For the residual surplus of this sort of [earthen] body, being present in the larger of the animals, is used by nature for protection and advantage (ἐπὶ βοήθειαν καὶ τὸ συμφέρον καταχρῆται ἡ

[39] See *GA* II.6, 744b11–27 and my discussion in Leunissen 2010a: 81–99.
[40] *PA* II.9, 655b4–15.

> φύσις), and [the surplus, which] flows of necessity to the upper region,[41] in some animals it distributes to teeth and tusks, in others to horns.

According to this explanation, the addition of horns in these larger animals is strictly speaking 'not necessary,' as they are not conditionally necessitated, but they are produced from excess residues that have come to be of material necessity in a way that is supposed to benefit the animal in question. In addition, it appears that nature can use these extra materials for the production of other defensive parts and not just for horns,[42] or perhaps can even decide not to produce such parts at all. This latter option is in fact suggested by Aristotle's earlier claim that nature only produces tool-like[43] parts such as horns in those animals who are strong enough to use them (*PA* III.1, 661b28–662a2):

> Of the tool-like parts that are for strength and protection, nature provides each of them only or more to those animals that are able to use them, and especially to the animals able to use them most – parts such as sting, spur, horns, tusks, and any other such part there may be. And since the male is stronger and more spirited, in some cases he alone has such parts, in other cases he has them more [than the female]. For as many parts for which it is necessary that the females have them as well, e.g., parts related to nourishment, they have, but they have them less; while those related to none of the necessities, they do not have. It is also on account of this that among the deer, males have horns, while the females do not.

In other words, because tool-like parts are 'not related to one of the necessary functions' and because nature only provides parts to those who are able to use them or use them best, in the case of deer, nature uses the excess earthy residues for the production of horns in the males but not in the females. For in females their presence would have been useless (or of lesser use),[44] and nature does nothing in vain. And in the many-toed animals, instead of producing horns, their formal natures produce teeth

[41] This material flows up – despite being earthy – because it is hot, as a result of the digestive process, and instead of it moving back down and being excreted, it is used for the production of horns; see *Somn.* 3, 456b20–28.

[42] For 'add-on' bony structures such as nails, hoofs, spurs, claws, and hair, see *PA* IV.10, 687b22–24 and 690a4–9; *PA* IV.12, 694a22–27; and *PA* II.14, 658b3–5.

[43] These earthy parts are moved 'from the outside' and do not produce sensations: see *DA* III.13, 435a11–435b4.

[44] Note that Aristotle makes a similar argument about why nature placed horns on the head in *PA* III.2, 663a34–663b12: had they been placed anywhere else on the body, their weight and presence would have been useless, so nature 'correctly' places horns where animals can use their weight to their advantage.

and claws for protection, and in camels and elephants, nature adds to their bodily size (*PA* III.2, 662b30–663a6).

Given this apparent 'flexibility' with which nature can provide protection to four-footed live-bearing animals, Aristotle's subsequent account of horned animals seems problematic (663a8–18):

> In those in which the projection of the horns is by nature useless (ἄχρηστος πέφυκεν), nature has added another means of protection, such as in deer [it added] speed, for the size and extensive branching of them [i.e. horns] is more harmful than helpful (μᾶλλον βλάπτει ἢ ὠφελεῖ) and [it] also [added speed] in antelopes and in gazelles, for standing up against some animals they defend themselves with their horns but from those that are wild and that are out to fight they flee, while in bison – for in these the horns are by nature bent towards one another – [nature added] the emission of excrement; for in this way they defend themselves when frightened.

The horns in these animals are "by nature" shaped and/or sized in a manner that makes them useless for protection, and in some cases they are even more harmful than beneficial for the animals that have them. The claim that the horns are shaped and/or sized in the way they are "by nature" seems to imply that this is the work of material necessity and that this is not something that can be altered by the teleological actions of the formal natures: so, while formal natures can redirect the flow of the excess materials into a different direction (*PA* III.1, 661b28–662a2), the growth of the resulting structures is determined by material necessity alone. Surely, then, given these useless or even harmful results, these animals would have been better off *without* those horns, and given the principle that nature does nothing in vain, the optimal teleological solution would have been for nature to remove these 'bad' horns or perhaps to redirect the flow of earthy residues toward some other use.[45] Perhaps the preservation of horns in gazelles and antelopes is teleologically justified by their occasional functionality, but no such explanation is available for the bison and deer. In the case of deer specifically, Aristotle recognizes that the males benefit from the seasonal shedding of the horns as a lowering of their load (*PA* III.2, 663b13–14: *epheleias kouphizomenon*) and that the females do not have horns exactly because they would be harmed even more than the males are by their presence (664a3–8):

> The females of deer do not have horns, but regarding teeth they are similar to the males, and the cause of this is that their nature is the same for both,

[45] On the suggestion that "nature can stop" a flow of extra residues "with a view to another use," see *PA* II.15, 658b14–25 and *PA* III.2, 664a1–3 and 8–11.

> namely horn-bearing.[46] But horns have been taken away (ἀφῄρηται) from the females for the reason that, while they are not useful (διὰ τὸ χρήσιμα μὲν μὴ εἶναι) either for the males, the latter are harmed less (βλάπτεσθαι δ' ἧσσον) on account of their strength.

In the case of female deer, then, nature takes away the horns, while preserving them in the males despite their uselessness and relative harm to them.[47] Interestingly, in his account of useless and harmful horns in deer and bison above, Aristotle side-steps the question of whether nature could or should have removed the horns and instead highlights the teleological ways in which nature *compensates* for the limited ability of these animals to defend themselves with their horns by *adding* alternative protective features to them, such as speed and the emission of excrement.[48]

The Bad Placement of the Windpipe in All Animals That Breathe Air

The last part I would like to discuss that involves a mistake that is due to the actions of formal natures is the windpipe, which Aristotle says is badly placed.

Aristotle discusses the windpipe in the context of his treatment of the function and locations of the parts in the neck in *PA* III.3. The windpipe connects the mouth with the lungs and thus serves the important function of breathing (i.e. cooling the area around the heart) but also vocalizes sound. It is located in front of the esophagus, which connects the mouth with the stomach and thus enables the function of nourishment (the esophagus is not *strictly speaking* necessary for nourishment, as it does not contribute anything to its function and is only required because the mouth does not lie directly next to the stomach). Since food needs to pass over the windpipe in order to get into the esophagus, the location of the

46 "Horn-bearing kind" presumably refers to their shared *material* nature, since this causes the excess earthy materials used for the growth of bony structures such as horns and teeth.

47 Cf. perhaps *HA* IX.32, 619a16–18 where Aristotle mentions a kind of eagle whose beak keep growing, perhaps similarly due to material necessity, given the earthy nature of beaks, which ultimately causes the bird to die of starvation.

48 Aristotle thus seems to engage in a kind of thought-experiment according to which nature "as a craftsman" realizes that the horns that grow in these animals of material necessity turn out badly and then decides to add extra features to their design to allow them to protect themselves. This is, of course, not a reflection of the *actual* causal procedure involved in the generation of these animals, but rather, by thinking of animals as products of art, put together by the teleological actions of formal natures in causal interaction with the operation of material necessity, Aristotle is better able to identify the causes and the reasons why animals have the features they have. I argue for this heuristic use of these thought-experiments in Leunissen 2010a: 119–135; 2012.

windpipe in front of it is functionally problematic, as Aristotle points out repeatedly: it "hinders" (*PA* III.3, 664b3–4: ἐμποδίζουσα) and "causes trouble for" (664b21: ἐνοχλεῖται) the reception of food, and when food drops into it, it causes choking and straining and painful coughs (664b5–6 and 664b30). In order to prevent the latter from happening, so Aristotle states, nature "has constructed the epiglottis" (664b21–22: ἡ φύσις … μεμηχάνηται τὴν ἐπιγλωσσίδα) and he praises this part for having been constructed "well" or "beautifully" (664b32: καλῶς).

None of the above is especially problematic for Aristotle's teleology, but he makes two additional claims about the windpipe that *in combination* suggest that nature in fact made a mistake in positioning it in front of the esophagus and that it subsequently produced the epiglottis as a remedy for its own teleological failure. For Aristotle insists that the positioning of the windpipe in front of the esophagus is due to nature: since it is a principle of teleology that more honorable parts should be placed in the more honorable locations of the body, and since the windpipe is functionally more honorable than the esophagus, the windpipe should be placed in front of the esophagus (*PA* III.3, 665a10–26). This is remarkable, because here (as in other uses of this principle) Aristotle states explicitly that this principle only applies "when nothing greater impedes" (665a23: ὅπου μή τι μεῖζον ἕτερον ἐμποδίζει), such that if functional needs require parts to be placed elsewhere, those functional needs trump the need to place that part in the most honorable location.[49] Indeed, in his preceding summary, Aristotle acknowledged that placing the windpipe in front of the esophagus was a mistake: the location is "bad" and nature had to 'remedy' the situation by constructing the epiglottis (665a7–8: τῆς ἀρτηρίας τὴν φαυλότητα τῆς θέσεως ἰάτρευκεν ἡ φύσις, μηχανησαμένη τὴν καλουμένην ἐπιγλωττίδα). One could argue that, given that it is possible for nature to make an epiglottis, it is therefore better – teleologically speaking – to keep the windpipe in the most honorable position and then add an epiglottis to preserve functionality.[50] However, this account would still form a notable exception to how this teleological principle is applied elsewhere in the *PA*, and I think that it would have been more consistent with Aristotle's overall explanatory program to leave teleology out of this account and to attribute the unfortunate placement of the windpipe *entirely* to necessity or to other constraints involved in the animal's design.

[49] Cf. *PA* III.4, 665b18–21 and *PA* II.14, 658a18–24.

[50] See Lennox 2001c: 269–272.

11.4 Weird Parts

There are also parts featured in *PA* that are weird or anomalous in such a way that Aristotle's insistence on explaining them teleologically also ends up showing weaknesses in this explanatory project. One such "weird"[51] part is the mouth in dolphins and sharks. While in fish, the mouth is placed straight across and towards the front of the head, in dolphins and sharks the mouth is placed *on the underside*, forcing them to turn belly up to grasp their food (*PA* IV.13, 696b25–27). Aristotle explains this odd 'fact' (which is, by the way, anatomically and behaviorally speaking false) as follows (696b27–34):

> It appears (Φαίνεται) that nature does this not only for the sake of the preservation of other animals (for during the turn the other animals are saved by the delay; for all such animals eat other animals), but also to keep them from pursuing their gluttony concerning food. For if food could be grasped easily, they would perish due to their rapid repletion. In addition, the nature of their snout is curved and narrow, making it unable to open widely.

In this explanation, two things stand out. First, Aristotle presents the mouths of dolphins and sharks not just as *different* in location, shape, and width from mouths in fish, which are here treated as the relevant comparison class, but also as *worse* in terms of serving their function as inlets for food, that is, their mouths make them *bad/slow* at catching food. Second, Aristotle's account here appears to be a – perhaps tentative[52] – *teleological reinterpretation* and *expansion* of the popular and non-teleological explanation recorded in *HA* VIII.2, 591b23–30. In the latter, Aristotle describes the need of dolphins and sharks to turn over on their backs to catch fish as the cause of smaller fish being preserved more often and mentions the popular counterfactual (591b28: δοκοῦσιν) that if this had not been the case, there would have been few small fish, because "the speed and the voracity of the dolphin seem incredible." Under this scenario, the preservation of smaller fish is a lucky but entirely accidental outcome. By turning the causal relationship between these two facts into a teleological one in his *Parts of Animals*, and thereby suggesting that nature *purposefully* designed a "bad mouth" in sharks and dolphins for the

[51] Aristotle sometimes labels parts as strange (*perittos*) or distinctive (*idios*) – e.g. the elephant's trunk (*PA* II.16, 658b33–35) and the bird's beak (*PA* IV.12, 692b15–19) – to highlight their extraordinariness or uniqueness; these terms do not necessarily carry any negative connotations.

[52] *PA* IV.13, 696b27: φαίνεται; Lennox 2001a: 341.

preservation of *other animals*, Aristotle notoriously strains his theory of natural teleology as involving natures doing what is best *for their own kind*.[53] The suggestion that nature *also* does this for the preservation of dolphins and sharks themselves seems contrived and *ad hoc*.

Also 'weird,' but in a different way, are breasts in male animals. Aristotle brings up breasts in *PA* IV.10, 688a11–689a3 in the context of a longer explanatory exposition that takes the external parts on the human body both as a guideline for the order of exposition, starting at the head and moving down to feet and toes,[54] and as the norm against which to compare the realization of similar parts in other blooded live-bearing animals. Aristotle starts with an explanation of breasts *in humans* (688a11–24):

> Between the arms in humans, and between the forelimbs in the other [blooded, live-bearing animals], is the so-called chest, and while in humans it is wide, as can be expected (for the arms – being attached to it on the side – do not prevent this space from being wide), in the other four-footed animals . . . this part is narrow. And because of this, four-footed animals do not have breasts (μαστοὺς) in that region; but in humans on account of the wideness of the area and there needing to be protection of the area surrounding the heart, because of that – with the area having become fleshy – the breasts become differentiated, and they are fleshy in men for the reason stated, but with regard to women nature uses them also for another function, as we say it often does: for it stores in there the food for the offspring.

According to this account, the primary function of breasts in humans is protection of the region around the heart, and this also explains why breasts in men are fleshy; in women, however, nature also uses breasts for a second(ary) function, which is the storage of nutriment for offspring. Now, in identifying protection as the main function of breasts, Aristotle reveals his (no doubt sexist) prioritization and normalization of the male body: the presence of breasts in men is explained by the function they perform in *all* humans, while their presence in women requires an additional teleological explanation to explain their 'special' use for feeding offspring. Aristotle's subsequent account of the differentiations of breasts in other live-bearing animals, however, bears strong evidence – of the kind that I think Aristotle should have been able to appreciate himself – that the primary function of breasts is in fact feeding offspring and that it is the

[53] See Johnson 2005: 208–209; Leunissen 2010a: 43–44.

[54] See *PA* II.10, 656a9–13; cf. also *HA* I.6, 491a14–26; *HA* I.15, 494a19–b1; and *HA* I.16, 494b18–21. See further Rapp (Chapter 13) this volume.

presence of male breasts that is anomalous and in need of a special explanation. For instance, it is the function of feeding offspring – and not protection – that explains how many and in what location breasts are present in live-bearing animals, and why in many animals, breasts are not even located near the heart. Moreover, Aristotle mentions male animals, such as horses, that lack breasts (688b30–33), whereas their female counterparts do possess them – the only case I have found in the *PA* where the female possesses the protective parts but not the male. And finally, Aristotle omits any discussion of nipples, the presence of which in male breasts should have struck Aristotle as odd or as "in vain," while their presence in females is necessary for the transfer of "stored food" to their offspring.[55] In this case, it is thus Aristotle's decision to center the teleological explanation on the male body and hence his failure to recognize the 'weirdness' of male breasts and nipples that produce the tensions and weaknesses in the accounts provided.

11.5 Conclusion

Aristotle's exhortation to biology invites us to see animals as functional – and therefore beautiful – pieces of 'the original art' designed by nature itself. The study of animals, he promises us, will fill us with a kind of knowledge that is unsurpassed in its width and depth and with a feeling of pleasure and wonder that provides its own kind of profit and compensation relative to the study of divine beings. And, for the most part, Aristotle's own explanations of the parts of animals as recorded in *PA* II–IV make good on this promise: even useless feet in squid, horns in deer, the position of the windpipe in front of the esophagus, badly shaped and positioned mouths in dolphins, and breasts in male live-bearing animals do little to undermine Aristotle's awe-inspiring teleological picture of the natural world. On the other hand, Aristotle's eagerness to explain even such 'bad' parts as products of nature constitutes – or so I have argued – a methodological weakness in his explanatory project and introduces explanations that are inconsistent with and seemingly *ad hoc* into what is otherwise mostly a very coherent and systematic account of why animals have the (differentiations of) parts they have. That teleology is ill-suited for explaining such parts is something already recognized by Theophrastus,

[55] Cf. also *PA* IV.11, 692a12 where Aristotle characterizes breasts exclusively as receptables – "a vessel as it were" – for milk. On the problems in Aristotle's account of breasts, see Lennox 2001c: 263–264.

who questioned the scope of natural teleology in general (*Metaph.* IX, 10a22–25; cf. 11b25–27) and identified parts like male breasts and horns of deer as 'in vain' and not due to teleology (10b7–16):

> And in addition, among living beings themselves, some things are like they are in vain (ὥσπερ μάταια), such as breasts in males and in females the emission, if indeed this does not contribute anything, and the growth of beards in some or in general the growth of hair in certain places; in addition, the size of horns such as those in deer to which belongs nothing of benefit, but to which is due damage through irritation, suspension, or covering of the eyes. And in the way in which some things happen by force or against nature, such as the way in which the heron copulates and the day-fly lives: and one could grasp quite a few other cases like that.

Accordingly, as Theophrastus suggests here, some of these parts are better explained by reference to necessity or chance as the primary cause, perhaps with teleology playing a secondary role. Furthermore, by extending – or perhaps overextending – his theory of natural teleology to such 'bad' features, Aristotle ends up making room for *genuine* teleological mistakes of nature, some of which then require further teleological actions to fix these mistakes. In the end though, Aristotle claims, we should not judge those parts – useful or useless, beautiful or ugly – by themselves; rather, we must look at the *whole animal* and its substantial being, and *at that level* the teleological actions of nature produce fully functioning, most beautiful creatures.

CHAPTER 12

Mind's Place in Aristotle's Science of Nature
Evidence of Parts of Animals I.1

Mary-Louise Gill

Introduction

Should mind (*nous,* νοῦς) be studied by Aristotle's science of nature or be reserved for first philosophy, the study of being *qua* being (τὸ ὂν ᾗ ὄν), sometimes called "theology"? We must first ask what notion of *nous* (νοῦς) might be relevant for natural philosophy, because Aristotle uses the word in more than one way in different places.[1] *Nous* can refer to a faculty of the human soul – mind or intellect – distinct from other psychic faculties, such as the perceptive and nutritive capacities, enabling us to think intelligible objects (νοητά). Unlike other psychic faculties, which are realized in material parts of the body, Aristotelian *nous* is realized in no special material part, such as the brain or the heart. *Nous* can also refer to separate divine substance, whose being is pure activity (ἐνέργεια) – thinking (νόησις) – and whose object of thought is thinking itself (*Metaph.* XII.9, 1074b34–35).[2] I translate the word as "mind" in my title, and sometimes as "intellect," because the key passage concerning *nous* in *Parts of Animals* I.1 mainly concerns human *nous* and asks whether it is part of nature. Divine *nous* is not out of the picture, however, because one section of the passage discusses the cause of cosmic order, arguably in reference to divine *nous.*

Even if we restrict our attention to the section on *nous* in *Parts of Animals* I.1, the question is much debated, with advocates on opposing

[1] Thanks to Ronja Hildebrandt for pressing me to confront at the outset this fundamental question. On the concept of *nous* in early Greek philosophy, Plato, and Aristotle, see von Fritz 1974 and Menn 1992, 1995.

[2] Aristotle's divine substance is sometimes thought to be without power (δύναμις), but Aristotle appears to attribute power to it at *Metaph.* XII.7, 1073a5–13. Fazzo 2016 cites textual considerations supporting the idea that the Unmoved Mover in *Metaph.* XII is *in* act (ἐνεργείᾳ), a mode of being typically contrasted with being in capacity (δυνάμει). On the status of divine *nous*, I remain neutral in this paper.

sides.[3] In this chapter, I shall go through the relevant section of *PA* I.1 and contend that Aristotle presents three arguments excluding mind from nature but all in the voice of an opponent. Then in a fourth argument he responds to the third, with implications for the second as well.

Before we turn to the main task, let me rehearse the reasons why one expects *nous* – divine *nous* as well as human *nous* – to be included in Aristotle's science of nature.

12.1 Reasons for Mind's Place in Nature

Aristotle identifies the theoretical sciences as each concerned with being, though they study being from different perspectives. Natural philosophy studies being *qua* changeable, arithmetic studies being *qua* numerable, while first philosophy (metaphysics) examines being itself simply *qua* being (*Metaph.* VI.1, 1025b7–10). Each science looks for the causes (αἰτίαι) and sources (ἀρχαί) of things studied from its own perspective. Because of its focus on changing things, natural philosophy seeks the causes and sources of change and defines nature (φύσις) as the inner source and cause of a thing's proper change and rest (*Phys.* II.1, 192b13–23). A thing's nature is responsible for its characteristic behavior. This definition of nature invites our main question: Does Aristotle regard mind as part of the nature of those natural beings equipped to think? Scholars on both sides of the debate agree that evidence in Aristotle suggests that it is.

In the first place, *De Anima*, Aristotle's investigation of the soul, bills itself as having great import for all truth, and especially for nature (φύσις), because soul "is a sort of source of living things" (ἔστιν γὰρ οἷον ἀρχὴ τῶν ζῴων) (*DA* I.1, 402a1–7). *De Anima* studies the soul as the cause and source of the main life functions, including nutrition, reproduction, perception, and locomotion, and contains two chapters on the intellect (*DA* III.4–5). Why is a treatment of the intellect featured in a physical treatise on the soul, if mind is not a suitable topic for natural science?[4]

[3] Those arguing that mind is not an appropriate topic for natural philosophy include Balme [1972] 1992; Charlton 1987; Frede 1992; Broadie 1998; and Lennox 1999, 2001a, 2009. Those defending mind's place in nature include Caston 1998 and Frey 2018.

[4] Cf. Broadie 1998: 169. Lennox 2009: 2–11, in discussing the methods appropriate to natural science, argues that *De Anima* is not a work of natural science. One reason: *Meteor.* I.1 lists the order of natural inquiry and does not mention the soul, or at least not explicitly. Instead, it mentions investigations that precede the one in *Meteorology* itself, starting with the study of the first causes of natural change (*Phys.*), continuing with the study of motions of the heavens (*Cael.*), followed by the transformations of the elements and generation and destruction in general (*GC*). *Meteorology* itself will deal with irregular but natural celestial and sublunary phenomena and will be followed by the

Second, although the intellect is not the psychic realization of any special material part of the human being (*DA* III.4, 429a24–27; cf. II.1, 413a5–7), various human organs are arranged as they are for the sake of the intellect, and for that reason explanation of those parts must appeal to the intellect.[5] It is because of our intellect that humans alone stand upright and travel on two legs (*PA* IV.10, 686a25–28); have hands as instruments to serve many purposes, including the arts (*PA* IV.10, 687a9–23); and have teeth, lips, and tongue that facilitate articulate speech (*PA* III.1, 661b14–15; II.16–17, 659b30–660a27). Human blood is thin and pure, conducive to intelligence (*PA* II.2, 648a2–11). If mind is not part of human nature and so not in the province of natural science, the special features and functions of those human parts would be left unexplained, or not fully explained, since they are as they are for the sake of our intellectual activity. As David Balme observed, before giving reasons why Aristotle excludes intellect from human nature, intellect is the chief feature differentiating human beings from other animals.[6]

Third, an analogue of human intellect operates in the cosmos as a whole, since the Prime Mover is an active intellect that thinks itself. *Phys.* VIII and *Metaph.* XII both give physical arguments for the existence of the Prime Mover. In *Metaph.* XII, Aristotle argues that the complex eternal circular motions of the heavenly bodies ensure the continuity and variety of generation and destruction in our sublunary realm. An unmoved mover, one for each heavenly sphere, ensures the continuity of the eternal motions (XII.6, 1071b3–11; XII.8). XII.7 argues that the Prime Mover causes motion as a final cause, as an object of desire (τὸ ὀρεκτὸν) and thought (τὸ νοητόν) (1072a24–27). The Prime Mover is in the first place the source of the eternal rotation of the outermost sphere, the sphere of the fixed stars (XII.7, 1072a23–26; 1072b3–10), but it is also the source of cosmic order and goodness in general (XII.10, 1075a11–25). The constant activity of the first mover guarantees that things continue to behave according to their natures for the good of one another and of the whole. Like human intellect, the Prime Mover has no material base, but the Prime

study of animals and plants. Many commentators assume that *De Anima* belongs with the biological works, and I agree. Aristotle merely gestures toward the biological works in a single sentence toward the end of *Meteor.* I.1 and in the final sentence of the last chapter of *Meteor.* IV. I take the little treatises included in the *Parva Naturalia*, which focus on joint activities of soul and body, to belong with the biological works as well. Thanks to Thomas Slabon for pressing me to think more about the status of *De Anima* as a work of natural philosophy.

5 See Broadie 1998: 168. Cf. Lennox 2019: 107–113 and in this volume the chapter by Rapp (Chapter 13).

6 Balme [1972] 1992: 89.

Mover is the preeminent final cause of the cosmos. As a final cause and first mover of the physical cosmos, the Prime Mover – divine *nous* – is indispensable to natural philosophy, because its existence grounds numerous physical results. That said, I add a caveat: At *Metaph.* VI.1, 1026a10–32, Aristotle says that if there is something eternal and unchanging and separate, it will be the subject matter of a theoretical science but not natural science. This claim suggests that, while divine *nous* is a first principle of natural science, and so appealed to in that science, the study of first principles and causes as such is reserved for first philosophy.[7] If we take seriously the claim in *Metaph.* VI.1, there can still be much, though not complete, overlap between the domains of natural philosophy and first philosophy.

Fourth, given the role of the Prime Mover in explaining motion in the cosmos in general, and the similarity of its activity with that of human intellect, it is not surprising that *De Motu Animalium* compares animal locomotion to the eternal motions of the heavens (*MA* 5, 700a31–32). The two sorts of motions differ chiefly in that the motions of the heavens are eternal, whereas those of animals have a limit (*MA* 6, 700b29–32). *MA* 6 looks for the source (ἀρχή) of animal motion (700b9–11). All animals move for the sake of something, and the limit for all their motion is the goal (οὗ ἕνεκα) of their motion (700b14–16). Aristotle identifies the sources of locomotion as intellect (νοῦς) and desire (ὄρεξις) (700b17–19) and claims that imagination and perception hold the same place as intellect (700b19–20), presumably by playing the same role in the teleological explanation of nonrational animal locomotion as intellect plays in human locomotion (cf. *DA* III.10, 433a9–12).[8] What first moves the animal, says Aristotle, is the object of desire and thought (*MA* 6, 700b17–24; *DA* III.10, 433b13–21). So, the ultimate source of motion is the object – the apparent good – grasped as good by intellect or imagination (cf. *DA* III.10, 433b10–13). An object conceived as good stimulates the animal's desire to pursue it. Animal locomotion is limited, because it ends once the goal is reached. Thus, *De Motu Animalium* and similar evidence in *DA* III.9 and 10 show that intellect plays a major role in Aristotle's account of human locomotion.[9]

[7] I owe my appreciation of this point to Ronja Hildebrandt. [8] Cf. Nussbaum 1978: 333–334.

[9] For an explicit description of the cosmic analogue of animal locomotion, see *Metaph.* XII.7, 1072a26–30.

These considerations suffice to show that study of the human mind is appropriate – and perhaps even necessary – for natural philosophy. Mind is a topic for natural philosophy because it figures in the explanation of various human parts and functions, including human locomotion.

12.2 Mind in *Parts of Animals* I.1

PA I.1 argues more than once that mind (νοῦς) is not a suitable topic for natural philosophy, and many scholars regard the evidence as decisive, despite the considerations just mentioned. In response to that challenge, I shall argue that the bulk of the section presents three separate arguments to exclude mind from nature, none of which Aristotle accepts. Then, in a fourth section, he replies to the third argument and indirectly to the second.

Leading up to the whole passage, *PA* I.1 has been discussing various methodological issues in the study of natural philosophy and argues that the final cause is prior to the efficient cause. In the prologue to our section (641a14–32), Aristotle contends that it is the task of the natural philosopher to study the soul more than matter. Though nature is spoken of in two ways, with reference to the matter and with reference to the 'being' (οὐσία), the latter is more important as both mover (ὡς ἡ κινοῦσα) and end (ὡς τὸ τέλος). Moreover, the matter is a nature because of the soul and not vice versa, since the matter is potentially the product, and the form makes the product (whether a part or the organism as a whole) the sort of thing it is. Aristotle leaves open the question whether the student of nature should investigate the whole soul or only part of it, but he does say that if natural philosophy studies only part of the soul, it should study that part of it that makes a living thing the sort of thing it is (641a21–23). Then he gives a series of reasons to exclude the intellect from nature.

Before I go through the text (641a32–642a1) section by section, let me call attention to some signposts marking out what I take to be its overall design: Aristotle presents three arguments against mind's place in the domain of natural philosophy, followed by his response to the third argument.

Section (1), the so-called Correlatives Argument, opens at 641a32: "Someone might raise a question (*aporēseie d' an tis*, ἀπορήσειε δ' ἄν τις), in view of what has just now been said, whether it belongs to natural philosophy to talk about all soul or about some <part of it>"; *tis* ("someone") could be anyone, Aristotle or someone else. Next, section (2), which I call "Nature as Moving Cause," begins at 641b4: "Or (*ē*, ἢ) is all soul not a source of motion...?" Most translators take the *ē*, which I translate as "Or" introducing a further worry, instead as a sort of adversative, allowing Aristotle to address a concern raised in section (1): Balme translates the *ē* as "The answer is that"; Lennox translates "However"; Peck: "All the same"; Ogle: "But perhaps."[10] Now Aristotle does sometimes state a puzzle and respond with a question opening with ἢ alone or with a negative.[11] The renderings in this passage strike me as over-translations, however, because section (2) on nature as moving cause does not respond to issues raised in the Correlatives Argument in section (1).[12] Section (2) very naturally reads as a second, independent argument for the same conclusion as section (1), that mind is not an appropriate topic for natural philosophy. Indeed, one might translate the opening line of section (2) as understanding the phrase *aporēseie d' an tis* (ἀπορήσειε δ' ἄν τις) from the start of the Correlatives Argument: "Or (ἢ) <someone might raise a question> is all soul not a source of motion...?" Whoever is speaking then states a strong conclusion: "So it is clear that (*dēlon oun hōs*, δῆλον οὖν ὡς) one should not talk about all soul, since not all soul, but some one or more parts of it, is nature" (641b8–10). If an opponent, rather than Aristotle, is speaking in section (2), then the opponent arrives at that conclusion.

Now consider the start of section (3) (641b11): "And further (*eti de*, ἔτι δὲ)." Aristotle regularly uses the phrase *eti de* to make a further point, so

[10] Balme [1972] 1992: 8; Lennox 1999: 3; 2001: 6; Peck 1968: 71; Ogle 1937: 1.998. Louis 2017: 17 in his French translation gives: "Peut-être bien," "It could well be that" (thanks to Eric Guindon for suggesting this translation of Louis' somewhat peculiar French). Pellegrin's French translation 2011: 103 leaves the ἢ un-translated.

[11] E.g. *NE* I.10 at 1100a31: "but let us go back to the first *aporia* we raised" and then at 1100b7 Aristotle begins his solution by saying: Or (ἢ), is following fortunes completely wrong?" Cf. *NE* V.9, 1136a23, where 1136b3 follows up with a question opening "Or is the definition not (ἢ οὐκ) correct?" I am grateful to Gabriel Richardson Lear for calling these passages from the *Nicomachean Ethics* to my attention and to Emily Kress for supplying me with numerous other examples to check. See also n. 14 below. Not all instances follow that pattern and there is no requirement to read ἢ οὐκ ... as a response to a previous difficulty. Aristotle typically responds to the phrase ἀπορήσειε δ' ἄν τις in his biological works with his solution, often introduced by the phrase αἴτιον δ' ὅτι ("the reason is that"): GA II.5, 741a6...a9; III.3, 754b20...b21; III.7, 757a14...a18; cf. PA III.7, 669b6–7.

[12] Lennox 1999: 2–5; 2009: 13–14 (cf. 2001a: 142–144) thinks that the Correlatives Argument presents a conditional argument leading to a counterfactual conclusion: if intellect were to be studied by natural philosophy, there would be no other branch of philosophy (in particular first philosophy), because natural philosophy would investigate everything, including all theoretical objects. According to Lennox, Aristotle rejects the conclusion mainly on the strength of the second, positive argument (2). But on that interpretation, simply the strength of the argument in section (2) and not its content allows (2) to respond to (1). For Lennox's interpretation of the second argument and my objections, see n. 20.

if he is the speaker in section (2), he should continue speaking in section (3).[13] But if the speaker in section (2) is an opponent, then an opponent is still speaking in section (3). In fact, I think we can hear the voice behind the argument in section (3), and if I am right, we should also read section (2) as an argument by an opponent, and if I am right that we should render the *ē* (ἢ) at the start of section (2) as "or" introducing a second objection to mind's place in nature, then the Correlatives Argument in section (1) likely expresses the voice of an opponent as well.[14]

Only in section (4), which I call "Nature as Final Cause," do we hear Aristotle's own voice. He begins at 641b23 with "But (*de*, δὲ) we say (*legomen*, λέγομεν) 'this for the sake of that' (τόδε τοῦδ' ἕνεκα) wherever some end appears toward which motion proceeds, if nothing impedes it." To be sure, the (*de*, δὲ) could be read as "and" and so as generalizing the previous conclusion. Therefore, this final signpost must await our interpretation of section (3) to which I take section (4) to respond. I now consider each section in turn.

12.3 Correlatives Argument

The first stretch of text about mind in *PA* I.1 argues that if natural philosophy studies mind and its objects, there will be no room for other branches of philosophy, such as first philosophy:

> 1. Someone might raise a question (ἀπορήσειε δ 'ἄν τις), in view of what has just now been said <viz. that the student of nature should study the soul more than the matter>, whether it belongs to natural philosophy to talk about all soul or about some <part of it>. For if it talks about all soul, no philosophy will be left besides natural science. For intellect (νοῦς) is of intelligible objects (τῶν νοητῶν). Therefore natural knowledge would be concerned with all things. For it belongs to the same knowledge to study

13 Another instance of ἔτι δὲ close at hand is 641b36, where the speaker in section (4) (whom I take to be Aristotle), having spoken at length about the role of the seed in generation, makes a further point (ἔτι δὲ) that the seed is in potentiality.

14 Thanks to Lawrence Evans for calling my attention to *NE* II.4, which presents a series of arguments similar to that in our passage. In *PA* I.1, starting with Ἀπορήσειε δ' ἄν τις (1105a17), followed by (1105a17) ἢ οὐδ' (a21), and then ἔτι (a26). Lesley Brown 2009, in her notes on the revised Oxford translation of the *Nicomachean Ethics*, reads ἢ οὐδ' as the start of Aristotle's response, a response completed in the section opening with ἔτι. The repetition of signposts in the two passages lends support to someone who wants to argue that Aristotle responds to the Correlatives Argument in sections (2) and (3); but since (1), (2), and (3), as we shall see, focus on independent considerations for excluding mind from nature, I find it more plausible to read them as three separate arguments for one conclusion – a conclusion Aristotle rejects.

intellect and intelligible objects, if in fact they are correlated, and the same discipline studies all the correlates, as in fact is the case with perception and perceptible objects.[15] (*PA* I.1, 641a32–b4)

Section (1) excludes mind from natural philosophy on the grounds that the same science studies both a faculty and its correlative objects, for instance, perception and perceptible objects. The objects of intellect presumably include all objects of investigation – not only elevated immaterial things but material things and mathematical objects as well. And so, if mind belongs in the domain of natural science, natural science will investigate everything, leaving no room for mathematics and first philosophy.

Let us take a moment to consider why this argument should not unduly concern Aristotle. There is much of it with which he agrees. In *DA* II.4, he states as a methodological rule that to understand a psychic faculty – he mentions the nutritive, perceptive, and intellectual faculties – we should understand its function (ἔργον), and to understand the function, we should study the correlative objects (415a14–22), and he mentions food (τροφή), the perceptible object (αἰσθητόν) and the intelligible object (νοητόν).[16] *DA* II.4 aims to clarify the nutritive soul, a causal capacity (δύναμις), and does so by examining the nutritive functions: growth, self-maintenance, and reproduction. In connection with those functions, he discusses the correlative objects, food – not a carrot or cabbage as such, which should be studied by botany—as nutriment, stuff used by the soul to feed the body and considered from the perspective of the function it serves. At least in the case of nutritive soul, food is more knowable to us than the psychic faculty we aim to understand, and so an investigation starts with the objects; yet those objects ultimately depend for what they are on the psychic faculty and function they serve. Therefore food, considered as food, is posterior to the function, though it is studied by the same expert, an expert in nutrition.[17]

In *Metaph.* VI.1, Aristotle commits himself to the existence of different sciences and distinguishes the three main theoretical sciences on the basis

[15] Translations are my own, though I have profited from consulting other translations, especially Lennox 2001a.

[16] Wedin 1988: 13 emphasizes this methodological rule, which he calls FFO (faculty/function/object), in his investigation of mind and imagination in Aristotle; he also stresses that the condition is not extensional, since an object of thought may be an object of desire as well, and so what counts as the object of a faculty depends on how the object is described.

[17] In Gill 2021a, I spell out in more detail Aristotle's method expressed in the passage on correlatives in *DA* II.4 and its application to the study of nutritive soul.

of their respective objects: natural philosophy studies things *qua* moveable and not separate from perceptible matter (1025b25–1026a6); mathematics studies objects that move but *qua* unmoved and separate from perceptible matter (1026a7–10); and first philosophy (also called theology, 1026a19) studies objects that are separate and unmoved (1026a15–16). Some scholars take *Metaph.* VI.1 to characterize first philosophy as 'special' metaphysics, focused exclusively on the highest objects, immaterial substances. But we need not, and in my view should not, take VI.1 that way. Aristotle discusses a universal mathematics that studies everything, and hence the very same objects as natural philosophy, but abstracted from changeable matter (1026a25–27). Similarly, first philosophy, though especially interested in eternal, separate, unchanging, immaterial objects, does not restrict itself to those. It also studies the same objects as physics does but from its own perspective, the one spelled out in *Metaph.* IV.1 and mentioned in the first and last lines of VI.1 (1025b3–4; 1026a23–32), namely, those same objects and their attributes *qua* being. This is not the place to defend a conception of first philosophy, but I believe that, unlike mathematics, which abstracts categorial content, especially quantitative content, from moveable things and studies that abstract content (numbers, lines, figures, magnitudes, and so on), first philosophy studies objects independent of categorial content, simply *as being*, and investigates how they are structured and related, what depends on what and why, and what is primary and on what grounds.[18] *Metaph.* VII and VIII investigate being in terms of predicative relations, and IX investigates them in terms of potential and actual being.

Aristotle allows room for both first philosophy and natural philosophy (as well as mathematics) to investigate the same objects, but in different ways. As stated above, *Metaph.* VI.1 indicates that first philosophy includes the study of things that are primary, separate, and unmoved – first causes and principles appealed to by natural philosophy but beyond its scope to investigate. Even so, first philosophy overlaps with natural philosophy in considering natural things *qua being*, leaving it to natural philosophy to study them as changeable and to mathematics to study them as numerable.

Aristotle would agree with the speaker in section (1) that different sciences deal with different objects, but he disagrees that those objects must be extensionally distinct, because in his view sciences can investigate different aspects of the very same objects. To avoid the outcome that natural philosophy is the only science, the objector needs to remove mind

[18] See Gill 2006. My understanding of first philosophy in Aristotle is indebted to Sefrin-Weis 2002.

from the domain of natural philosophy.[19] Notice that Aristotle agrees with the opponent on all but one main point, that the objects of different disciplines must be extensionally disjoint and hence that mind must be banished from natural philosophy. We shall see this general pattern repeated in sections (2) and (3).

12.4 Nature as Moving Cause

Section (2) argues that mind is not a source of motion and so not part of nature:

> 2. Or (ἦ) is all soul not a source (ἀρχή) of motion, nor all the parts, but the source of growth is something also in plants, and of alteration the perceptive part, and of locomotion something else and not the intellectual part (τὸ νοητικόν)? For locomotion is present in other animals too, but thought (διάνοια) in none. So it is clear that (δῆλον οὖν ὡς) one should talk not about all soul, since not all soul, but some one or more parts of it, is nature (φύσις). (*PA* I.1, 641b4–10)

Section (2) presents a second argument excluding mind from the domain of natural philosophy, this time focusing on the sources of various sorts of motion. The source of growth is nutritive soul, which we share with plants, as well as other animals. The source of alteration is perceptive soul, which we share with many lower animals. Intellect is not a source of locomotion, because we share locomotion with other animals, and none of them has intellect. Animal locomotion should therefore be explained by some lower psychic faculty humans share with other locomotive animals. Since natural philosophy studies things *qua* moveable, and the source of motion for natural things is their nature, and intellect is not a source of motion, intellect is not part of nature, even human nature. The passage reaches a strong conclusion: "So it is clear that (δῆλον οὖν ὡς) one should talk not about all soul, since not all soul, but some one or more parts of it, is nature (φύσις)" (641b8–10). If Aristotle is presenting this argument on his own behalf, he excludes intellect from the province of natural philosophy.

[19] We need not identify particular opponents, but they could be radical Platonists committed to a two-worlds ontology according to which knowledge and opinion deal with distinct objects – knowledge (ἐπιστήμη) with eternal, unchanging being (Forms), perception and opinion (δόξα) with material becoming. Individuals called Friends of the Forms in the Battle of Gods and Giants in Plato's *Sophist* appear to hold such a view. They want to keep *nous* in the intelligible realm (*Sph.* 246a4–249d8) but face difficulties on that score. See n. 22.

But why would Aristotle make such an argument, given the conflict with *MA* 6? Recall that *De Motu Animalium* discusses the role of intellect in locomotion and mentions perception and imagination by the way, as filling the same role as intellect in the explanation of the motion of nonrational animals. The passage also conflicts with *DA* III.9–10 for the same reason.[20] Intellect plays a prominent role in human locomotion, because intellect conceives an object as a practical good, and that object so conceived arouses desire, which in turn moves us.

Furthermore, as previously noted, Aristotle appeals to intellect later in *Parts of Animals* to explain various human parts. In some of those passages, he speaks of human nature and clearly with reference to intellect. For instance, in explaining why we stand upright, he says that humans are the only animals to do so, because human nature (τὴν φύσιν αὐτοῦ) or being (τὴν οὐσίαν) is divine (θείαν); moreover, thinking would be difficult if there were a mass of body weighing us down from above, for weight makes thought (τὴν διάνοιαν) and the common sense (τὴν κοινὴν αἴσθησιν) hard to move (δυσκίνητον) (*PA* IV.10, 686a27–31).[21] In explaining why human beings have hands, he insists that humans have hands because we are the most intelligent (διὰ τὸ φρονιμώτατον). Hands are special instruments – and nature (ἡ φύσις), like a reasonable human being, always distributes each organ to one that can use it (*PA* IV.10, 687a9–12). The fact that part of human nature is divine is insufficient reason to exclude it from the domain of natural science, since the scientist needs to appeal to it to explain certain bodily arrangements.

I have argued that the second objection to *nous* in *PA* I.1 does not square with Aristotle's treatment of intellect elsewhere or even in *Parts of Animals* itself. If I am right that this section presents an opponent's position, we should construe *dēlon oun hōs* (δῆλον οὖν ὡς, "So it's clear that") as the opponent's conclusion. Again, Aristotle agrees with the

20 Lennox 1999: 4–5 argues that in *DA* III.9–10 intellect is neither a necessary nor a sufficient condition for human locomotion. It is not a sufficient condition, because desire is also required. It is not a necessary condition, since humans can act irrationally. But Aristotle could surely reply that when humans behave irrationally they are behaving like other animals, on the basis of imagination (φαντασία), conceiving of some object as desirable and pursuing it. I see no evidence that Aristotle ceased to regard mind as a necessary condition for distinctively *human, rational* action. I agree with those who stress that Aristotle is speaking of *practical* wisdom (φρόνησις). Practical wisdom is the intellectual virtue of the λογιστικόν, the calculative part of rational soul (*NE* VI.1, 1139a11–15 and VI.5, esp. 1040b20–28). Both practical and theoretical wisdom are acquired states (ἕξεις) of the rational soul, gained through learning (II.1, 1103a14–18).

21 Thanks to Sophia Connell for calling to my attention the word δυσκίνητον ("hard to move"). See Johnson 2021: 23–24; Connell 2021b: 205; and Christof Rapp's contribution (Chapter 13) to this volume.

speaker's argument, with one main exception. He agrees that faculties of soul are topics for natural philosophy insofar as they are part of an organism's nature, defined in *Physics* II.1 as an inner source of motion and rest (192b13–27), but apparently, unlike his opponent, he regards intellect as part of human nature as a source of locomotion.[22]

12.5 Cause of Order

Section (3) does not explicitly mention *nous* but speaks of some unnamed cause abstracted from natural things, which we possess from the universe:

> 3. And further (ἔτι δὲ), natural philosophy cannot study any of the things from abstraction (τῶν ἐξ ἀφαιρέσεως), since nature makes all things for the sake of something (ἕνεκά του). For (γὰρ) it appears that, as art is in the products of art (ὥσπερ ἐν τοῖς τεχναστοῖς ἐστιν ἡ τέχνη), so some other source (οὕτως . . . ἄλλη τις ἀρχή) and cause of such a sort (αἰτία τοιαύτη) is in the things themselves (ἐν αὐτοῖς τοῖς πράγμασιν), and we possess it – just like hot and cold – from the universe (ἐκ τοῦ παντός). Hence it is more reasonable that the heaven has come to be by such a cause, if it has come to be, and is on account of such a cause than mortal animals. At any rate that which is ordered and definite is much more apparent in the heavens than around us, and the fluctuating and chance are more so in the mortal sphere. But some people say that each of the animals is and comes to be by nature (φύσει), whereas the heaven, in which no chance and disorder appears, is constituted in such a way by chance and spontaneity. (PA I.1, 641b11–23)

The first sentence of section (3) excludes things "from abstraction" (*ex aphaireseōs*, ἐξ ἀφαιρέσεως) from the domain of natural philosophy on the ground that nature makes everything for the sake of something (641a10–12). We initially expect the items excluded to be objects of mathematics, since Aristotle often uses the phrase *ex aphaireseōs* in connection with them and treats them as separate from changing and perceptible matter, though they are not actually separate from physical objects. Yet why mention mathematical objects here? The rest of the passage, starting with *gar* (γὰρ, "for") at 641b12, suggests that certain abstract objects are

[22] Again, we need not pinpoint a specific opponent, though this opponent also sounds like an extreme Platonist, indeed a Friend of the Forms in the Battle of Gods and Giants in Plato's *Sophist*, this time with a different bee in his bonnet. This individual excludes all *kinēsis* from the realm of what is really real. The Eleatic stranger makes trouble for him by pointing out that if he excludes *kinēsis* from the realm of being, he will render the Forms unknowable, since our knowing them has an impact on them in relation to us. Moreover, by removing *kinēsis* from the realm of being, *nous* itself will be excluded from that realm (*Sph.* 248a4–249b7).

being excluded from natural philosophy because, though causes of order, they do not act for the sake of ends as Aristotelian nature does.

Let me summarize the passage briefly: to ground his claim that natural philosophy focuses on things that act for the sake of something and not on abstracted things, the speaker compares some cause to art: as art is in the products of art, so a similar cause is in the things themselves, and we possess it, as we possess hot and cold, from the universe. Hence it is much more reasonable, he says, that the heaven came to be from such a cause, if it came to be, than that perishable animals did, since the universe is more ordered and limited than things around us, which seem to be more due to chance. The speaker contrasts his position with that of some other people, who think that the reverse is true: animals come to be by nature, whereas the heaven comes to be by chance and spontaneity. The speaker rejects that view, saying that there is no chance and spontaneity in the heaven.

Maybe Aristotle is comparing nature to art, as he often does, to make the point that both act for the sake of something. Yet why does he say that we have the relevant cause, as we possess hot and cold, "from the universe"? This passage in fact seems reminiscent of an earlier passage in *PA* I.1, where Aristotle states Presocratic positions about efficient causes. He said that the first natural philosophers mainly investigated the material cause but then tried to explain how the whole was generated from the matter by appeal to efficient causes: strife or love (Empedocles), mind (νοῦς) (Anaxagoras), or spontaneity (perhaps Democritus) (*PA* I.1, 640b4–8).[23] Since the speaker's opponent in section (3) attributes the generation of the cosmos to spontaneity, perhaps the main speaker in section (3) attributes such generation to *nous*. Anaxagoras treats *nous* as totally separate from the cosmic soup it differentiated, though portions of it are in some things. *Nous* for Anaxagoras could count as something "from abstraction" (DK 59B11, 59B12).

Section (3) is also highly reminiscent of the cosmological passage in Plato's *Philebus* (27c3–31b1), in which Plato's Socrates argues that *nous* is the cause of the cosmos against some opponents who attribute the heavens to chance.[24] Those opponents appear to be the same as those mentioned in the final sentence of section (3), people also mentioned in *Phys.* II.4, there described as believing that nature (*phusis*) and mind (*nous*) are responsible for animals and plants, whereas the heavens came about by chance

[23] See Lennox 2001a: 136, though he doubts that the third is Democritus.

[24] This connection is also noticed by Sedley 2007: 195.

(196a24–b5).[25] Moreover, both section (3) and the cosmological passage in the *Philebus* claim that material stuffs and their properties come from the universe. The *Philebus* is more explicit: we get our paltry, feeble portions of fire from pure, powerful fire in the universe. That passage can clarify section (3) in *PA* I.1.

In the *Philebus*, Socrates and Protarchus have been debating what constitutes the good life, with Protarchus championing pleasure, Socrates intellect (νοῦς). Socrates then persuades Protarchus that neither the life of pleasure without intellect nor the life of intellect without pleasure suffices for the good life and that the prize should go instead to a life combining both (*Phlb.* 20b1–23b3). Taking as settled that the best life combines pleasure and intellect, Socrates asks which ingredient in the mixed life is responsible for its goodness, so as to confer second prize to it. The next two sections of the dialogue – the fourfold division of beings (23b5–27c2) and the cosmological section (27c3–31b1) – together present themselves as deciding the question in favor of intellect (23b5–9, 27c3–6). The fourfold division distinguishes four kinds – the limit, unlimited, mixture of the two, and cause of the mixture – and awards priority to the cause ahead of the other three (26e1–27c2). The cosmological section then argues that *nous* is the cause (28a4–7, c1–9, 31a7–10), whereas pleasure belongs to the unlimited kind. Socrates thereby implies without explicitly concluding that *nous* deserves second prize in the contest.

In a little more detail, Socrates undertakes to show in the cosmological section, against some opponents who regard the cosmos as the product of chance, that the cause is *nous* (*Phlb.* 30c2–8). Not just any *nous* but divine *nous*, which creates and governs a beautiful cosmos. Thanks to divine *nous*, and not to mere chance (28d5–9), we have an ordered cosmos, with sun, moon, and stars, and revolutions of the whole heaven. Moreover, we get our impure and feeble fire down here from the pure and powerful fire in the universe (29b6–11), and likewise with the other stuffs that compose us, and also soul (30a1–8). Things around us are impure and have a power unworthy of their nature (φύσις) (29b6–10).

Grant the similarities between the cosmological section of the *Philebus* and section (3) of our passage in *PA* I.1, and the *Philebus* sheds considerable light on our passage. Most important, though section (3) does not explicitly mention *nous*, we can surmise that *nous* – indeed divine *nous* – is

[25] Monte Johnson called my attention to the passage in *Phys.* II.4. For discussion of it, see Charlton 1970: 105; Johnson 2005: 108; and Sedley 2007: 191. Opinions differ as to who the opponents are, but Democritus is a strong contender.

the relevant cause of cosmic order and is among the things "from abstraction" not studied by natural philosophy.[26] If we accept that parallel, we can make sense of the peculiar claim in section (3) that "as art is in the products of art, so some other source (ἄλλη τις ἀρχή) and cause of such a sort (αἰτία τοιαύτη) is in the things themselves, and we possess it – just like hot and cold – from the universe (ἐκ τοῦ παντός)" (*PA* I.1, 641b12–15). Art is an efficient cause of its products and external to them in the craftsman, though both cause and product belong to the same physical realm. Art is *in* its products in the attenuated sense that the form in the artisan's mind (say the blueprint of a house) is realized in the materials that compose the house produced. What Aristotle characterizes as "some other source and cause of such a sort" is analogous. Divine *nous* is an efficient cause analogous to art and is external to – indeed utterly separate from – its products. We get our *nous* – *nous in* us, such as it is – from without, from the universe.[27] One final comparison between the *Philebus* and section (3): the speaker in section (3) says that chance is more likely in our realm than in the cosmos. Though Socrates does not make that point explicitly in the *Philebus*, he claims that the things around us are impure and their power unworthy of their nature. The pure nature of fire exists apart from us in the cosmos, and it and other pure natures sustain their paltry counterparts around us. Our human *nous*, though sustained by divine *nous*, is a feeble, impure thing whose power falls far short of its pure cause, divine *nous*.

Let me conclude this section by indicating points in section (3) with which Aristotle would agree based on evidence elsewhere. First, he would agree that divine *nous* is the preeminent cause of the universe and that it is separate from the world around us, including the heavenly spheres, and that *nous* is an efficient cause, the first unmoved mover of cosmic motion (*Phys.* VIII.10, 267b6–9; *Metaph.* XII.7, 1073a3–5). He would also agree that human *nous* comes in from outside and is divine (*GA* II.3, 736b27–29). Although his statement could express the view I have

[26] Sedley 2007: 194–203 discusses our passage in *PA* I.1 together with *Physics* II.4–6 on chance and the cosmological passage in Plato's *Philebus*. Whereas he thinks our passage in *PA* I.1 expresses Aristotle's own view and indeed his commitment to cosmic teleology, I take the opening two sentences to exclude divine *nous* from natural philosophy because natural philosophy studies final causes, and this divine *nous* (reminiscent of Anaxagoras' *nous*) is solely an efficient cause.

[27] Socrates does not explicitly speak of *nous* in our realm in the cosmological passage, though he does say that the cause is in everything and furnishes our bodies with soul (*Phlb.* 30a9–b7). Earlier in the dialogue he denigrates his own *nous* to divine *nous* (*Phlb.* 22c1–7). Sedley 2007: 195 calls attention to a fascinating passage in Xenophon, *Memorabilia* I 4.8 (Marchant 1979), where the speaker Aristodemus asks Socrates if our *nous* derives from cosmic *nous*.

attributed to the speaker in section (3) of our passage, it could instead express a more mundane idea, that we are born with perceptual capacities enabling us to acquire *nous* through education without being born with it as a separate psychic power.[28] In addition, he would agree that things around us are less orderly than things in the heavens: whereas the heavenly bodies in the superlunary realm move eternally in circles, things in the sublunary realm move in straight lines and are perishable (cf. *Metaph.* IX.8, 1050b6–28; XII.10, 1075a11–25). Not everything in our realm can be explained teleologically, and much is due to material necessity, as Aristotle states more than once in *PA* I.1 (639b21–26; 642a1–3; 642a31–b4) and reminds us often in his biological works.[29]

But Aristotle would disagree with the speaker in section (3) that things in our realm get their natures from cosmic natures and that human beings get their *nous* from cosmic divine *nous*. Instead, living organisms in our realm get their natures from their parents. This is the topic of section (4).

12.6 Nature as Final Cause

We turn finally to section (4) of our text in *PA* I.1. I take this passage to express Aristotle's own view and to respond directly to section (3) and indirectly to section (2), while also pointing ahead to the end of *PA* I.1. I argued on the basis of a similar passage in Plato's *Philebus* that the unnamed cause of cosmic order is divine *nous*, an efficient cause. We have portions of it from the cosmos, though our *nous* falls far short of pure, divine *nous*. Now, in a significant correction of the speaker in section (3), the speaker in section (4) claims that we speak of a *final* cause ("this for the sake of that") wherever there is an *end* toward which motion proceeds, if there is no impediment, and he identifies the end as a thing's *nature* (φύσις):

> 4. But (δὲ) we say (λέγομεν) "this for the sake of that" (τόδε τοῦδε ἕνεκα) wherever some *end* (τέλος) appears toward which the motion proceeds, if nothing impedes it. Therefore it is evident that there is some such thing, the very thing we also call a *nature* (φύσις); for it is not the case that some chance thing comes to be from each seed, nor some chance seed from a chance body, but this from that. So the seed (τὸ σπέρμα) is a source (ἀρχή)

[28] Cf. Burnyeat 2002: 70 n. 111. For a thorough discussion of this topic, see Connell 2021d.

[29] *Meteor.* IV.1–11 focus on material necessity, and the biological works mention many outcomes that come about both for the sake of something and from necessity. E.g. *PA* IV.11, 692a1–8; IV.12, 694a22–b12; *GA* II.1, 731b18–732a11; II.4, 738a33–b4 and 739b20–33; II.6, 743a36–b5 and b16–18; IV.8, 776b31–33; V.8, 789b2–15.

> and productive (ποιητικὸν) of what comes to be from it. For these things are *by nature* (φύσει); at any rate they grow from seed. But prior to that is that of which the seed is a seed. For the seed is *becoming* (γένεσις), whereas the end is *being* (οὐσία). Yet prior to both is what the seed is from. For the seed is a seed in two ways, *from* which and *of* which. For in fact it is a seed of that from which it came, e.g., from a horse, and a seed of what will be from it, e.g., of a mule, though not in the same way, but of each in the way mentioned. And further (Ἔτι δὲ) the seed is in potentiality (δυνάμει); and we know how potentiality (δύναμις) is related to actuality (ἐντελέχεια). (*PA* I.1, 641b23–642a1)

The first two sentences recall the arguments in both sections (2) and (3). In section (2), the speaker mentions the causes of three sorts of motion – growth (change of quantity), alteration (change of quality), and locomotion (change of place) – but says nothing about substantial generation, its source (*archē*, ἀρχή) and end (*telos*, τέλος), topics central to section (4). Section (3) does speak of generation, but the speaker attributes generation of the cosmos to separate divine *nous*, an efficient cause separate from things in our realm, and describes generation in our mortal realm as haphazard, with much due to chance. Against that speaker, Aristotle contends in section (4) that nature is an end, the goal for the sake of which generation occurs. Generation in our realm is orderly, not haphazard: "It is not the case that some chance thing comes to be from each seed, nor some chance seed from a chance body, but this from that" (641b26–28).

The seed, he says, is the source and productive of what comes to be from it (641b29–30). The seed is becoming (*genesis*, γένεσις), whereas the end, that for the sake of which the generation occurs, is being (*ousia*, οὐσία), an end Aristotle has already identified as nature (*phusis*, φύσις) (641b23–26). This nature is substantial form as distinct from the whole offspring consisting of matter and form, which exists "by nature" (φύσει) (641b29–30). The organism exists by nature in virtue of having an inner source of its proper motion and rest: a nature (cf. *Phys.* II.1, 190b8–193a9).

But how is nature a *final* cause, an *end*, that for the sake of which the generation occurs? Aristotle's definition of nature in *Phys.* II.1 helps us understand how nature is an *efficient* cause of motion and rest but leaves obscure how nature is a final cause, the form of the emergent offspring. His discussion of seed at the end of the passage helps to answer the question. The seed (identified as the *source* and as *becoming*) can be considered in two ways, in relation to the parent *from* which the seed came and in relation to the offspring *of* which it is the seed. The seed connects parent and offspring, enabling the parent to replicate its kind, an oak tree an oak tree via an acorn, a human being a human being via the σπέρμα (*sperma*, "seed"), understood as the first combination of male and female factors. Aristotle has prepared us for this discussion earlier in *PA* I.1, 640a19–33, in a passage taking Empedocles to task for saying that features of creatures, such as the spine, are due to accidents that occur during their formation. No, says Aristotle: Empedocles did not grasp that the seed must have a *capacity* (δύναμις) of a certain sort and that what produced the seed (τὸ ποιῆσαν) is prior to it both in account (τῷ λόγῳ) and in time (τῷ χρόνῳ). Section (4) ends with a reminder of that earlier passage, saying that the seed is in potentiality (δυνάμει) and reminds us that we know how potentiality (*dunamis*, δύναμις) is related to actuality (*entelecheia*, ἐντελέχεια) (641b36–642a1).

Besides gesturing back to the earlier discussion in *PA* I.1, the end of section (4) also gestures toward Aristotle's detailed discussion of potentiality and actuality in *Metaph.* IX, and especially IX.8, where he argues that actuality is prior to potentiality in several ways. Actuality is prior to potentiality in account, because a potentiality is defined with reference to the actuality it is a potentiality for (*Metaph.* IX.8, 1049b12–17): an acorn is potentially an oak tree because it will turn into an oak in appropriate conditions. As for priority in time, although the acorn is prior in time to the mature oak it can become, prior to the acorn is the mature oak that produced it (1049a17–1050a3): the parent as producer is prior in time to the seed and uses the seed as an instrument in its production. The parent's nature is a first efficient cause of generation. Note that Aristotle calls three things *sperma* ("seed") in different contexts: the male semen, the female menstrual blood, and the fetation, the combination of male and female factors. None of these entities is ensouled, but the semen and fetation are vehicles that carry the soul principle, which will eventually be internalized once the heart or analogous organ is formed in the developing creature.[30]

We can understand that the formal nature of the parent is the first efficient cause of the offspring; the parent causes motion in virtue of its form, using seed (in all three senses) as its tool.[31] Yet we still need to understand how the form of the offspring operates as a *final* cause of organic development.

[30] See Gill 1989: 227–34 and texts cited. For detailed recent discussions, see Henry 2019 and Gelber 2022.

[31] See *GC* I.7, 324a24–b4, on grades of efficient causes; see also Gelber 2020 and Kress 2023 on tools of the soul.

Metaph. IX.8 discusses a more fundamental way in which actuality is prior to potentiality than priority in account and time: priority in being (οὐσίᾳ) (1050a4–1051a3). Although Aristotle distinguishes several ways in which actuality is prior to potentiality in being, most relevant for understanding section (4) of *PA* I.1 is the first, the priority of the mature organism to an immature specimen:

> But surely <actuality is prior to potentiality> also in being (οὐσίᾳ), first because things later in generation (τῇ γενέσει) are prior in form (τῷ εἴδει) and being (οὐσίᾳ). For example, a man (ἀνὴρ) is prior to a boy (παιδὸς) and a human being (ἄνθρωπος) to a seed (σπέρματος), for the one already has the form, whereas the other does not (τὸ μὲν γὰρ ἤδη ἔχει τὸ εἶδος τὸ δ' οὔ). Second, <actuality is prior to potentiality in being> because everything that comes to be marches toward a source (ἀρχὴν) and end (τέλος), since that for the sake of which is a source (ἀρχὴ γὰρ τὸ οὗ ἕνεκα), and generation is for the sake of the end (τοῦ τέλους δὲ ἕνεκα ἡ γένεσις). And the actuality (ἡ ἐνέργεια) is an end, and the potentiality (δύναμις) is acquired for the sake of that. (*Metaph.* IX.8, 1050a4–10)

Here Aristotle claims that the adult male is prior in being to the prepubescent boy and a human being to a seed, because the one has realized the relevant soul principle missing in the other – an individual merely on the way to having that form. The seed has a potentiality for the form, and the form will eventually be actualized in the course of organic development. As Aristotle will put the point at the end of *PA* I.1 (642a6–13) also stated earlier (639b21–640a10, a33–b4), the boy in the one case and the seed in the other are conditionally necessary (*ex hupotheseōs*, ἐξ ὑποθέσεως) as stages in the development of the mature specimen, stages required if the form is to be realized. Nature as final cause explains how the whole material body and each of its parts come to be organized as instruments, for the sake of that end: the organism's soul.[32]

Intellect is not mentioned explicitly in section (4). Aristotle uses as his example a horse as parent and mule as offspring, animals without *nous*.[33] What about human beings? Does human nature include *nous*? As we have noted, Aristotle says in *GA* II.3 that *nous* comes in from outside. Certainly, *nous* is not present in the seed, but neither is nutritive soul or perceptive soul present in actuality. These emerge in the course of organic development. Nonetheless, the seed (fetation) has the potentiality to come into the

[32] Cf. *PA* I.5, 645b14–20.

[33] Lennox 2001a: 146 observes that Aristotle's example of the mule is odd, because it is an *unnatural* generation, but probably Aristotle chose the example simply to make sure that reference to the offspring is not confused with reference to the parent.

form, and in the case of humans the form that differentiates us from other creatures is our rational capacity. The seed has the *dunamis* enabling the organism to acquire *nous*, if development occurs unimpeded in appropriate conditions.

Recall the discussion of motion in section (2) of *PA* I.1. Not only did the speaker ignore substantial generation, he also ignored the most distinctive motion of humans – thinking – the activity that distinguishes humans from other animals and moreover distinguishes the mature human from infants and youths not yet able to engage in their natural adult activities, procreation and thinking. Whatever the role of intellect in locomotion – the main issue in section (2) – intellect is the capacity equipping human beings to achieve their end and fully be what they are, rational animals. Intellect is certainly part of human nature, the source of our most proper and divine human behavior, contemplation. Not only that. Mind also affects our lower functions, including nutrition and procreation by unifying them into a coordinated system: we do not graze like cattle, we have candlelit dinners; we do not simply copulate, we court the person we love.

Parts of Animals need not discuss human intellect in its own right, because intellect has no corresponding material part, and material parts are the main topic of *Parts of Animals*. Aristotle leaves examination of *nous* itself to such works as *De Anima*, *De Motu Animalium*, *Metaphysics* XII, *Nicomachean Ethics*, and *Posterior Analytics* II.19. Even so, as we have seen, *Parts of Animals* does discuss parts whose structure and function are determined by our rational nature, such as our upright stance and consequent two-footed mode of locomotion, and our possession of hands as a tool of tools (687a2–688a11) for artificial production. Mind is part of human nature and therefore a vital topic to be studied by natural philosophy.[34]

[34] I gave a first version of this chapter at Western University, Ontario, in 2007, as a comment on a paper that became Lennox 2009. On that first occasion, I focused only on the first and second arguments in *PA* I.1. More recently, I gave versions of the expanded version at conferences on Aristotle's biology in Williamstown (2018) and Cambridge (2019), at Stanford and Chicago (2020), and at Barbara Sattler's Kolloquium in Bochum (2021). I am grateful to all those audiences for valuable questions, and especially thank Fei-Ting Chen, Sophia Connell, Ronja Hildebrandt, Emily Kress, Samuel Meister, and Thomas Slabon for detailed written comments.

CHAPTER 13

The Human Body and the Role of Erect Posture
Aristotle's Parts of Animals IV.10

Christof Rapp

13.1 The Project and Structure of *PA* IV.10

At the beginning of *PA* IV.10, Aristotle announces an examination of the (nonuniform, external) parts of blooded, live-bearing animals – thus resuming, as he says, a previous discussion of parts surrounding the head and the neck (presumably a back-reference to the agenda of *PA* II.10 to III.3). For one paragraph he dwells on the head and neck again. All sanguineous animals have a head, he says, and nature has assigned three functions to it. Above all, the head exists for the sake of the brain and is hence located in a place opposite to the heart – for reasons mentioned elsewhere (*PA* II.7), namely that the brain has to cool down the heat coming from the heart. Second, the head has to accommodate those senses that require a certain mixture of blood, a certain temperature and a certain amount of tranquillity that is needed for the accuracy of perception.[1] Third, nature has equipped the head with those parts that are needed for the intake of food. The head, he says, is a suitable position for the mouth, and he suggests in a counterfactual thought experiment that it would be strange both if the stomach were above the heart and if the mouth together with the stomach were below the heart.[2] The neck exists for the sake of the

[1] In general, perception (or the transmission of the perceptual changes from the peripheral to the central organ) is impaired or even erased by random motions in the blood. For the more accurate modes of perception, the purer blood in the head therefore provides an ideal surrounding (see, e.g., *PA* II.10, 656b3–7).

[2] Admittedly, both alternative anatomic configurations would take some getting used to, but it is unclear how these meager hints are meant to conclusively exclude any alternative position of the mouth or the stomach. In light of what follows in this chapter, one might think that a stomach located above the heart would add to the unwelcome amount of weight that presses upon this important organ. But why couldn't the mouth be located close to the stomach in its actual position beyond the diaphragm? Since Aristotle is inclined to associate the function of food intake with the 'up' of an animal body (see Section 13.3), it is clear that this alternative location of the mouth would have unwelcome consequences for his project of mapping the dimensions of the animal body. Still, commentators are puzzled by the argument he gives to rule out this second alternative possibility: the length of the body, he says, would become great, and the mouth (if this is indeed, as Lennox 2001a:

windpipe and its function is to protect the windpipe together with the esophagus. All this is summarized in a very condensed way (686a5–24). When Aristotle turns immediately afterwards to the parts 'that come next,' namely to the trunk and the forelimbs, it becomes clear that he intends to go through the various parts from the top to the bottom (in the case of human beings) or from the front to the rear (in the case of quadrupeds).[3] In principle, the structure of the chapter follows this pellucid ordering principle: from the head and neck to the thorax and forelimbs, with the hand as the peculiarly human variant of the forelimbs, then to the chest with the nipples or mammae, followed by the belly down to the so-called end of the thorax, the pelvic region, which serves the excretion of both dry and moist residues and is at the same time used for the reproductive organs. The chapter terminates, in accordance with the same anatomic logic, with remarks on the (rear) legs and, finally, on the feet.

Arguably, this is the chapter's main structure; however, if we describe the chapter only like this, we will miss the main attraction and, most notably, the common thread of the chapter beyond the top-down (or front-to-back) structure. Triggered by the first mention of forelimbs early on in the chapter, Aristotle points out that human beings have hands and arms instead of front legs and that this is due to their erect posture. He immediately connects the erect posture with human beings' alleged divine nature, exhibited in their intellectual capacities. From this point on, the comparative anatomy of *PA* IV.10 primarily aims at accounting for the anatomic differences between human beings and other blooded, live-bearing animals (mostly quadrupeds) with regard to their external, nonuniform parts. While it is common for Aristotle's biological writings to start with the examination of human beings for heuristic or pragmatic reasons (*HA* I.6, 491a19–23; *PA* II.10, 656a7–10), since human anatomy is more familiar to the human researcher, the present chapter does not deal with human anatomy just for pragmatic reasons but rather displays a genuine interest in the peculiar shape and orientation of the human body. All the distinctive features of the human body discussed in the course of this

316 suggests, the grammatical subject of this particular sentence) would be too distant from the moving and concocting origin, i.e., the heart (*PA* II.10, 686a16–17); apparently, this makes sense only if Aristotle thinks of an extension of the lower abdomen (see Kullmann 2007: 689: "Verlängerung des Unterkörpers durch Versetzung des Mundes nach unten"), namely in accordance with the given length of the oesophagus. However, this would presuppose – without any necessity – that the esophagus would have the same length that it actually has with the mouth integrated in the head.

[3] Compare the listing of external parts in *HA* I.15, 494b1ff.

chapter are ultimately traced back to the erect posture (or so it will be argued in Section 13.4). In light of this explanatory structure, the erect posture of humans, as (part of) the explanans of all these peculiar anatomic features, constitutes the hidden thread of this chapter in that it connects the discussions of the particular parts of the human body. And while the erect posture is used to account for these other distinctive features of the human body, this posture itself is said to facilitate the intellectual capacities of human beings. How exactly this is supposed to work turns out to be the main challenge posed by this chapter. It is relatively uncontroversial, though, that it is a peculiar proportion between the bulk of the upper and the bulk of the lower part of the animal body that, according to this chapter, provides the missing link between upright posture and gait, on the one hand, and the quality of thinking or intelligence, on the other. And it is also owing to this important role of the bodily proportions that the chapter discusses the various external parts of blooded, live-bearing animals with a view to these animals' overall shape and posture. The project and purpose of the chapter can hence be characterized as follows. Obviously, it discusses external bodily parts of the various blooded, live-bearing animals with a view to what is most distinctive of the human body (namely its upright posture) and relates most of the highlighted differences to the overall orientation and posture of human and nonhuman animals.

13.2 Erect Posture and the Divine Nature of Human Beings

Let us proceed to the chapter's central argument. It is presented in several chunks of text (**[T1]**, **[T2]**, **[T3]**, **[T5]**), the first of which provides the general outline that will be filled out with more details in the subsequent paragraphs.

> **[T1]** Humankind, however, instead of forelimbs and forefeet has arms and what are called hands. For it alone of the animals is upright, on account of the fact that its nature and substantial being are divine; and the function of that which is most divine is thinking and being intelligent (διὰ τὸ τὴν φύσιν αὐτοῦ καὶ τὴν οὐσίαν εἶναι θείαν· ἔργον δὲ τοῦ θειοτάτου τὸ νοεῖν καὶ φρονεῖν).[4]

It is peculiar to human beings that they have arms and hands.[5] The connection between the possession of hands and upright posture is that

[4] *PA* 686a25–29, Lennox's translation 2001a slightly altered.

[5] Actually, Aristotle says that they have *what is called* or *so-called hands* (τὰς καλουμένας ... χεῖρας). The Greek word χείρ is the usual designation for hands – why "so-called" then? Perhaps the

for animals that stand and walk uprightly there is no need and no function for front legs,[6] so that nature was free, as it were, to assign a different anatomy with different functions to the human forelimbs (otherwise they would have been without function) and to equip human beings with hands instead of (front) feet. Here, as in other passages, Aristotle seems to assume that the upright posture is strictly distinctive of human beings, even though he admits later in the chapter (689b31–690a4) that apes, having neither a tail (like biped animals) nor fleshy buttocks (like quadrupeds), hold a sort of middle position.[7]

Next, the text asserts a causal or explanatory connection between human beings' upright posture (explanandum) and, roughly speaking, their intellectual capacities (explanans), even though it is far from clear, at first, how and why the latter is supposed to provide an explanation for the former. As for the intellectual capacities involved, Aristotle speaks of *noein* and *phronein* (νοεῖν καὶ φρονεῖν).[8] Both words have a range of different nuances and connotations, and given that Aristotle uses two words, one might wonder how exactly they are meant to be distinct. The first word, *noein*, might plausibly be taken to refer to the activities of thinking and reasoning; saying of an animal that it engages in thinking or reasoning presupposes that it possesses *the part of the soul* by which one thinks and reasons or *the capacity* to think or to reason, both of which, the part of the soul and the capacity, might be called *nous* (νοῦς). Even though there are uses of this word in Aristotle that apply to both human and nonhuman animals, the present context with its interest in peculiarly human characteristics seems to require the exclusive use, that is, the sense of having *nous* that only applies to human beings. The second word, *phronein*, can also be used to refer to various modes of thinking; thus, it cannot be excluded that both terms together are used as a hendiadys, in

qualification reflects the fact that the Greek word χείρ can also refer to the paws of certain animals, while people have come to use it in an exclusive sense for the human hands: what are called 'hands' in human beings.

6 See *PA* 687a5–7.

7 What about birds in this context? After all, birds like penguins or ostriches seem to have a similar upward orientation. In *De Incessu Animalium* 11, 710b17–711a7, Aristotle argues that birds in general cannot have an upright posture. Also, Aristotle might be more interested in the relatively stable orientation of terrestrial beings, while the orientation of aerial and aquatic animals is more volatile and more difficult to account for.

8 In writing "thinking and being intelligent" as a translation I follow Gregoric's suggestion; for his discussion of the Greek words *noein* and *phronein*, see Gregoric 2007: 93–96. Gregoric aims at showing (prompted by the overall topic of his monograph) that each of the two verbs is connected with one of the two capacities that are mentioned in the following lines (see [**T2**.A]), namely the first to *dianoia* and the second to κοινὴ αἴσθησις ('common sense').

order to cover the whole range of thinking broadly conceived. However, if one looks for a semantic difference and a division of labor, one might assume that *noein* is rather on the side of theoretical thinking, while *phronein* is connected with practical thinking and practical wisdom (*phronêsis*),[9] or, to take up another familiar Aristotelian distinction, that the latter is used for discursive thinking, while the former is preferably used for nondiscursive thought. However, it might also be significant that in the further course of this chapter, Aristotle keeps using cognates of *phronein*, namely *aphronesteron* and *phronimôtaton*, in the comparative and in the superlative, referring to different degrees of a property that needs to be connected with *phronein*. Arguably, animals that are better able to *phronein* might be called 'more intelligent' and animals that are less capable of thinking 'less intelligent.' If this is the relevant context, one might wish to understand *phronein* either as the activity on account of which animals are more or less intelligent or as the property of being intelligent itself.

Another difficulty posed by this section is the connection between divinity and human intelligence. Commentators usually refer to *Nicomachean Ethics* X.7–8,[10] where Aristotle points out that complete happiness, consisting in uninterrupted activity of the intellect, is beyond human nature, but rather belongs to the divine sphere, an element of which is also present in human beings (*EN* X.7, 1177b26–28). Plausibly, this might be taken to mean, first, that through their intellect human beings partake in the divine sphere in the sense that it is the intellect that connects the divine and the human domain, and second, that the life and the unhindered activity of the divine unmoved mover is a paradigm or role model – though unreachable – for human happiness. So, the difficulty is not that there would be no comprehensible connection between the human intellect and what Aristotle takes to be divine; the difficulty is rather to pinpoint why the theorem that human beings have intellect – an idea that is almost omnipresent in Aristotle's oeuvre – is introduced with the help of the (less frequently mentioned and less scientifically minded) thesis concerning the alleged divine nature of human beings. Does the allegedly divine nature of human beings boil down to saying that they have intellect? Is this just a rhetorically augmented way of saying that human beings have a rational nature, that is, a nature that involves the capacity to think? Such a deflationary reading is conceivable; however, it is striking

[9] Something like this seems to be presupposed in Ogle/Barnes' translation (= Barnes 1984): "to think and to be wise."

[10] See, e.g., Lennox 2001a: 318; Kullmann 2007: 690.

that Aristotle first stresses the divine nature (*phusis*) and essence (*ousia*) of human beings and only then mentions the human capacity to think – as a characteristic that seems, according to this passage, to be grounded in and derived from their divine nature. Elsewhere Aristotle seems happy to connect human beings' erect posture directly to divinity without the explicit mention of the connecting middle term, as it were, namely human intelligence. Before we move on with the argument of *PA* IV.10, we should take a look into this context.

13.3 The Human Posture in the Cosmological Context

It seems that Plato and Aristotle share the idea that the upper part of human beings is directed toward the upper part of the universe and that this orientation indicates, to put it cautiously, a special dignity or a special degree of compliance with the order of the cosmos as a whole. Plato's *Timaeus* puts this thought as follows:

> Now we ought to think of the most sovereign part of our soul as god's gift to us, given to be our guiding spirit. This, of course, is the type of soul that, as we maintain, resides in the top part of our bodies. It raises us up away from the earth and toward what is akin to us in heaven, as though we are plants grown not from the earth but from heaven. In saying this, we speak absolutely correctly. For it is from heaven, the place from which our souls were originally born, that the divine part suspends our head, i.e., our root, and so keeps our whole body erect.[11]

Clearly, this is a different genre of text and not written in the scientific spirit of Aristotle's biological treatises; still, it includes claims that are pertinent to the Aristotelian attempts to make sense of the human posture. Not unlike Aristotle, the *Timaeus* passage presupposes that the erect posture of humans is unique among sublunary living beings. The reference to the most sovereign part of the soul makes clear that the human intellect is meant to play a central role in the account to be given. Since, however, Plato locates this part of the soul – unlike Aristotle – in the top part of the human body, the head, the rest of the account follows quite smoothly. Having postulated a kinship between human beings, or, to be precise, the most sovereign part of the human soul, introduced as "god's gift," and what resides up in heaven (i.e. god or the divine), the argument makes use of the 'attraction of like by like' principle; this attraction is strong enough

[11] Plato, *Timaeus* 90a, translation by Zeyl/Cooper, see Cooper 1997.

to pull up the human body away from the earth and to make it stand uprightly, so that the head points to the heaven and is closer to it than the lower parts of the same body. The same idea is finally illustrated and amplified by the picture that through their heads (and, thus, through the most sovereign part of their souls) human bodies are rooted, as it were, in heaven, where the souls originally come from, just like plants are rooted in the earth.[12]

How does this relate to the interpretation of *PA* IV.10? Owing to his cardiocentric stance, Aristotle clearly cannot use exactly the same pattern of argument (for either thinking does not have a proper location at all, assuming that there cannot be a location for the incorporeal *nous*, or thinking takes place in the region where the soul and the common sense are located, which would require the chest, not the head, be the part that is closest to heaven). Still, the passage from the *Timaeus* offers a model for explaining the erect posture through the human intellect and, more than that, brings in a cosmological perspective in that it turns the human posture into a matter of the human body's position relative to the cosmic directions. And this latter kind of approach is also taken by Aristotle – though of course without Plato's peculiar premises about how the various parts of the soul are distributed to several bodily regions.

In Aristotle, the background to this cosmological approach, as it were, is his theory of dimensions and the distinction between cosmic dimensions on the one hand and the functionally defined dimensions on the other. The main source for this theory is Aristotle's *De Incessu* 4, 705a28–b8, where he points out that living beings have an up and a down and that this distinction is not merely defined by their position relative to the earth and the heavens. This means that the up for animals is defined as the place from which food is distributed, while the down is defined as the place where the distribution of food, that is, its digestions, ends (so that up and down in this functional sense become connected with the functions of the nutritive soul). That this functional sense of up and down does not always correspond to what we usually take to be up and down, namely 'down' as what touches ground and 'up' as what points to the heavens, becomes most obvious in the case of plants, since their functional up, the place where they draw nourishment, can be found in their roots, which are positioned in the ground and thus mark the 'down' with reference to the position relative to the earth and the heavens.

12 For a thorough discussion of these claims, see Gregoric 2005.

> Furthermore, all living beings have the up and the down, for the up and the down are found not only in animals but also in plants. Now, this distinction is one of function, and not merely of position relative to the earth and the heavens. For that from which food and growth is distributed is the up, while that to which it is distributed and in which it ends is the down. The former is some kind of origin; the latter is an end. The up is an origin, although in the case of plants it might seem that the down is more appropriate as origin. But that is because plants do not have the up and down in the same position as animals. With respect to the universe plants do not have the up and down in the same way as animals, but with respect to their function they do. The reason is that roots are the up for plants.[13]

Plants, which inhabit the lower region of the *scala naturae*, have their functional 'up' where animals usually have their functional 'down,' thus reversing the order of the directional sense of 'up' and 'down' relative to the heavenly sphere and the earth. Not being aligned to the cosmic and natural directions might be taken as a (further) indication for plants' inferior rank among living beings; by the same token, being fully aligned to the directions of the universe can be seen as an indication of an animal's superior rank in the *scala naturae*.[14] For the modern reader, this might be a somewhat strange criterion for ranking living beings; it is, however, exactly the principle that Aristotle highlights in the following section:

> In human beings more than in any other animal the upper and the lower parts of the body are determined in accordance with what is naturally upper and lower: in other words, upper and lower in man correspond with upper and lower in the universe itself . . .. Of course, in all animals the head is up above with regard to the creature's own body; but, as I have said, man is the only animal which, when fully developed, has the head above in the sense in which "up" is applied to the universe.[15]

In other, nonhuman animals the head is up above in relation to the rest of their bodies (which might be taken as an indication that they are superior to plants), but only the orientation of human beings is fully aligned to the directions of the cosmos, with their heads pointing to the uppermost sphere. Therefore, the orientation of the human body fully corresponds to the *natural* up and down, which suggests that other beings that fail to match the natural directions are inferior in this respect. Aristotle even goes

[13] *De Incessu Animalium* 4, 705a28–b6, translated by A. Falcon. For a thorough discussion of this passage, see Dimas 2021.

[14] This thought is well put by Dimas 2021: 136: "So the degree to which the up of a species inhabiting the sublunary region is aligned with that of the universe tells us something about its nobility. The animal species whose functional up is best aligned with the positional up is indeed the noblest."

[15] *Historia Animalium* I.15, 494a26–b1, Peck's translation slightly altered.

one step further by associating this *natural* orientation with the thought that they partake in the divine sphere:

> Such is humankind; for of the animals known to us either humankind alone, or humankind most of all, partakes of the divine. So both because of this and because the shape of the external parts of humankind is most familiar, one ought to speak about humankind first. For straight away the natural parts are disposed according to nature in this kind alone, that is, what is above for humankind accords with what is above for the whole cosmos; for humankind alone among the animals is upright.[16]

It is not perfectly clear in this passage whether human beings' kinship to the divine is meant to be grounded in their erect posture or in some other peculiar characteristic (the section summarizes the thought that human beings alone partake in living well); however, the reference to the erect posture that accords with the directions of the cosmos directly follows upon the remark about the divine nature and is clearly meant to ground the paradigmatic, standard setting role of human bodies. Aristotle's interest in relating the functional and the directional sense of 'up' and 'down' provides the background for the attempt of **[T1]** to associate human beings' divine nature with their upright position. In light of this cosmological context, one might be inclined to say that "man's divinity is not a consequence of his intellectual capacities alone, but has something to do with his spatial orientation,"[17] even though it is one of the main ambitions of *PA* IV.10 to prove that intellectual capacities of human beings and their upright posture or orientation are inextricably connected.

13.4 Thought, Weight, and Bodily Proportion

Let's go back to *PA* IV.10's main train of thought! Owing to their divine nature, it is human beings' function to think and to be intelligent; however, this, namely thinking and being intelligent, is not easy, Aristotle says,

> **[T2]** [A] when much of the body is pressing down from above, since the weight makes thought and the common sense averse to change (τὸ γὰρ βάρος δυσκίνητον ποιεῖ τὴν διάνοιαν καὶ τὴν κοινὴν αἴσθησιν). [B] For this reason, when their weight and bodily character become excessive, it is necessary that their bodies incline towards earth, so that for stability nature placed forefeet beneath the four-footed animals, instead of arms and hands.

[16] *PA* II.10, 656a7–13, Lennox's translation slightly altered.
[17] Freudenthal 1995: 63. See also Preus 1987.

> For it is necessary that all those able to walk should have two hind limbs, and such animals become four-footed because their soul is unable to bear the weight.[18]

Here, Aristotle connects thinking and upright posture by employing a partly anatomical explanation. If humans were not upright, then the capacity to think or the activity of thinking[19] would be impaired or even obstructed by a weight pressing down from above. This remains somewhat obscure as long as we do not assume that the mentioned weight would press upon a specific bodily region or upon specific bodily parts that are in one way or another crucial for thinking. According to standard Aristotelian doctrine, the heart is such a bodily part, for the heart is said to be the place where the central or common sense and the soul are located. Thinking would become 'averse to change' and thus, probably, less versatile, less versed or accomplished, when this particular bodily region is jammed under a heavy weight.

Maybe this is the right moment to do away with a fundamental objection. In *De anima* III.4, Aristotle claims that *nous*, thinking/understanding, does not require a specific bodily organ and that the plasticity of *nous*, which is supposed to be receptive to any thinkable content whatsoever, would be impaired by the materiality of the organ. How is it possible, then, one might object, that thinking is impaired by a weight that presses upon a particular bodily organ? There is a lot one could say about this. In the most condensed way, the answer to this objection should mention two points. First, 'thinking' does not univocally refer to the mental operation that is the subject matter of *De Anima* III.4–6; it might also be used in a broader sense for several kinds of intelligent activities that involve sense perception, memory, or conative states, all of which are clearly bodily. Intelligence in the sense of learning from past experience involves thinking about experiences that are somehow stored in the thinker's body. Second and more importantly, the thesis of *De Anima* III.4 does not imply that thinking takes place without any body or bodily organ. Most notably the famous theorem that thinking requires *phantasmata* links thinking with the bodily storing and transmission of residual contents of sense perception. Hence, in order to make sense of the claims of [**T2**.A], it suffices to assume that if the bodily organs that are

[18] *PA* 686a29–b2, Lennox's translation slightly altered. "Averse to change" for the Greek *duskinēton* has been suggested by Gregoric 2007: 92.

[19] Actually, he speaks of thinking *and* the common sense; on the role of the common sense in the context, see Gregoric 2007: 96–98.

typically involved in the processing of *phantasmata* or any other contents of memory are impaired, the kind of thinking that is dependent on them (and the development of intelligence that involves thinking) gets equally impaired or obstructed.[20]

Even if this much is granted, one might wonder how all this is related to upright posture. How is the amount of weight that presses upon the animal's heart related to different postures? The argument is constructed as though upright posture would imply little to no weight and horizontal posture excessive weight. However, it is not the case that an upright posture as such would guarantee zero weight pressing from above. Imagine a circus elephant who has learnt to stand on his hind legs only: There is a considerable weight pressing on this heart both in the horizontal and in the upright posture; at best, the difference between the two positions is a matter of degree. Text **[T2.B]** builds some kind of bridge to the issue of posture, but instead of directly comparing the effective forces in upright and horizontal postures, Aristotle uses the supposed explanatory power of his model of excessive weight to account for the occurrence of quadruped animals. Excessive weight explains not only why in some animals the central organ, where the soul is located and which is crucially involved in many higher cognitive activities, gets impaired, but also why the same animals are quadruped in the first place. The argument in [B] takes on a natural-historical tone, strongly reminiscent of Plato's *Timaeus*, pointing out that some animals start to bend toward earth due to excessive weight and that nature at this point provided front legs in order to stabilize their position. All this leads us away from the topic of thinking.[21]

In the passages that immediately follow upon **[T2]** it turns out that the references to weight must be read against Aristotle's theory of bodily proportions that he starts unfolding immediately after the quoted passage.

[20] For the broader background of this question, see van der Eijk 1997 and Connell 2021b; for more detailed discussions of the physiology of thinking in *PA* IV.10, see Coles 1997 and Gregoric 2007: 83–98.

[21] Like all commentators, I am perplexed by the final remark in section **[T2.B]**, that in quadrupeds the soul is unable to bear the weight. One would have expected that the rear legs in quadrupeds are unable to bear all the weight alone. The idea that the soul, or perhaps the bodily region where it is located, plays a mechanical or quasi-mechanical role for the stability of the body is not as outlandish as it may seem, if we consider the argument of *De Motu Animalium* 9, where Aristotle argues that for the soul to move the body it is crucial that it be located in the middle of the body. Still, it remains unclear how this could elucidate the existence of quadrupeds. Alternatively, one might want to relate this obscure remark to the thought of section **[T2.A]** that certain cognitive soul functions are impaired under certain bodily conditions; in this latter sense, the soul could not bear the weight because its functioning would be impaired or obstructed.

In fact, he is only interested in one such proportion, namely the relation of the upper to the lower part of a body, where the upper part is defined as the thorax that extends from the head to where excretion takes place and the lower part as the rest of the body (see also *HA* II.1, 500b26–30). If we imagine, for example, a hedgehog standing on its rear legs (after all, this is the position that, according to Aristotle, *HA* V.2, 540a3–4; *GA* I.6, 717b29–31, they use to mate, in order not to sting each other with their prickles), it becomes immediately obvious that, according to the given definition, its upper part is disproportionally big and bulky compared to its lower part, only consisting of the hedgehog's skinny and short legs. Aristotle describes animals that exhibit this disproportion as "dwarf-like" (*nonōdes*); it turns out that all animals, quadrupeds, fish, birds, etc., except fully developed adult human beings are "dwarf-like" in this sense, even human children.[22] Only adult human beings are said to be well-proportioned in this sense, since in a human being the legs together with the pelvic region are of a similar size and mass as the thorax, and not disproportionally small as in the case of the hedgehog. This very rough scheme is then used to provide the piece of explanation that we found to be missing in the account given in **[T1]**.

> **[T3]** [A] And because of this all animals are less intelligent than human beings. [B] And in fact among human beings, children compared with adults, and among adults in their prime the naturally dwarf-like, are deficient in the possession of reason (τῷ τὸν νοῦν ἔχειν ἐλλείπουσιν) even if they have a surplus of some other potential. [C] And a cause of this deficiency, as was said before, is that the origin of the soul (ἡ τῆς ψυχῆς ἀρχή) is, in very many animals, averse to change and bodily.[23]

In [A] it is confirmed that it is here, in the proportion of the upper to the lower part, that we have to look for the explanation of why human beings are more intelligent than all other animals. Before he elaborates on this

[22] The portion of animals that, on this account, does *not* qualify as 'dwarf-like' thus becomes amazingly small; in *On Longevity* 4, 467a32–33, Aristotle even claims, curiously enough, that male animals in general are 'more dwarf-like' than their female counterparts (which phenomena does he want to capture in the first place? Men's proverbial broad shoulders? Men with a paunch or with a flat butt?). Does this mean that women are more intelligent than men? (see Connell 2021e: 8) This would be a highly welcome compensation for some of Aristotle's weird claims about women in the *Politics* and elsewhere. Aristotle himself, however, does not draw this conclusion. According to thesis (i), however, as defined further down in Section 13.4, he would be committed to admitting that people with a proportionally lighter trunk would be better *disposed*, at least, *to develop* intelligence; the significance of these differences is limited, however, by thesis (ii), according to which it is the upright position (exercised by male and female humans alike) that primarily matters for the disposition to develop intelligence.

[23] *PA* 686b22–686b28, Lennox's translation slightly altered.

thought in [C], Aristotle adds a brief remark on human children and adult human beings with 'dwarfism' in [B], demonstrating that the explanatory model chosen is supposed to account not only for different degrees of intelligence across different species of animals but also for different degrees of intelligence among human beings. This might work for small children, but it is less clear where Aristotle gets his views about human adults with smaller stature; he hastens to add that they may have other capacities abundantly.[24] In [C], Aristotle refers back to the brief account given in **[T2.A]** ('as was said before'), and the main message seems indeed familiar by now. This time, however, he does not refer to 'thinking' explicitly but to 'the origin/principle of the soul,' which is a bit misleading,[25] since it seems to refer to something that is the principle of the soul, while in fact it is the soul that is the principle. He might mean 'the place from which' (= origin) the soul moves and steers the body and all of its life-related functions. Such an origin, again, is the heart. If this region becomes "averse to change" and "bodily" (*sōmatēdēs*), it causes a deficiency in the animal's intelligence. This does not go much beyond the argument provided in **[T2]**; however, at this point of the argument the reader is supposed to understand that this kind of affection is – in some way or another – the effect of the disproportional size and bulkiness of the upper part of an animal body and that the same sort of disproportion is found in animals that have no erect posture.

It is time to summarize the argument thus far. Having a divine nature, human beings' function includes the capacity to think and to be intelligent. For this reason, human bodies need to have an overall constitution and orientation that facilitates, rather than impairs, this function ("need to have" in the sense of "hypothetical necessity"). What must such a suitable, intelligence-facilitating body look like? Since the region where the intellectual capacities are primarily located, namely the central region of the trunk, must not be strained by excessive weight and bodily matter, their bodily proportion needs to be such that the upper part is lighter (i.e. less 'bodily,' having less weight and less bodily matter) than the lower part. How is this connected with an animal's posture? When an animal's upper part has excessive weight, the upper bodily part of this animal species tends to lean toward the earth (so that these animals need front legs), while animals with permanent upright posture need to have a lower part that is

[24] In *De Memoria* 2, 453a32–b3, Aristotle points out that by the same token people with 'dwarfism' (*hoi nanōdeis*) have poor memory.

[25] See also Althoff 1992: 73 n. 46.

heavier,[26] for otherwise it would require some permanent effort to maintain this position.[27]

> **[T4]** So in order that it may easily carry the upper parts, which are light, nature, taking that which is bodily from the upper parts, added the weight to the lower parts. This is why it made buttocks, thighs, and calves fleshy.[28]

Having a lower part of the body that is heavier is a precondition for the upright posture (or, after all, for maintaining the upright posture more easily, for adult human beings with 'dwarfism' (*nanoi*) are "dwarf-like" (*nanōdē*) but still have an upright posture). In such a body, the upper part, that is, the trunk, is less fleshy and less heavy, which facilitates the development of higher intelligence. That much is clearly implied by the text.

Still, one might be puzzled which of the following two theses Aristotle ultimately wishes to defend: (i) Thinking and intelligence are impaired by a (disproportionally) heavy and bulky upper part, which, concomitantly, would make it difficult to maintain an upright posture. (ii) Thinking and intelligence is impaired in the case of quadrupeds by the weight that presses on the heart *due to their horizontal orientation* (while not impaired in animals with upright posture). Thesis (i) is implied by the reconstruction of the argument just given and also by Aristotle's claim that people with "dwarfism" are less intelligent (since it makes use of the principle "the (proportionally) heavier the upper part, the less intelligent the animal"). Thesis (ii) seems to be in the background of a formulation in **[T2**.A], namely that much of the body is pressing down *from above* (πολλοῦ τοῦ ἄνωθεν ἐπικειμένου σώματος), which seems to appeal, not to the (proportional) weight and bulkiness of the upper part as a whole but to the position of the heart relative to the greater part of the body's weight. According to thesis (ii), the question is not whether the heart is located in a (proportionally) lighter or heavier trunk but whether the greater part of the weight is below the position of the heart or whether it presses from above; and this again might be a question of the posture *alone*. At least, Aristotle does not clearly distinguish between (i) and (ii), and perhaps he is even happy to endorse a combination of both. This, the combination of (i) and (ii), would imply, among other things, that if an animal assumes a horizontal posture due to its disproportionally heavy upper part, its horizontal orientation will double the thought-impairing load in that (not only the upper part is proportionally heavier by nature, but) the main part of

[26] *De Incessu Animalium* 11, 710b5–8. [27] *De Incessu Animalium* 11, 710b9.

[28] *PA* IV.10, 689b12–15, Lennox's translation slightly altered.

the weight won't be below the place where the soul is located. Maybe this combination of (i) and (ii) has other interpretative advantages as well. If Aristotle took (i) in isolation, he would have to invoke the (proportional) weight and bulkiness of an animal's upper part to account for different degrees of intelligence in different species of nonhuman animals; on this account, for example, the thinking of a greyhound with a slim body would be more volatile or flexible than the thinking of a bulky elephant, but Aristotle never uses the explanatory model of *PA* IV.10 to that end. The decisive factor for the question of intelligence rather seems to be whether an animal has an upright posture or not, and this kind of reasoning is better accounted for by thesis (ii). Aristotle's remarks about people with 'dwarfism' draw on (i) when these people are said to be less intelligent; however, he does not remotely question that they have human intelligence, probably because, after all, they keep an upright posture, which according to (ii) is not only a concomitant of the lighter trunk but is decisive for a body in which the heart is not impaired by too much weight and thus facilitates the development of intelligence.[29]

To sum up, the human body is suitably shaped for facilitating intelligence, in that nature has allocated a greater share of weight to the lower part. This is why the upper part is lighter and less 'bodily' and why human beings can maintain their upright posture easily. The upright posture again is the reason why the central organ that is crucial for perception and thinking is high up in the body and is hence not pressed down by the weight of the trunk.

13.5 The Uprighting Power of Heat

[T2] and **[T3]** together with the theorem about different bodily proportions seem to provide a more or less self-contained and complete explanation for how the upright posture is connected with intelligence (as suggested in Section 13.4). However, **[T3]** is directly followed by an argument that brings in an entirely different route:

> **[T5]** And further, as the heat which rises becomes less and the earthen material becomes greater, the bodies of animals grow smaller and many-footed, and finally become footless and stretched out on the earth. Proceeding in this way a little, even their origin is below, and the part

[29] For another application of this see n. 22.

> corresponding to the head is in the end unable to move and perceive, and a plant comes to be, having the above below, and the below above.[30]

Again (as already in **[T2.B]**), Aristotle shifts to a natural-historical narration that takes human beings as the starting point and the gold standard, so that the generation of quadrupeds and then footless animals appears as the result of a 'degenerative process' (Lennox 2001a: 320), which in the end – following the *scala naturae* downwards – leads even to plants.[31] Since Aristotle does not endorse any kind of evolutionary theory – much less a 'reverse' evolution from human beings to plants – one cannot help but thinking that this is an allusion to and, indeed, an imitation of the degenerative chain of incarnations provided in Plato's *Timaeus* 91d–92c.[32] In Plato, this process is driven by people who neglect philosophy and hence fail to develop the rational part of their souls that is located in the head; as a consequence of this lifestyle, the part of the soul located in the chest becomes predominant and their bodies incline toward earth, so that god provides an increasing number of supporters, that is, legs and feet, under their more and more mindless bodies up to the point that their bodies become legless and fully stretched out along the ground. By imitating this sort of narrative, Aristotle is able to highlight how he, as opposed to Plato, is able to offer a physiological-anatomic explanation for the differences in the posture of various animals. Also, he inverts the explanatory direction, for while in the Platonic version the neglect of using the rational powers, that is, a defective lifestyle, gives rise to the series of degenerated bodies, Aristotle's account is interested conversely in the anatomic preconditions for an unimpaired development and exercise of the rational powers.

Right at the beginning of **[T5]** Aristotle brings in a new thought, namely that "the heat which rises becomes less and the earthen material becomes greater." It stands to reason that this text suggests a certain correlation between the decrease of heat and the increase of earthen material and that the increase of earthen material is meant to be concomitant to the phenomenon mentioned in the immediately preceding **[T3.C]**, namely that the heart becomes more 'bodily.' Assuming that this is so, the seemingly new explanatory element in **[T5]**, that is, the reference to heat and earthen material, would not be independent from the proportion

[30] *PA* 686b28–34, Lennox's translation slightly altered.

[31] On the functional up and down of plants, which inverts the cosmic directions, see Section 13.3.

[32] See Lennox 2001a: 319; Gregoric 2005: 191f.

theory summarized in **[T3]**. But how do these two pieces of explanation fit together? There are, it seems, two possible ways to proceed.

First, there is abundant evidence that for Aristotle the rising heat plays a crucial role in the upward growth of animals. For a body to become upright, it must grow upwards, and in order to grow upwards it needs more heat than other animals, probably because it is natural for heat to move upwards. Therefore, there is also a direct connection between heat and upright posture.

> This is also why human beings alone among animals are upright; for the nature of the prevailing heat produces growth from the middle according to its own movement.[33]
>
> For heat promotes growth, and a profusion of blood is a sign of heat. And further, the bodies of those that are hotter are more erect, which is why humankind is the most erect of all the animals.[34]

The message is clear: the bodies of those that are hotter are more erect. Is this an alternative or even competing explanation for the upright posture of human beings?[35] Rather than competing, it might be seen as a complementary explanation, for, in addition to the teleological explanation of why human beings need to have a body with upright orientation (because they have superior intelligence), it offers a material cause for why some bodies grow upwards, while others do not. The extra portion of heat in these bodies is required – if not for other purposes – because the body is to grow upwards and to assume an upright posture.

Second, it is tempting to think that heat shows up in this passage because elsewhere, for example, in *PA* II.2 and 4, heat is also mentioned in the physiological account of thinking and character, which is obviously related to the topic of our chapter. Strictly speaking, however, *PA* II does not arrive at the simple formula 'the hotter – the more intelligent'; rather it is a fine-tuned mixture of purity, thinness, and a certain temperature of blood that gives animals a sharper sense-perception, a better intelligence, or a more courageous character.[36] So if there is a connection between the amount of heat in the human bodies and humans' capacity to think – beyond the fact that heat is among the material causes for their upright posture – it has to be a more subtle one. The most promising path is probably this: If the rising heat diminishes the earthiness of an animal's body (or, after all, the earthiness of its upper part), the heart or the

[33] *PA* II.7, 653a30–32, Lennox's translation slightly altered.

[34] *PA* III.6, 669b3–6, Lennox's translation slightly altered.

[35] See Kullmann 2007: 694.

[36] See *PA* II.2, 648a2ff. and II.4, 650b19ff.

neighbourhood where it is located will also be less earthy and, perhaps, this is the sense in which the 'origin of the soul' can become less bodily (according to **[T3.C]**). Earthiness, the presence of earthy matter, in Aristotle's view, clearly affects and impairs the quality of the upper sense organs and of the blood.[37] Less earthiness is hence better for thinking and, indeed, human beings are said to have less earthy matter than other animals.[38] It remains to account for how earthiness and heat are supposed to be related. In general, earthy matter in an animal's body is the residual substance that remains unconcocted and the process of concoction is more complete, that is, leaves less residue, the hotter the body is. What about the peculiar role of the *rising* heat? Perhaps the point simply is that an animal that is not erected by heat becomes "earthbound"[39] and hence has more earthy matter. Or perhaps the rising heat makes it more difficult for earthy or "fleshy deposits to gather in the upper and middle regions of the body."[40]

None of these options, of course, is mentioned or alluded to in our passage **[T5]**. It is uncontroversial that heat is meant to be responsible for the upward growth. It is most likely that heat affects the quantity or location of earthy matter within the body and that this, again, is the background for the claim in **[T3.C]** that the origin of the soul becomes more bodily. Aristotle emphasizes twice that the central region becomes more 'averse to change' (due to the weight **[T2.A]** or in accordance with increased earthiness **[T3.C]**). Apparently, this is Aristotle's preferred explanatory path in *PA* IV.10. The idea seems to be, at any rate, that thinking becomes less versed or accomplished when the bodily part that is most intimately connected with this activity is less versatile. Even though there is no explicit hint to the above-mentioned theory of blood mixture from *PA* II.2 and 4, one might see a parallel to this theory in that thick blood, which contains more earthy fibers and thus tends to solidify more easily (and is hence, most probably, more averse to change), is similarly said to be less suitable for thinking.

13.6 The Upright Posture as Explanans

After a bit more than one Bekker page (approximately a fifth of the total length of *PA* IV.10), Aristotle thinks to have shown why some living beings are biped, some quadruped and some without feet and why human

[37] For a discussion of some pertinent passages, see Freudenthal 1995: 48ff. [38] *GA* II.6, 745b20.
[39] Lennox 2001a: 320. [40] Coles 1997: 309.

beings alone have an upright posture (687a2–5). Up to this point, the peculiar human posture was the explanandum; it was explained by the humans' divine and intelligent nature together with the claim that such a nature requires a body with an upright orientation (see Sections 13.2–5). Once this has been established, Aristotle resumes his original program of discussing the external parts of sanguineous, live-bearing animals from the top to the feet. In this discussion, he continues to highlight the distinctive features of the human body and it seems that all these distinctive features are ultimately traced back to the erect posture: the shape of human hands, the broadness of the human chest, the position of the nipples, the lack of a tail, the bulkiness of the lower extremities and the buttocks, and finally the size of the feet. This means that from this point on (i.e. from 687a5), the erect posture is no longer the explanandum but appears regularly in the explanans of the position and characteristics of the mentioned parts.

Let's start with the feet and work our way upwards! Among quadrupeds, Aristotle distinguishes solid-hoofed, split-hoofed, and many-toed animals. Owing to the size and the amount of earthen matter, nature has allocated proportionally more matter to the hoofs of solid-hoofed animals than to their horns and teeth (690a5–9). Human beings have the proportionally largest feet – with good reason (*eulogōs*), for they alone stand upright, so that the (two) feet have to bear all the weight (690a26–32). For the same reason – because the feet have to guarantee stability of the erect posture – it is useful that the toes are rather short, and shorter than the foot's unsplit part; also, shorter toes are less vulnerable than lengthy ones, and it is also good that the human feet are split into toes, for if they were not, everything that hurts a part of the foot would at once hurt and impair the entire foot (690a29–b7).

Speaking of the animals' lower region, Aristotle addresses the tail and the bottom (689b1–12); while most quadrupeds have a tail, human beings do not, but have fleshy buttocks, comfortable to sit upon, which proves to be useful given that human beings, unlike many quadrupeds, cannot stand on their feet all day long. Also, human beings have fleshy legs, in particular fleshy thighs and shanks, while the corresponding parts of other live-bearing animals are often bony and spiny. All this contributes to the typical human proportions (see Section 13.4) and "the only reason, as it were, for all of this" (689b10–11) is the fact that, of all animals, human beings alone have an upright posture.

The chest is located between the arms or between the forelimbs. The human chest is broader though than in quadrupeds. Why is this so? For the human body, the sidewards-attached arms provide no obstacle for the breadth of the chest, whereas in quadrupeds the chest has to be narrow in order to facilitate the forward extension of their forelimbs when they walk and change place (688a10–17), so that, again, it is the upright posture of human beings that explains the difference. Moreover, this is also the reason why it is impossible for nonhuman animals to have their nipples on the chest between the forelimbs; for, obviously, this would be an impediment for progression (688a28–31). For human beings, by contrast, nothing (in particular no forelimbs required for progression) impedes the nipples from being attached to the chest, where, as Aristotle notoriously insists, they contribute to the protection of the region of the heart in male and female animals alike. In female animals alone has nature assigned the additional function of breastfeeding to the chest and the nipples (688a19–25).[41]

The philosophically most challenging discussion in this context is dedicated to the role of human hands.[42] That there is a peculiar connection between the upright posture, the use of arms and hands, and the development of higher intelligence is all but far-fetched; even in modern evolutionary biology the so-called Freeing of the Hands Hypothesis is considered to pinpoint a contributing factor to the bipedalism of some ancestors of human beings.[43] What matters for the finer-grained comparison between Aristotle and the modern point of view however is the causal direction. Aristotle does not argue, as the modern advocates of the "Freeing of the Hands Hypothesis" would do, that human beings assume an upright posture in order to get their hands free. According to Aristotle, human beings have arms and hands in the first place, *because* they have no need for front-legs in order to stabilize the position of the thorax. Therefore, nature was free to assign a different function (and, accordingly, a differently shape) to the human forelimbs (see Section 13.2), meaning that, within the Aristotelian framework, naturally grown animals are expected not to possess parts that have no function whatsoever, so that the forelimbs of human adults – given that they are not needed for stabilization and progression – were available to perform some other useful function. Hands are, according to Aristotle, an instrument for (the use of)

[41] With this move, Aristotle gets rid of the burden of explaining why male animals that are, as is well known, unable to breastfeed their offspring have nipples at all, which seems to be an objection to the teleological-heuristic postulate that "nature does nothing in vain." Aristotle's student and fellow researcher Theophrastus by contrast claims (*Metaph.* 10b7–9) that nipples of male animals are purposeless.

[42] See also the contribution by Emily Kress, Chapter 14 in this volume.

[43] See Niemitz 2010. Having the hands free enabled the ancestors of human beings, among other things, to reach out for food and to carry food, to carry infants, to throw things, etc.

instruments,[44] that is, an unspecialized, universal instrument that is useful for a wide range of different purposes (687a19–23) and enhances human beings to execute many different skills or arts (*technai*). Accordingly, the shape of human hands is such that "they can grasp and hold everything" (687b4–5). From Aristotle's perspective, then, the possession of arms and hands is explained by human beings' upright posture, not the other way around.

Similarly, Aristotle neither accepts Anaxagoras' view that human beings are intelligent and, indeed, the most intelligent animals, *because* they have hands[45] (which would not be patently implausible in light of Aristotle's view that hands are like universal instruments – if one only adds the idea that such unspecialized instruments might trigger the development of all kinds of capacities that are required to make intelligent use of them). It is reasonable to assume, says Aristotle, that it is the other way around, namely,

> [**T6**] that it is because they are most intelligent that human beings are given hands. For the hands are instruments and nature, like an intelligent human being,[46] always apportions each instrument to the one able to use it. Surely it is more fitting to give flutes to the flutist than to provide the ability to play flutes to one who has them; for nature has provided the lesser to the greater and superior, not the more honourable and greater to the lesser. So if it is better thus, and nature does, among the possibilities, what is best, it is not because they have hands that human beings are most intelligent, but because they are the most intelligent of animals that they have hands.[47]

That human beings possess hands is thus explained by their intelligent nature (and, in another way, by their upright posture, which is similarly grounded in their intelligent nature), for only intelligent beings are able to deal with and deserve to possess such subtle instruments. The peculiarly human capacities do not flow from the instruments they happen to have; rather they have such subtle instruments owing to their peculiar capacities. Against the picture provided, for example, in Plato's *Protagoras* (321c), which suggests that human beings are, in Aristotle's words, "not well constituted" but have a deficient nature,[48] since they are "barefoot, naked and without weapons for defense" (687a23–26), Aristotle builds on the

[44] See also *De Anima* III.8, 432a2–3. [45] Listed in Diels/Kranz as fr. 59A102.

[46] Does this formulation reveal "a bit of humour" (Lennox 2001a: 320) or is it a metaphorical expression of the rationality of nature (Kullmann 2007, 697)? See also Aristotle's metaphor that nature acts like a good housekeeper in *GA* II.6, 744b11–27.

[47] *PA* IV.10, 687a9–18, translation by Lennox.

[48] Or are 'Mängelwesen' as the German philosopher Arnold Gehlen put it.

idea of the multifunctional use of human hands. While other animals have only one weapon or one kind of protection and cannot get rid of the singular kind of protection they happen to have, human beings can have many weapons and many kinds of protections, for the multifunctional hand can become "a talon, claw, horn, spear, sword and any other weapon or instrument" (687b2–4). What people might conceive as a deficient equipment of human beings turns out to be a true advantage, for the absence of specialized weapons and protections is more than compensated by the possession of a nonspecialized instrument, which, again, can only be used by an intelligent being.

CHAPTER 14

The Human Model
Polymorphicity and Scientific Method in Aristotle's Parts of Animals

*Emily Kress**

Introduction

Parts of Animals II.10 makes a new beginning in Aristotle's study of animals. In it, Aristotle proposes to "now speak as if we are once more at an origin, beginning first with those things that are primary" (655b28–9).[1] This is the start of his account of the nonuniform parts of blooded animals: parts such as eyes, noses, mouths, etc., as opposed to uniform parts like blood and flesh.[2] *PA* II.10 proposes a new strategy for studying these parts: "one ought to speak about the human kind first" (656a10).[3] Beginning "first" with the "primary" things thus amounts to beginning with humans (655b28–29).[4]

Why does Aristotle think this strategy is appropriate for his project? One answer is that it reflects a fundamental anthropocentrism. Lloyd, for instance, has raised the possibility that the basic assumptions that guide Aristotle's biology may to some extent reflect the anthropocentrism of his culture, where the relation of humans to animals was "a preoccupation of popular beliefs."[5] In this chapter, I develop an interpretation that both

* For astute comments and suggestions, I thank Sophia Connell, Jessica Gelber, Kosta Gligorijevic, Robbie Howton, Giouli Korobili, Jim Lennox, Mariska Leunissen, Diana Quarantotto, and Michael Ziegler. I am also appreciative of the thoughtful conversations at the Cambridge conference and to any interlocutors I may have inadvertently omitted here. Ancestors of some of these ideas were presented (by Skype) at a workshop on the *PA* at the Munich School of Ancient Philosophy.

1 Translations of the *PA* are from Lennox 2001a, with modifications. The Greek is from Bekker 1831. Translations of other texts are influenced by the *ROT* and Loeb editions.

2 II.10 begins with blooded animals' external nonuniform parts (on the head). III.3 transitions (eventually) to their internal parts, then to bloodless animals' internal and external parts. IV.10 returns to blooded animals' external parts. Cf. Lennox 2001a: 220, 254, 292–293, 305, 314; Gotthelf [1987b] 2012: 158–164.

3 The method is new to *PA* but also used elsewhere: *HA* I.6 (491a19–23) and *GA* II.6 (737b25–27).

4 As Gligorijevic has emphasized to me. Cf. Quandt 1983: 368.

5 Lloyd 1983: 42. Lloyd emphasizes the human's exceptional status (26–35) and the role of hierarchical assumptions (35–37); cf. n. 23. For criticism, see Lennox 1985: 308–315. On hierarchy and anthropocentrism, cf. Pellegrin 1986: 91ff.; Müller 2019: 119–120; Clark 1975.

builds on and challenges this suggestion. I do this by investigating how Aristotle thinks this strategy works in theory and in practice: his justifications for adopting it and its interaction with his scientific commitments. I argue that he adopts it in part because he thinks (possibly wrongly) that humans are such that his scientific concepts apply to them in a particularly clear way. (This too is a form of anthropocentrism: one that might be resisted by arguing that humans are not the best illustrations of these concepts – that he ought to look elsewhere for models with the features he desires.)[6]

More specifically, Aristotle holds that starting with humans helps establish the causal explanations of the parts of other animals, particularly when they are recalcitrant. What makes humans suitable for this role is a special teleological relation between their parts and the ends they serve: in humans, he supposes, this connection is particularly tight (in a way I explain below). Consequently, Aristotle thinks he can use humans to illuminate which sorts of features tend to be for the sake of which ends and then extend the results of this inquiry to the parts of the other animals.

14.1 The Human Model

The context in which Aristotle introduces his new strategy provides important clues to how he understands its purpose. *PA* II.10 begins by identifying three particularly important nonuniform parts: that "by which they receive nourishment," that "by which the residue departs," and "a third part present in all animals [that] lies midway between the two most necessary parts, within which is their origin of life" (655b29–37). However, not all living creatures have all these parts. Whereas most animals (at least, the "complete" ones) have all three, plants "have no place for useless residue" (655b32–34). More generally, plants do not have as many distinct parts as animals do: it is "of the nature of plants, being immobile, not to have many forms of the non-uniform parts (*polueidēs tōn anomoiomerōn*); for few actions require the use of few instruments (*organōn*)" (655b37–656a2). For this reason, "we should study the visible character (*ideas*) of plants independently" (656a2–3).[7] While this observation allows Aristotle to set plants aside – as outside the scope of his treatise – its emphasis on the number of "forms" of the parts puts him

[6] I thank Connell, Leunissen, and Korobili for helpful questions.
[7] On the "prospects for a unified study of living beings," see Falcon 2015: 84–85.

in a position to make the distinctions that will structure his study of animals.

Indeed, Aristotle soon introduces a similar contrast among animals: between those that are more and less "polymorphic in visible character." This contrast provides one reason for adopting his new method:

> But those things with perception in addition to life are more polymorphic in visible character, and some of these more than others. And there is generally greater variety among those whose nature partakes not only of living but, in addition, of living well.[8] Such is the human kind; for of the animals known to us (*gnōrimōn*) either the human kind alone, or the human kind most of all, partakes of the divine. So both because of this and because the shape (*morphēn*) of the external parts of the human kind is most familiar (*gnōrimon*), one ought to speak about the human kind first. For straight away the natural parts are disposed according to nature in this kind alone, that is, what is above for the human kind accords with what is above for the whole cosmos; for the human alone among the animals is upright.[9] (656a3–13)

Living creatures thus fall on a spectrum of being more or less "polymorphic" (*polumorphoteran*) in their "visible character" (*tēn idean*); or, in the earlier formulation, their natures vary in whether they have "many forms" (*polueidēs*) of their nonuniform parts.[10] Humans, Aristotle thinks (possibly wrongly!), are most polymorphic, then other animals, then plants.

This polymorphicity clearly looms large in justifying Aristotle's claim that we should discuss humans first. But what is it? And why is it important? A first step is to note that *PA* II.10 does not merely observe that living things differ in this way. It also offers an explanation: they differ because of differences in the actions they perform (cf. I.5, 645b28–29). Whereas plants perform "few actions" (and so need "few instruments," II.10, 656a2), animals and especially humans lead more complex lives.[11] Humans, apparently, are "among those whose nature partakes not only of

[8] Cf. *DA* III.12, 434b22–29, 435b4–25; *Sens.* 1, 436b12–437a3; with Howton 2019; Leunissen 2010: 57–75; and Gotthelf 2012e: 53–55.

[9] Cf. *Juv.* 1, 468a4–13.

[10] While the two formulations use different vocabulary, they should be equivalent: the second formulation's reference to the *idea* ("visible character") picks up the observation that it is a consequence of the first formulation that we should study the *idea* of plants separately (656a3, 4).

[11] Cf. *GA* I.23, 731a24–9 and *De Caelo* II.12, 292b2–10 (we have the "most actions"). Thanks to Gligorijevic. On our multiplicity of actions and intermediate status in the cosmos, see Rapp 2019; cf. Osborne 2007: 117–122.

living but, in addition, of living well" (656a6). This is why they are more polymorphic: because they live well and perform more complex actions, humans have more parts of different sorts with which to do so.[12] Relatedly, IV.7 notes that some animals lack a "many-parted" (*polumeres*) body because of their "actions," whereas "those partaking of many motions are in need of many instruments" (683b4–8).

In explaining the polymorphicity of parts ("instruments") by reference to actions, II.10 extends a line of thought begun in II.1. There, Aristotle observed that the "uniform parts are for the sake of the non-uniform," in part because of facts about the "functions and actions" of the latter (646b11–13). These facts hypothetically necessitate further facts about the composition of the parts that perform them: "Since the actions and movements present both in animals as a whole and in their non-uniform parts are polymorphic, it is necessary for their components to have distinct capacities; for softness is useful for some things, hardness for others," etc. (646b14–8).[13] Nonuniform parts thus have the complexity they do because it is (hypothetically) necessitated by the complexity of the actions they perform. Sometimes, their complexity consists in being composed of multiple (uniform) components with different capacities: bones, sinews, flesh, etc. (646b18–27). Sometimes, it involves a "polymorphic" shape, as in some viscera (646b32–4; cf. 647b33).

This context helps us understand *PA* II.10's strategy. *PA* II.1's idea is that the complexity of nonuniform *parts* is hypothetically necessitated by the complexity of their functions; II.10 simply extends this observation to the creature as a whole: if a *creature* has complex functions, then that creature – its visible character (656a4) or its nature (655b37–656a1) – is more polymorphic in its nonuniform parts. While II.10 is less explicit, part of the idea may be that polymorphic natures have more nonuniform parts of distinct kinds (by contrast with plants' "few instruments," II.10, 656a2, and by analogy to the "distinct capacities" of the multiple components of nonuniform parts, II.1, 646a14–8). Alternatively, again by analogy to II.1, it may also be that (collectively?) the parts have a more polymorphic "shape" (646b32–34).

[12] The exact nature of this complexity is difficult, but see Leunissen 2010a on the distinction between living and living well and the suggestion (with reference to 2.10's *scala*) that "'well' indicates a more complex performance of the being's life functions" (62). At the conference, Korobili offered an insightful analysis of this complexity, seeing "formal" complexity as due in part to "material complexity." Cf. Karbowski 2019: 224–230 on functional and morphological complexity in a political context.

[13] Cf. Tipton 2013: 96–97.

This, according to II.10, is what humans are like: polymorphic in their uniform parts, because of the complexity of their actions. This is one reason Aristotle gives humans methodological priority: "both because of this and because the shape of the external parts of the human kind is most knowable, one ought to speak about the human kind first" (656a8–10).[14] In context, "this" picks up the claim that the human kind is polymorphic, a point just justified – in parallel to II.1 – by appeal to the actions it engages in (656a5–7), justified in turn by its participation in the divine (656a7–8).[15] It is because, Aristotle thinks, humans are polymorphic and because their parts are most knowable that we should discuss them first.

14.2 Why Does Polymorphicity Matter?

My question concerns Aristotle's first reason for starting with humans: our supreme polymorphicity. What connection, exactly, does he see between polymorphicity and methodological priority? Why not work our way up to humans, beginning with simpler creatures? A natural thought is that starting with humans is an organizational device. But what kind of device is it, and what makes it helpful?

One possibility is that humans provide a convenient table of contents for a treatise.[16] Aristotle might have thought that if humans possess a large variety of parts (if not *all*), enumerating them yields a comparatively thorough list. From this point of view, there is nothing special about human parts as such: any long, accessible list would do. Another possibility is that he thinks that starting with humans ensures that the organization of our inquiry reflects the organization of the natural world. Leunissen observes that "nature provides all kinds of hierarchies in its organization from which Aristotle can draw so as to find normative principles for the organization of his own writings in those cases where didactic or

[14] Contrast the order for the other external parts (*PA* 4.6, 682a31–33) and generation (*HA* V.1, 539a2–8).

[15] I agree with Lennox 1999 that the claim that we participate in the divine justifies the claim that we live well, which explains why we are polymorphic.

[16] While I am not sure the view is ever put in exactly these terms, this formulation provides a useful contrast to the other options. But note Gotthelf's observation that "the list is taken from man" ([2006] 2012: 160) and occasional observations that the method does not involve discussing humans as often as one might think or for the reasons one might have guessed, e.g. Lennox 1999: "while . . . the anatomy of mankind is regularly discussed, it is equally true that mankind is not at center stage" (8). Quandt 1983 discusses Aristotle's use of "one serialized list or another" to organize his biological treatises, though "sometimes he abandons one list for another" (361). For additional discussion: Kullmann 1999: 115–117; and Kullmann 2007: 448.

conceptual concerns yield no particular preference for discussing one natural entity or its attribute before another."[17]

While both suggestions capture something important, there is evidence that Aristotle thinks he has further reasons for adopting this organizational device. *PA* II.10 gives two justifications: greater knowability and *polymorphicity*. By contrast, when *HA* I.6 introduces a similar method, it claims only that the human is most knowable to us (491a19–22).[18] The *PA* and *HA* have different projects: roughly, the *PA* offers causal explanations for the facts reported in the *HA*.[19] While the first suggestion explains why polymorphic animals provide a thorough list, it does not explain why starting with them is particularly germane to the *PA*'s explanatory project. The second suggestion does better. It can allow that hierarchically ordered expositions have explanatory benefits: perhaps they reveal which parts of a given explanation should be "assumed for now" and which should be argued for here.[20]

However, we may go further: does polymorphicity have anything to do with the project of identifying the causes of animal parts? There is evidence that Aristotle thinks it does: that an organization that begins with polymorphic animals helps us construct causal explanations and present them in an illuminating way – one that makes manifest how this part is for the sake of that function.[21] Such an organization is not only convenient or in line with the hierarchy of the universe – though Aristotle probably thinks it is – but is also, in his view, uniquely suited to achieving understanding of the causal structure of the phenomena.

This perspective develops Carbone's observations that Aristotle often takes a visual and spatial approach to morphology and that the human body can play a role in identifying the "body plans" central to that thinking, as well as the idea that there is sometimes a "coalescence" between morphological and teleological concerns.[22] I build on this

[17] Leunissen 2017a: 177; cf. Leunissen 2018. Gotthelf and Falcon 2018 comment on the *GA*'s idea that "the maximum possible (sc. the most complex) outcome, namely the production of a human being, is the natural place to start" (25).

[18] Though see *HA* IX.1, 608b4–8: the "traces of the characters" – differences between males and females – are clearer in some animals and especially in humans, since its nature is most "completed." Thanks to Gligorijevic.

[19] See Lennox 2001a: 223; 2012.

[20] Cf. Leunissen 2017a: 169. Falcon 2015 also emphasizes explanatory concerns.

[21] I thank Jessica Gelber for suggestions about formulation.

[22] Carbone 2016: 27. Carbone notes that the fact that polymorphicity follows from functional complexity "means that the organism's shape needs to be explained teleologically, but it also means, from a methodological point of view, that studying the body plan provides the evidential support for a teleological explanation of the body's anatomy" (18). While Carbone's discussion is

approach, fleshing out the special role of polymorphicity and what *makes* it effective for identifying causal explanations. The answer appeals to an important feature of polymorphic animals: an especially tight correlation between a part's features and the end they serve. This makes it easier to illuminate *which* features are for the sake of *which* end. This explanation can be used to shed light on the causal explanations of the parts of other animals, particularly where they are recalcitrant: where they perform a given function but in peculiar ways.[23]

14.3 Identifying Causes

Understood in this way, *PA* II.10's methodological innovation is meant to play an important role in Aristotle's implementation of his research program. That program, laid out in *PA* I, centrally involves division and explanation: "first to divide the attributes associated with each kind that belong in themselves to all the animals, and next try to divide their causes" (I.5, 645a36–b3). Following Gotthelf and Lennox, I understand it as follows. One begins by identifying the *widest class* of animals possessing a given feature. This might be a generic (like feathers in birds) or an analogical (like eye protection across kinds) unity. One divides this class according to *differences* in how animals in it have the feature. Then, one

wide-ranging, its main focus seems to be visual, topological, and morphological thinking, including its role in explanation, via "la correspondance … entre la position des parties … et la fonction" (Carbone 2011: 16). Humans come in most explicitly via their role in identifying the *body plans* essential to that thinking, which then facilitate comparative anatomy (Carbone 2016: 19–21; Carbone 2011: 16, 49–50, 77ff., 107–108). The emphasis is less on the nature of polymorphicity and what *makes* it valuable for Aristotle's method (beyond the general relation between multiplicity of functions and of parts and how this makes identifying differences relevant to explanation, 78; cf. 50–51, 107, 158). I offer an explicit account of how polymorphic animals' teleological structure lets Aristotle use *their* parts as models (especially in terms of hypothetical necessity), in ways that do not always refer to body plans. To the extent that connections between body plans and teleology are particularly manifest in polymorphic animals, I am offering a new argument for Carbone's account of the role of topological reasoning and what makes it work. My account also explains Aristotle's use of the human model for identifying hypothetical necessity relations in parts serving multiple functions. Wilson 2009 briefly mentions the view that "in the biology hierarchies are explanatory and supply cause"; Aristotle's tendency is "to explain the varieties as deviations from a norm" (85); cf. Wilson 2013. Cf. Stavrianeas 2018: 52, 66 on hierarchical concepts that allow evaluations across species.

[23] I am sympathetic to claims that *analogy* has a methodological role, e.g., Henry 2014; cf. Lennox 2005b: 95–97. Lloyd 1983 notes how "Aristotle's interpretations of the role and function of various of the parts of the lower species of animals are influenced or even determined by doctrines derived from his study of the higher animals" (37; see 37–40). Meyer 2006 treats the human as a reference point for analogy and starting with it as a condition for zoology (27–28), helpful for understanding the functions of other animal parts (33). On the nature of analogy, see Frey in this volume (Chapter 8).

identifies the widest class that possesses each difference. In this way, one prepares to give corresponding causal *explanations* of the parts (639a15–b6, 644b1–7).[24] Nature, after all, produces "a differentiation of this part for the differences of its operation" (662a23–24). Explaining a part thus requires identifying the cause common to the parts in the wider class, as well as the cause of its variation.

This is the task of the *PA*. Importantly, it does not merely identify these causes but also sketches the reasoning behind its identifications. In Gotthelf's words, it offers "systematic discussions – arguments, actually – aimed at *establishing* what the cause (generally, the function) of the part in question is": "attempts to establish the definitions of these parts."[25] I will argue that Aristotle takes polymorphicity to be useful in constructing some such arguments.[26] The reason is that some parts are recalcitrant: it is hard to identify their causes, what they explain, and what this means for other features such parts might have.

My central example will be *PA* II.16's discussion of organs for smelling. Aristotle began investigating the sense organs in II.10, starting with humans. There, he noted that animals have nostrils in order to smell, and that in animals (like humans) with nostrils (rather than other means of smelling), their function is smelling "by means of breathing" (657a6–7; cf. *DS* 5).[27] This common cause explains certain features, including position (*PA* II.16, 656b26–657a11). *PA* II.11 turns from human sense organs to those of the "other animals," whose "sense-receptors are also well situated ... in relation to each one's proper nature" (657a11–12). *PA* II.16 tackles their organs of *smell*, turning from the common part and cause to the *differences*. There are few, aside from being positioned in the snout in long-jawed animals (658b27–33). Aristotle's account of organs of smell is thus in keeping with his program of identifying common and differentiated causes, starting with humans.

The remainder of II.16 introduces two hard cases. The second is certain "channels" for smelling:

[24] Gotthelf [1997] 2012: 193 n. 18; cf. Gotthelf [1997b] 2012: 206–208; Lennox [1987a] 2001b, cf. Leunissen 2010a: 114–115.

[25] Gotthelf [1987b] 2012: 164 and n. 35, citing Bolton 1987 and applications in Freeland 1990; Wilson 2009.

[26] Cf. Carbone 2011: 57–60, 65, 69, 77ff., 106, and *passim* on the usefulness of body plans for grasping differences among parts and in identifying them and their functions.

[27] Some animals smell through gills, pipes, or the mid-section (659b13–19). We might capture a common and differentiated *cause* by distinguishing kinds of smelling (through breathing vs. other means). Gelber 2015 understands "references to organisms' habitats" similarly, "as references to the specific or determinate way they have their vital capacities" (284); cf. Stein 2018: 44–45.

> The birds, the snakes, and the other blooded egg-layers among the four-footed animals have the channels of the nostrils in front of the mouth, but they do not have them clearly articulated so much so that one would say they are nostrils,[28] unless on account of function; but the bird, at any rate, has them in such a way that one would not say it has a nose at all.[29] (659a36–b4)

Lest there be doubt, Aristotle soon states that "in the beak they have channels for smell, but are unable to have nostrils" (659b13). His verdict is thus that birds have neither "nostrils" nor "noses" – even though their beaks have "channels" for smelling.[30] My question concerns what "arguments" should establish such an account of beaks and channels: how to identify the appropriate causal explanation for them – one whose implications reveal why birds are "unable to have nostrils" and why their channels in their beaks (or, perhaps, the part that contains the channels) do not count ("unless on account of function").

Aristotle's remarks here are puzzling: he looks to be identifying a common part – a "nostril" – and gesturing toward a common cause – a "function" – that would justify (*dia*) labeling it as such (659b2–4). However, he ultimately denies that the part is a nostril (b13). How can he support this claim? What, in the final analysis, makes a *part* be of a given type – say, a *nostril*?

In general, Aristotle inclines toward the view that nonuniform parts are defined by their functions. *PA* I.1 claims that nonuniform parts differ in virtue of their capacities: "About the non-uniform parts such as face, hand, and foot, one should say in virtue of what each of them is such as it is, and in respect of what sort of capacity" (640b21–22; cf. *GA* I.18, 722b31–32). Saying what such a part is just is to say what capacity it essentially has. But activity (*energeia*) is prior in being to capacity (*Metaphysics* Θ.8 argues).

28 After Louis 2017's "nettement distinct au point qu'on" (58 and 58 n. 2). See Düring 1943: 152–153 for explanation of the difficulty. Lennox 2001a offers: "They do not have any clearly differentiated nostrils to speak of" (43). One *desideratum* is to avoid contradicting 659b13. Also note: having "channels of the nostrils" is compatible with not having "nostrils" (*HA* II.12, 504a21–23).

29 After Louis 2017: 58 and 58 n. 3, supplying "has them" (cf. 659a2, 3). Lennox 2001a offers: "The *bird*, at any rate, has nothing one would call a nose."

30 Might it be that "nostril" (unlike "channels," 659b1–2, 12; cf. 405a21–23) labels a part common at a lower level of generality that includes humans but not birds, so that saying a part might be called a "nostril" on account of function is to extend that label, loosely, to a more general level that includes birds? I agree there is "looseness," but work needs to be done to tease out the source; see below. One difficulty is that a "loose" extension should be to a *more general* level than the one that defines "nostrils" (cf. 658b28–29), but that function appears to be *smelling through breathing* (657a4ff.). Beak channels, however, *do* let birds *smell through breathing* (659b13–19); surely this is why they are called nostrils on account of function. I thank Leunissen for questions; cf. n. 57.

Thus, just as we cannot say what a capacity is without saying what activity it is for, so too we cannot say what a nonuniform part (defined by its capacity) is without saying what it is for.

PA I.5 adds that what they are for is a "certain action": "Every instrument is for the sake of something, and each of the parts of the body is for the sake of something, and what they are for the sake of is a certain action (*praxis*)" (645b14–16). For example, "sawing is not for the sake of the saw, but the saw for sawing; for sawing is a certain use" (645b17–19). Similarly, "whenever there are actions that are for the sake of other actions, the things whose actions they are differ in the same way that their actions do" (645b28–29) and vice versa (29–32). While the exact structure of these arguments is difficult, their direction is clear. As Charles puts it, in holding that "priority between actions determines priority between organs," I.5 appears to be taking actions to be "definitionally prior to organs."[31]

Aristotle's idea is that a part is defined by its function – what it does: its characteristic *action* – and not what it is made of or its (non-functionally defined) shape (cf. *PA* IV.12, 694b13–14). An object is a saw because it cuts wood, which hypothetically necessitates that shape.[32] A part is a hand because it grasps, which hypothetically necessitates fingers of a certain shape. The function makes it what it is. *PA* I.1's critique of Democritus reinforces the point.[33] Democritus held that "it is by virtue of its configuration and color that each of the animals and their parts is what it is" (640b29–30), where this involves looking to "shape" and "figure" (640b29–35; cf. 640b34).[34] Aristotle's complaint is that this approach does not identify the *cause* – especially the final one (641a11–13): the "work" the part "is able to do" (641a1–2).

The underlying idea may (if borne out by *PA* II–IV) be that nonuniform part types are defined by *the functions that hypothetically necessitate* (and in this way causally explain) their other features. Two parts are of the same type (at some level of generality) if and only if they have the same hypothetically necessitating function (at that level of generality).[35] The upshot is a question about how to fit together *PA* II.16's claims that beak channels can be called "nostrils" on account of their *function* and that they are not nostrils. Are beaks an exception to Aristotle's usual functional account of part types – or does that account, properly understood, show

[31] Charles 2000: 313–314.

[32] I am sympathetic to the view that the *functions* are matter-involving (Charles 2009).

[33] The context is definitional: 640b24.

[34] For useful discussion and a different view, see Carbone 2016: 8–9. Cf. De Ribera-Martin 2018.

[35] Cf. Stein 2018: 38. The brain is a tricky case: Lennox 2021: 238ff.

why beak channels are not nostrils? And how, more generally, does Aristotle go about illuminating the causal explanations of recalcitrant parts like these?

14.4 Polymorphicity and Teleology

In the remainder of this chapter, I argue that Aristotle's approach to such parts – whose explanation is not straightforward – is to appeal to what he thinks he knows about the parts of the most polymorphic animals: humans. This strategy works because being polymorphic is a consequence of – and so a good clue as to – the *teleological* relation between the functions humans perform and their parts' features. As we saw in *PA* II.1 and II.10, complex lives hypothetically necessitate complex bodies: ones with a variety of nonuniform parts. Moreover, according to I.1 and I.5, each such part is defined by its function. Humans (Aristotle thinks) are therefore animals with many functionally defined parts, which have the hypothetically necessitated features appropriate for human life.

Aristotle's insight is that where there is a *variety* of such parts, they can be "specialized": defined in terms of a single function or functions whose requirements do not conflict (are not too "dissimilar").[36] He develops this idea in *PA* IV.6's account of insects that use a single part for defending themselves and for taking in nourishment:

> And it is better, where possible, not to have the same instrument for dissimilar uses, but rather the defensive one most sharp, and the one that is to be a tongue spongy and able to draw in nourishment. For where it is possible for two things to be used for two functions without impeding each other, nature is unaccustomed to making things as does the coppersmith who, to economize, makes a spit-and-lampstand; but where this is not

[36] I present specialization as a consequence of polymorphicity, with polymorphicity strictly concerning only the number of parts, since this seems to fit 683a19–23 best (cf. 683b4–8). (Cf. the *Politics*' discussion of the implications of the *number* of citizens for whether offices should be doubled up: IV.15, 1299a30–b10). A complex life hypothetically necessitates polymorphicity, *because* it facilitates specialization and its functional advantages. I am nonetheless sympathetic to a stronger view (cf. IV.10, 687a19): that II.10's "few"/"many" parts are *types*, defined by ends/functions, and that the tokens that best exemplify (*apoteloito kallista,* 1252b3–4) such functional definitions are specialized ones. On this view, the parts of polymorphic animals still count as especially *distinct* (first and foremost in type), by analogy to the "distinct capacities" of the components (II.1, 646b17) of nonuniform parts with polymorphic actions and in contrast to the "conjoined" capacities (IV.5, 682a11) of stinger-tongues. On either view, polymorphic animals exhibit less doubling up of "dissimilar" functions (IV.6, 683a20). Thanks to Lennox for questions.

possible, nature makes use of the same thing for multiple functions. (683a19–26; cf. 661a15–29, 682a8–12, 682b36–83a3)[37]

Animals with more distinct parts are better off: they have distinct parts for nourishment and for defense.[38] They avoid the negative consequences that sometimes follow when "dissimilar" functions are doubled up in a single part: their tongue can have just the sponginess appropriate for the nourishment humans need, without compromising it for protective sharpness.[39] After all, "nature makes the instruments to fit the function" (694b13–14).[40]

The idea is that where there are more (distinct) parts, these parts need not combine "dissimilar" functions and so need not be subject to trade-offs. The result is that the relations of hypothetical necessity between a part's function and features can be "tighter": such parts can have not just features that make it possible for them to perform their functions but also a degree ("most sharp") or quality of them that is particularly appropriate for doing so, perhaps in animals of this sort.[41] Moreover, they need not have "extra" features required for one function but not a dissimilar one: they need not be *both* sharp and spongy. If the features of nonuniform parts are those hypothetically necessitated by their ends, this is to say that such parts have all and only the end-promoting features appropriate for the animal they belong to.

The animals that have more parts, II.10 suggests, are the polymorphic ones. In claiming that humans are especially polymorphic, Aristotle is thus making space for the view that they need not "have the same instrument for dissimilar uses" – that their parts can be tightly adapted to and specialized for human needs. This view raises difficult questions. On the one hand, perhaps it need not require Aristotle to suppose (falsely) that

[37] Cf. the *Politics*' claim that each tool is made best when it is made for a single use (1252b1–5). Still, in *small* cities – like lower animals? – several offices may be combined in one person (1299b1–10). Thanks to Gligorijevic. For discussion, see Karbowski 2019: 224–227; Gottlieb and Sober 2017: 258–260; Leunissen 2011: 43. Cf. *Juv*'s claim that user and used should be distinct in capacity and place (IV, 469b1–2); thanks to Korobili.

[38] Humans can choose their weapons (*PA* IV.9, 687a19–b9). Because hands *adapt* to all these instruments in virtue of their ability to hold them, they are "many" instruments (687a21). Beaks do not adapt.

[39] "Sometimes": human tongues are for speech as well as flavor perception (II.17, 660a1). This doubling is not problematic, as the same features – softness and detachment – are serve both (660a17–23); IV.6, 683a19–26 rightly emphasizes "dissimilar" functions. Cf. Leunissen 2010a's "secondary teleology." My point is that for polymorphic animals, these bad consequences are less likely, which makes them a model even where doubling occurs.

[40] For application to human society, see Karbowski 2012: 342–345; 2019: 225.

[41] I take it that an animal's way of life (*bios*) partly determines this; see n. 65.

polymorphic animals excel at everything they do,[42] but only that their parts are *finely tuned* to their needs, without *interference* from "dissimilar uses." At the same time, this points to a further question: Whether parts that combine "dissimilar" functions really *are* poorly tuned: isn't this just the tongue these animals need? And here we might well resist: perhaps Aristotle should simply say that in parts like these, the relation between feature and function is clearest to *us*, and no more.

Difficulties aside, the present question is this: *Do* these ideas play a role in Aristotle's method? There is reason to think they do: that the reasoning that establishes the causal explanation of beaks takes human parts as models and that their fine tuning facilitates this. Before we can get there, however, we should recall what makes beaks so puzzling. Aristotle has claimed that birds do not have their channels so "clearly articulated so much so that one would say they are nostrils, unless on account of function" (*PA* II.16, 659b2–4). Apparently, being more clearly "articulated" would allow the channels in their beaks to count as nostrils, for reasons beyond those of "function." What might Aristotle mean?

The opening of *PA* II.16 provides a clue: in animals "with long jawbones that become progressively narrower the part consisting of the nostrils is, as far as is possible, actually present in what is called the 'snout' (*rungchei*), while in the rest the nostrils are more articulated from the jaws" (658b20–33). In long-jawed animals, there *is* a "part consisting of the nostrils" – a nose (cf.659b4) – "present in" the snout. In other animals, however, "the nostrils are *more* articulated from the jaws" (658b32–33). The implication is that in long-jawed animals there is *less* articulation of nose and nostrils from snout: they are present but blend in. The situation is worse in birds: they have a snout/beak (*rungchos*) but no distinct nose – let alone nostrils – on it.

The question is how articulation fits into Aristotle's approach to defining parts: is it a new criterion, or an aspect of an old one? Lennox suggests that "Aristotle is unwilling to define 'nostrils' in *purely* functional terms." Instead, "a nostril must be a separately differentiated [i.e. articulated] part, such as a human nose" (Lennox 2001a: 237). Thus "the bird, at any rate, has them in such a way that one would not say it has a nose at all" (659b4). These remarks are along the right lines: what is needed *is* a nose, which is

[42] We smell poorly (*DS* IV, 440b31–441a2; cf. *DA* II.9, 421a9–13). Still, some smells support brain health and are special to humans (444a3–4), where animals smell through breathing *because* this carries smells to the brain (444a19–25). Here, humans do excel at smelling through breathing (the focus of *PA* II.16; cf. *GA* V.2, 781b17–19). Thanks to Diana Quarantotto and Sophia Connell for questions.

more articulated. However, I will argue, this need not mean that a part's defining features are not "functional" or that there are no connections between articulation and teleology. Indeed, *Physics* II.8 suggests there must be: "being for the sake of something" is present in plants, but "is less articulated" (II.8, 199b9–10).[43]

More specifically, Aristotle appeals to articulation because as parts become more articulated, the connection between their features and end can get tighter, in the way sketched above.[44] The reason is that articulated parts are *distinct* from others, where – *PA* IV.6, 683a18–36 suggests – distinct parts can be specialized.[45] The upshot is that barely articulated parts lack this tight connection. (By contrast, parts with tight connections – as we find in polymorphic animals – may offer good models. See Section 14.5.)

Relations between articulation and teleology emerge in *PA* II.17's account of unarticulated crocodile tongues (660b33; cf. IV.11, 690b23–26) and minimally articulated fish tongues (660b20, 661a2–3).[46] Less articulated tongues sometimes appear not to be "separate" tongues (660b22–23, and unarticulated ones are "fused" to the lower jaw [660b28]). The first lesson is thus that *articulated* tongues are distinct and separate. Moreover, limited articulation and separation are connected to limited *function:* "Just as the tongue's use is slight, so too is its articulation" (660b18–20). In fact,

> since the perception of the taste found in flavors is for the sake of nutrition, this part is tongue-like, though not in every part equally but mostly in the tip.[47] Because of this, in the fish only the tip is separated. (661a3–6)

Perhaps the tip is most perceptive, or it most of all has "tongue-like" features suitable for perception. Either way, this is why "only the tip is

[43] *DA* II.1: their "organs" are "simple" (412b1–2).

[44] On articulation, see Quarantotto 2019, especially 350–351. I am sympathetic to her idea that a unitary substrate is "progressively articulated" into animal parts (351). Cf. Quarantotto 2022 for the claim that as embryos get more complex, "they become systems articulated into parts with a *fixed* specialization." Because more articulated parts are more unified (they are *one* part) and more distinct from other parts in a complex organization, they can have more characteristic specialization.

[45] Cf. *HA* IV.8: some animals have "very manifest" sense organs, when they have "the place of the eyes and of hearing distinguished" (533a18–21); such animals have *ears* rather than mere channels (a21–2). I take this idea to be pre-causal: in the *PA*, Aristotle can add that to lack *ears* is to lack a *part* sufficiently *specialized* to have features hypothetically necessitated by the function of ears (cf. 657a19–20).

[46] Cf. *HA* IV.8, 533a25–28 on "indistinct tongues." On articulation and function more generally, *PA* II.16, 659b29–30, III.3, 667a8; *GA* IV.6, 775a2.

[47] Alternatively, supply, with Düring, *tēn d' aisthēsin*: the tongue-like part has perception, but mostly in the tip (by analogy to *HA* II.1, 492b27) (156; see Louis 52 n. 3 for an alternative text with a similar meaning; cf. Peck 201–202).

separated": articulation and separation promote its function, directly or by allowing features suitable for it. Softness would be good (660a17–19), but fish tongues are in a "spinous" place (660b16, 24–25). Indeed, unarticulated tongues have a "lame character" (660b25). Articulation may thus promote function, insofar as articulated parts are *distinct* from the place – a "spiny" one – where they are located, which can allow them features that differ from those of the place and that are tuned to their function. In other words, articulated parts better specialize.

This picture of articulation gains support from the *De Incessu*, which builds on *PA* II.10's claim that "the natural parts are disposed according to nature in this [human] kind alone" (656a10–12; cf. *HA* 494a26–b1)[48] and connects it with articulation. Here too, articulated parts are *distinct* (cf. 706a20–21): *IA* 4 notes that left and right are "articulated more in some than in others," particularly in those with "instrumental parts" like feet or wings (705b21–22, 24–25; cf. 706a25–26).[49] Other animals may have a functional left and right (705b25–26) but less clearly (705b29) – because they lack right and left "instrumental parts" and so make the "distinctions" (*dialēpseis*) with their body (705b25–26). Moreover, such distinct parts have other appropriate features, such as being positioned "according to nature" in certain places (706a11–12; cf. 706a21–22, 706b3–10), being "detached" and easily moved (706a18–24).

14.5 The Causes of Bird Beaks

Here are the lessons so far. First, highly articulated[50] parts tend to be distinct, and so more able to have appropriate features for parts of their

[48] On the placement of our parts and/or their role as models, see Lennox 1985, 314–315; 2019, 107–111; Meyer 2006: 30–32; Osborne 2007 (on "pedagogical order": 124 n. 38). When II.10 insists the shape of our external parts is more knowable (656a9–10), I take it that it is so both to us and by nature (*Physics* I.1 184a16–18). (In *HA* I.6, 491a21–22, emphasis is on knowability to *us*.) Falcon 2018 offers a similar reading of *HA* I.15's claim that these dimensions are "confounded" in nonhuman animals (494a32): the issue is not that "these functional dimensions are not present" but that they are all "in the same place"; we are not necessarily *psychologically* confounded (49–50). Cf. Wein draft, and on the superiority of human stature more generally, see Rapp in this volume (Chapter 13).

[49] I have benefited from Lennox 2001c: 268–271; 2010e; Carbone 2011; and especially Corcilius 2021. Corcilius also emphasizes the difference between "functionally determined bodily differentiations" and whether those differentiations are "parts of the body that are set apart"; he notes that these "match exactly the number of points that is necessary and sufficient for accomplishing their functional task" and discusses the normative status of the human in connection with its parts' articulation. Cf. Carbone 2016 on "the *convertibility* of the teleology principle and of the morphology principle" (28).

[50] As Giouli Korobili reminds me, the three parts possessed by all complete animals – cf. *PA* II.10 – are "more articulated" in some (*De Juv.* I, 468a17; cf. IV, 469b1–2 on difference in capacity and

type. Beak channels, in contrast, are not clearly articulated; indeed, this appears to explain why they are not nostrils. Second, polymorphic animals (like humans, Aristotle supposes) have more distinct parts, which can likewise be more specialized. This, I suggest, is what Aristotle thinks makes them good models: because their features can be just the ones that serve their functions, articulated to the appropriate degree for human life, starting with them illuminates exactly how those features serve that function. By comparing beaks to such parts, Aristotle can identify those features and functions – and then show why the resulting causal explanation rules out the articulation that would make beaks noses.

Aristotle does just this in *PA* II.16. Indeed, his next move – after drawing our attention to the channel's lack of articulation – is to make such a comparative claim:

> The bird, at any rate, has them in such a way that one would not say it has a nose at all. This is a consequence of the fact that instead of jaws it has what is called a beak. (659b4–6)

This comparative claim – that the beak is "instead of" *jaws* – is Aristotle's justification (*touto de sumbebēken, hoti*) for thinking birds lack noses. In speaking of one part being "instead of" another, Aristotle is often considering whether they are analogues (645b6–10);[51] his idea here is that the beak is analogous to *jaws.*[52] Indeed, III.1 claims it is a mouth: "The beak, as it is called, is a mouth; for birds have this instead of lips and teeth" (662a33–35).[53] *PA* IV.12 adds: "In the birds there is a bony beak in place of teeth and lips" (692b18–19). That the beak is a mouth – analogous to jaws, teeth, and lips – is in one respect not surprising. *PA* II.10 identified the part "by which they receive nourishment" as necessary (655b29–31). Whatever else Aristotle says about beaks, it must accommodate the facts that birds need such a part and that the beak is the obvious candidate.[54] My question is simply what implications this fact has for what beaks are like – and how Aristotle goes about drawing this out.

place); *On Sleep* likewise identifies "three distinct places" (456a2–3). In her commentary, Korobili argues that *De Juv.*'s three "parts" are "sections of the body." These considerations support connections between articulation and distinctness (Korobili 2023).

[51] On analogy, see Henry 2014; Leunissen 2014; Wilson 1997, 2000 (contrasting it with the *scala*); Lloyd 1996. For caution about the expression, see Wilson 1997: 347; 2000: 70. All I require is that the uses I focus on here have parallel hypothetical necessity relations.

[52] To my knowledge, Aristotle never claims the beak is "instead of" a nose or nostrils.

[53] *HA* II.12 notes that birds have an unusual *mouth*, in that they have a beak but no teeth or lips; they also have no nose, but only the channels of the nostrils (504a19–22; cf. *HA* IV.8, 533a22–24).

[54] Thanks to Mariska Leunissen and Robert Howton for emphasizing that this point functions as important background.

Aristotle appears to be arguing that we are justified in denying that birds have noses (and thus nostrils) because their beak is a mouth, given what that mouth is like.[55] This reasoning will work if it entails that beaks are such that they cannot be (even analogous to) noses: that their characteristic features both make them mouths and also prevent them from being noses (and their channels nostrils). This last point is important: the spit-and-lampstand indicates that a part might be of two types (stingers and tongues). Why *does* Aristotle think beaks are not noses – what are *these* mouths like, such that they can't be noses *too*?

The key move emerges in Aristotle's account of what makes beaks mouths – and the human-first reasoning that supports it. A few lines later, he follows up his claim that beaks are "instead of jaws" (659a5) with the claim that they are "in place of teeth and lips" (659b23). Together, both claims support the claim that beaks are mouths, which contain (parts analogous to) teeth, lips, jaws, etc. (661a34–35, 662a20ff.). Aristotle then adduces a thought experiment – which harkens back to *PA* II.10's method of starting with humans:

> For the birds, as we said, for nourishment and strength their beak is bony. It has been joined together into one, in place of teeth and lips, just as if someone who had removed the lips from a human being were both to fuse the upper teeth together, and separately the lower teeth, and then were to draw them both out to a point; in fact this would already be a bird-like beak. (659b21–27)

The thought experiment identifies the human part whose components can be transformed into a "bird-like" beak.[56] This part is a *mouth* – specifically, lips and teeth – and not a nose. It explains how that transformation is achieved: by removing extra bits and accentuating the shapes of the remaining ones.

[55] Thanks to Robert Howton, Mariska Leunissen, and Diana Quarantotto for raising useful questions about this claim's role. I think I agree with Leunissen and Quarantotto that Aristotle's project is to show how having a "mouth" *of the sort birds do* – not just any mouth – means they cannot have articulated nostrils. I believe Leunissen also denies that Aristotle here shows that beaks are not like the spit-and-lampstand (or elephant trunk): that their mouthlike features prevent them from being parts *defined by* the nostril function. My main point, however, is how human parts help Aristotle illuminate the beak's causal structure.

[56] Kullmann 2007 also connects this thought experiment with "die vergleichende Morphologie des Aristoteles, der vom Typus des Menschen aus den Bau anderer Tiere zu verstehen sucht" (479). Carbone 2011 takes it to *rely on* the "l'analogie fonctionnelle des parties concernées" and emphasizes its "topologique" aspect (121; cf. 139–146). Wilson 2000 notes it as a puzzling case for the distinction between differing by analogy and by the "more and less" (59). On thought experiments, cf. Leunissen 2010a.

Why does Aristotle appeal to humans *here*? Notice the claim this reasoning illustrates: that "for nourishment and strength their beak is bony." This remark echoes a similar one above. Immediately after noting that the fact that the bird lacks a nose "is a consequence of the fact that instead of jaws it has what is called a beak" (659b5–6), Aristotle added:

> And these things are so because nature has constituted the birds in this way . . .. In order, then, that it may be useful for both physical strength and nourishment, the beak they have is bony; while it is narrow on account of the smallness of their head. And in the beak they have channels for smell, but are unable to have nostrils. (659b6–13)

Here too, Aristotle offers a causal (*aitia de toutōn*, 659b6) explanation of the beak's mouth-like nature, notably, its boniness.[57] The explanation refers to the birds' "nature" and identifies the final cause it aims at: "physical strength and nourishment" (659b9–10; cf. 659b22). Nourishment and strength, importantly, are common and differentiated functions of *teeth* (661b1–2) and sometimes *mouths*, which are constituted by the teeth (661a34–36) and are for nourishment and sometimes strength, respiration, and speech (662a20ff.).

The upshot of the thought experiment (and, generally, the comparative claim) is thus that the final cause that makes beaks *as they are* is analogous to that of human mouths – and that what it explains is likewise analogous. Indeed, its first result is that the main *explanandum* is (as at 659b6–13[58]) teeth-like *boniness* (659b22).[59] For the human parts "fused" and "drawn out" – teeth (659b24–5) – are *bony* (663b29–664a3; cf. 692b18–19). The second result concerns the explanation of those bony features: it is like that of teeth, where boniness promotes "protection" and "advantage" (663b29–664a3; cf. 655b2–13). This reveals what makes comparisons to the human so valuable. The thought experiment shows how these features and final causes "hang together" through transformation. These features are essential to the beak, if it is to be defined by the ends of strength and nourishment.

The resulting causal analysis can, moreover, be used to justify the claim that "in the beak they have channels for smell, but are unable to have nostrils" (659b12–13). A beak like *this* cannot have nostrils "clearly articulated" out of it; it cannot specialize in what nostrils do, for the

[57] For the demonstrative character of this explanation, see Kullmann 2007: 474–475.

[58] That claim that the small head explains the narrowness explains the part's differentiation.

[59] I am sympathetic to Wilson 2000's suggestion that "analogy involves not only identity of function, but also material and structural similarities" (69). Lennox has also emphasized this to me.

features it needs for its mouth-like function (e.g. boniness) don't allow for features that promote smelling. *Those* features might include material with a different one of *PA* II.1's "distinct capacities" (646b14–18): cartilage.[60] Parts are cartilaginous "where there is an advantage in the hard part being soft and pulpy ... as is the case with the ears and nostrils; for brittle appendages are quickly broken" (II.9, 655a29–32.) A nostril should be "easily moved" (*HA* I.11, 492b14); cartilage allows more "fluid" movement (II.9, 655a24). Moreover, longer, more protruding channels also promote smelling (*GA* V.2, 781b8–10). Just as ears capture far off movements (781b15–16), nostrils help us smell from a distance (781b7–13). Thus, just as an ear is added to channels because it secures the movement of air through the channel (781b24–26), a nostril is presumably added to channels because it secures the movement of air in smelling.[61] Here we have a more refined statement of features that promote the nostril function. Beaks, however, don't have these features, being bony and brittle rather than "soft and pulpy" (II.9 655a29–32) and having channels that do not protrude but are rather near the eyes (II.13, 657b18–19) and mouth (II.16, 659b2).

All in all, while it is *possible* to smell through a beak channel, beaks and their channels lack features that promote this end *as* noses and their nostrils do.[62] The reason is that beaks are defined by nourishment and strength, and so they have features hypothetically necessitated by these ends (of the sort appropriate for a bird's way of life: *PA* IV.12).[63] These features are not compatible with further features that promote smelling – but, the thought experiment reveals, they could not be given up without compromising the beak's mouthlike function. While it is *sometimes* possible for parts to have features that promote multiple functions, most[64] of the relevant features are not options *here:* the beak's lack of nasal articulation is inevitable. In this way, looking to humans clarifies the relations of hypothetical necessity in the beak. It does so because the comparative specialization of polymorphic parts makes these relations especially tight, allowing us to see which features serve which function – what, for instance,

[60] Birds have channels but lack ears, for "they do not have the sort of matter from which ears may be formed" (II.12, 657a19–20). Cf. n. 47.

[61] Cf. *Sens.* 4 on air in smelling: 443a2.

[62] Thanks especially to Leunissen and Quarantotto for helpful questions about this section.

[63] Beaks are for nourishment and overpowering, differing according to different ways of life, e.g., a swamp-dwelling life (*PA* IV.12, 693a10–19; cf. 662a35–b); the relevant "articulation" is tuned to *bios.* Cf. Lennox 2010a.

[64] Excluding channels.

boniness does. When Aristotle extends this analysis to beaks, he can spell out its implications for a part's *other* features.

While my focus is Aristotle's use of human models, the present analysis may also reveal why he denies that beak channels are "clearly articulated" enough to count as nostrils, "unless on account of function" (659b2–4). He may be distinguishing *performing* a "function" (birds *do* smell by breathing) from being a *part* that essentially does so (a nostril).[65] If to be defined by a function is to have features that function explains, meeting a minimum threshold of such features – beyond simply having channels – may be necessary for being a *part* of that type. Because less articulated channels are less distinct from the beak's mouth-like features, they may struggle to pass that threshold. They are somewhat like the functional left and right in animals that lack right and left "instrumental parts" (705b21–29; cf. 709b13–14): they perform the function, but lack the *parts*.[66]

14.6 Concluding Remarks

This is the methodological value of polymorphicity: where there are more parts, Aristotle thinks, there can be a tighter relation between their ends and features. Our accounts of that causal structure can also serve as models for the explanations of other parts. In *PA* II.16, this approach allows Aristotle to identify the implications of beaks' causal structure for their other features and functions. But while this is the most vivid example, there are others.[67] *PA* II.13 remarks that hard-skinned animals lack eyelids and "in place of this safeguard . . . have hard eyes, as if seeing through a fused eyelid" (657b29–35).[68] Aristotle's illustration – "as if seeing through a fused eyelid" – echoes II.16's thought experiment. And here, too, humans are in view: "the human most of all" (II.13, 657a36, b1–2) carries out the eyelid function – a "quick and membranous operation" (657b33) – by *blinking* (657a37).

[65] Beaks thus differ from elephant noses, which *are* nostrils (659a15), and from stinger-tongues, where there is a trade-off (but not a defeat) between the features required by the ends.

[66] Thanks to Leunissen and Quarantotto for questions and suggestions. Corcilius 2021 also emphasizes articulation of *parts*. *PA* 646b25 calls nonuniform parts, including nostrils, "instrumental." Cf. Lennox 2001a: 182.

[67] Cf. II.14, 658a11–31, 658b2–10.

[68] Hard eyes are in an analogical unity "safeguard for the eye" (II.13, 657a26). Cf. Gotthelf [1997b] 2012d: 194–195; 2012: 210. For other such parts, see *PA* II.13, 657a35, 657a37, 657b16.

PA IV.12 also presents an interesting series of comparisons: "While in elephants there is a trunk in place of hands, and in some of the insects a tongue in place of a mouth, in the birds there is a bony beak in place of teeth and lips" (692b15–19). Two of these comparisons implicitly feature humans: II.16 compared beaks to human mouths, and hands are distinctively human parts (IV.9, 687a2–b9).[69] Insect tongues are trickier. Earlier, Aristotle compared these tongues – which also function as stingers and are "intermediate between a sting and a tongue" (679b7–8) – to trunks (II.17, 661a15–29; IV.6, 682b36–83a3; cf. IV.5, 678b22–23), which also function in several ways (II.16, 659a20ff.[70]). That is, he compared them to parts themselves compared to human parts. Moreover, "the so-called sting is inside the mouth, as if the capacities of the tongue and lips were conjoined and possessed together" (IV.5, 682a10–12). The strategy is to examine a transformation of *separate* parts to make sense of the causal structure – the capacities – of *this* one.

The broader lesson, however, is that Aristotle takes humans to be models not only for causal explanations of difficult parts but also for how to *inquire* into such parts at all: for the importance of comparative anatomy more generally.[71] And here we might have an opening, if we wish to resist his anthropocentrism. Taken seriously, his interest in the tightness of the relation between end and features ought perhaps to have led him to look elsewhere – as he seems to begin to do for insect tongues.

[69] Cf. Meyer 2006: 32–33. [70] Cf. Gotthelf 1997b. [71] Cf. Tipton 2013: 148 n. 10.

References

Adler, A. 1933 (ed.). *Suidae Lexicon*, vol. III. Leipzig: K. G. Saur.

Althoff, J. 1992. *Warm, Kalt, Flüssig und Fest bei Aristoteles. Die Elementarqualitäten in den zoologischen Schriften*. Stuttgart: Steiner.

Anagnostopoulos, A. 2022. "*De Generatione et Corruptione* II.7," in P. Dimas, A. Falcon and S. Kelsey (eds.) *De Generatione et Corruptione II: Translation and Essays*. Cambridge: Cambridge University Press, 151–177.

Appel, T. A. 1987. *The Cuvier-Geoffroy Debate: French Biology in the Decades Before Darwin*. Oxford: Oxford University Press.

Arabatzis, G. 2009. "Michael of Ephesus on the Empirical Man, the Scientist and the Educated Man (*In Ethica Nicomachea* X and *In De Partibus Animalium* I)," in C. Barber and D. Jenkins (eds.) *Medieval Greek Commentaries on the Nicomachean Ethics*. Leiden: Brill, 163–184.

Balme, D. M. 1962. "Γένος and Εἶδος in Aristotle's Biology," *The Classical Quarterly* 12.1: 81–98.

[1972] 1992. *Aristotle's de Partibus Animalium I and de Generatione Animalium I (with Passages from II.1–3)*. Oxford: Clarendon Press.

1987a. "The Place of Biology in Aristotle's Philosophy," in Gotthelf and Lennox (eds.), 9–20.

1987b. "*Aristotle's Use of Division and Differentiae*," in Gotthelf and Lennox (eds.), 69–89.

1987c. "Aristotle's Biology Was Not Essentialist," in Gotthelf and Lennox (eds.), 291–312.

1991. *Aristotle, History of Animals, Books VII–X*. Cambridge, MA: Harvard University Press.

1992. *Aristotle, De Partibus Animalium I and De Generatione Animalium I (with Passages from II.1–3)*. Oxford: Clarendon Press.

Barber, C., and D. Jenkins (eds.). 2009. *Medieval Greek Commentaries on the Nicomachean Ethics*. Leiden: Brill.

Barker, E. 1957. *Social and Political Thought in Byzantium: From Justinian I to the Last Palaeologus*. Oxford: Oxford University Press.

Barnes, J. (ed.). 1984. *The Complete Works of Aristotle*. The Revised Oxford Translation, 2 vols. Princeton: Princeton University Press.

1993. *Aristotle's Posterior Analytics*, 2nd ed. Oxford: Clarendon Press.

Bekker, I. 1831. *Aristotelis opera*. Berlin: Academia Regia Borussica.

Berger, F. 2012. "Die Textgeschichte der *Historia Animalium* des Aristoteles, Aristophanes von Byzanz und die zoologische Sylloge des Konstantinos Porphyrogennetos," *Rursus* 7: 1–13.

Bolton, R. 1987. "Definition and Scientific Method in Aristotle's Posterior Analytics and Generation of Animals," Gotthelf and Lennox (eds.), 120–166.

Bouras-Vallianatos, P. 2019. "Galen in Byzantine Medical Literature," in P. Bouras-Vallianatos and B. Zipser (eds.) *Brill's Companion to the Reception of Galen*, Leiden: Brill, 86–110.

Broadie, S. 1998. "Νοῦς and Nature in *De Anima* III," *Proceedings of the Boston Area Colloquium in Ancient Philosophy* 12: 163–176.

Brown, L. (ed.) 2009. *Aristotle,* The Nicomachean Ethics. Trans. D. Ross. Oxford University Press.

Browning, R. 1990. "An Unpublished Funeral Oration on Anna Comnena," in R. Sorabji (ed.) *Aristotle Transformed: The Ancient Commentaries and Their Influence*. Ithaca, NY: Cornell University Press, 393–406.

Bubb, C. 2019. "The Physiology of *phantasmata* in Aristotle: Between Sensation and Digestion," *Apeiron* 52.3: 272–315.

2020. "Blood Flow in Aristotle," *Classical Quarterly* 70.1: 137–153.

2024. "Aristotle and the Medical Tradition on Sleep and Food," in H. Bartos and V. Linka (eds.) *Aristotle Reads Hippocrates*. Leiden: Brill, 169–187.

Burnyeat, M. F. 1992. "Is an Aristotelian Philosophy of Mind Still Credible? A Draft," in Nussbaum and Rorty (eds.), 15–26.

2002. "De Anima II 5," Phronesis 47.1: 28–90.

2004. "Introduction: Aristotle on the Foundations of Sublunary Physics," in F. de Haas and J. Mansfeld (eds.) *Aristotle, On Generation and Corruption Book 1 Symposium Aristotelicum*. Oxford: Clarendon Press, 7–24.

Bywater, I. 1924. *Aristotle: Poetics. The Works of Aristotle*, vol. XI. Oxford: Oxford University Press.

Carbone, A. L. 2011. *Aristote illustré: Représentations du corps et schématisation dans la biologie aristotélicienne*. Paris: Classiques Garnier.

2016. "The Axes of Symmetry: Morphology in Aristotle's Biology," *Apeiron* 49: 1–31.

Caston, V. 1998. "Aristotle on the Relation of the Intellect to the Body: Commentary on Broadie," *Proceedings of the Boston Area Colloquium in Ancient Philosophy* 12: 177–191.

Cerami, C. 2015. *Nature et Sagesse: Les rapports entre physique et métaphysique dans la tradition aristotélicienne. Recueil de textes en hommage à Pierre Pellegrin.* Louvain: Peeters.

Cerami, C., and A. Falcon (eds.). 2014. "Continuity and Discontinuity in the Greek and Arabic Reception of Aristotle's Study of Animals," *Antiquorum philosophia* 8: 35–56.

Charles, D. 1988. "Aristotle on Hypothetical Necessity and Irreducibility," *Pacific Philosophical Quarterly* 69: 1–53.

2000. *Aristotle on Meaning and Essence*. Oxford: Oxford University Press.

2009. "Aristotle's Psychological Theory," *Proceedings of the Boston Area Colloquium in Ancient Philosophy* 24: 1–49.

2010. *Definition in Greek Philosophy*. Oxford: Oxford University Press.
2021. *The Undivided Self: Aristotle on the "Mind-Body Problem."* Oxford: Oxford University Press.
Charlton, W. 1970. *Aristotle's Physics II, III*. Oxford: Clarendon Press.
1987. "Aristotle on the Place of Mind in Nature," in Gotthelf and Lennox (eds.), 408–423.
Clark, S. R. L. 1975. *Aristotle's Man: Speculations upon Aristotelian Anthropology*. Oxford: Clarendon Press.
Code, A. 1997. "The Priority of Final Causes over Efficient Causes in Aristotle's *PA*," in W. Kullmann and S. Föllinger (eds.) *Aristotelische Biologie. Intentionen, Methoden, Ergebnisse*. Stuttgart: Franz Steiner Verlag, 127–144.
2010. "An Aristotelian Puzzle about Definition: Metaphysics Z.12," in Lennox and Bolton (eds.), 78–96.
2011. "Commentary on Devereux," *Proceedings of the Boston Area Colloquium in Ancient Philosophy* 26: 197–209.
Cohoe, C. (ed.). 2022. *De Anima: A Critical Guide*. Cambridge: Cambridge University Press.
Coleman, W. 1964. *George Cuvier, Zoologist*. Cambridge, MA: Harvard University Press.
Coles, A. 1997. "Animal and Childhood Cognition in Aristotle's Biology and the Scala Naturae," in Kullmann and Föllinger (eds.), 287–324.
Cooper, J. M. 1985. "Hypothetical Necessity," in Gotthelf (ed.), 151–167. Reprinted in Cooper 2001: 130–147.
1987. "Hypothetical Necessity and Natural Teleology," in Gotthelf and Lennox (eds.), 243–274. Reprinted in Cooper 2001, 130–147.
(ed.). 1997. *Plato: Complete Works*. Indianapolis; Cambridge: Hackett.
2001. *Knowledge, Nature, and the Good*. Princeton: Princeton University Press.
Connell, S. M. 2016. *Aristotle on Female Animals: A Study of the Generation of Animals*. Cambridge: Cambridge University Press.
2018a. "Aristotle's Explanations of Monstrous Births and Deformities in Generation of Animals 4.4," in Falcon and Lefebvre (eds.), 207–223.
2018b. "Mariska Leunissen (2017) *From Natural Character to Moral Virtue in Aristotle*," in *Mind* 127.507: 938–946.
(ed.). 2021. *The Cambridge Companion to Aristotle's Biology*. Cambridge: Cambridge University Press.
2021a. "The Female Contribution to Generation and the Nutritive Soul in Aristotle's Embryology," in Lo Presti and Korobili (eds.), 63–84.
2021b. "Thinking Bodies: Aristotle on the Biological Aspects of Human Cognition," in P. Gregorić and J. L. Fink (eds.) *Encounters with Aristotelian Philosophy of Mind*. London: Routledge, 223–248.
2021c. "Animal Cognition in Aristotle," in Connell (ed.), 195–210.
2021d. "'Nous Alone Enters from Outside': Aristotelian Embryology and Early Christian Philosophy," *Journal of Ancient Philosophy* 15.2: 106–138.
2021e. *Aristotle on Women: Physiology, Psychology, and Politics*. Cambridge: Cambridge University Press.

2024. "Aristotle on Memory and Emotion in Human and Non-human Animals," in D. Spatharas and G. Kazantzidis (eds.), *Memory and Emotions in Antiquity*. Berlin: De Gruyter, 129–152.

Corcilius, K. 2021. "The Architecture of Locomotive Bodies," in Falcon and Stavrianeas (eds.), 141–164.

Corcilius, K., and P. Gregoric. 2013. "Aristotle's Model of Animal Motion," *Phronesis* 58: 52–97.

Coughlin, S. 2021. *Hephaistus' Cook. A Collection of Autobiographical Passages from the Works of Michael of Ephesus and Other Witnesses to His Life*. www.academia.edu/38778680/Michael_of_Ephesus_autobiographical_passages_with_brief_notes_and_English_translation_revised_Nov_2021_.

Cramer, J. A. (ed.). 1836. *Anecdota Graeca*, Vol. III. Oxford: Oxford University Press.

Crowley, T. J. 2008. "Aristotle's So-called Elements," *Phronesis* 53: 223–242.

2013. "*De Generatioine et Corruptione* 2.3: Does Aristotle Identify the Contraries as Elements?," *Classical Quarterly* 63: 161–182.

Cunningham, A. 1999. "Aristotle's Animal Book: Ethology, Biology, Anatomy, or Philosophy?," *Philosophical Topics* 27.1: 17–41.

Curd, P. 2007. *Anaxagoras of Clazomenae: Fragments and Testimonia*. Toronto: University of Toronto Press.

Denyer, N. 2000. "Never Will and Cannot," *Proceedings of the Aristotelian Society, Supplementary Volume* 74: 163–178.

De Ribera-Martin, I. 2018. "External figure (*Schêma*) and homonymy in Aristotle," *International Philosophical Quarterly* 58: 389–406.

Depew, D. 1997. "Etiological Approaches to Biological Aptness in Aristotle and Darwin," in Kullmann and Föllinger (eds.), 209–227.

Deslauriers, M. 1991. "Plato and Aristotle on Division and Definition," *Ancient Philosophy* 10: 203–219.

Devereux, D. 2011. "Aristotle on the Form and Definition of a Human Being: Definitions and Their Parts in *Metaphysics* Z.10 & 11," *Proceedings of the Boston Area Colloquium in Ancient Philosophy* 26: 167–196.

Devereux, D., and P. Pellegrin (eds.). 1990. *Biologie, Logique et Metaphysique chez Aristote*. Paris: Éditions du Centre National de la Recherche Scientifique.

Diels, H. 1951–1952 (cited as DK). *Die Fragmente der Vorsokratiker*, 3 vols., 6th ed., revised by W. Kranz. Leipzig: Weidmann (Original work published 1903).

Dimas, P. 2021. "De Incessu Animalium 4. Aristotle's Conception of Dimension," in Falcon and Stavrianeas (eds.), 119–139.

Dimas, P., A. Falcon, and S. Kelsey (eds.). 2022. *Aristotle's De Generatione et Corruptione: Translation and Interpretative Essays*. Cambridge: Cambridge University Press.

Dorandi, T. 1999. *Antigone de Caryste. Fragments*. Paris: Les Belles Lettres.

2017. "La ricezione del sapere zoologico di Aristotele nella Tradizione paradossografica," in Sassi, Coda, and Feola (eds.), 59–80.

Dow, J. 2015. *Passions and Persuasion in Aristotle's Rhetoric*. Oxford: Oxford University Press.

Drossaart-Lulofs, H. J. 1957. "Aristotle's ΠΕΡΙ ΦΥΤΩΝ," *Journal of Hellenic Studies* 77.1: 75–80.

1965. *Nicolaus Damascenus: On the Philosophy of Aristotle*. Leiden: Brill.

Duckworth, W. L. H. (trans.), M. C. Lyons, and B. Towers (eds.). 1962. *Galen On Anatomical Procedures. The Later Books*, Cambridge: Cambridge University Press.

Düring, I. 1943. *Aristotle's De Partibus Animalium: Critical and Literary Commentaries*. Göteborg: Elanders boktryckeri aktiebolag.

1980. "Aristotle's Method in Biology. A note on *De Part. An.* I 1, 639b30–640a2," in S. Mansion (ed.) *Aristote et les problèmes de méthode. Communications présentées au Symposium Aristotelicum tenu à Louvain du 24 août au 1er septembre 1960*. Louvain: Éditions de l'Institut Supérieur de Philosophie, 213–221.

Duffy, J. M. 1992. *Michaelis Pselli Philosophica Minora, Vol. I, Opuscula Logica, Physica, Allegorica, Alia*. Stuttgart and Leipzig: Teubner.

Ebrey, D. (ed.). 2015. *Theory and Practice in Aristotle's Natural Science*. Cambridge: Cambridge University Press.

Falcon, A. 2005. *Aristotle and the Science of Nature: Unity without Uniformity*. Cambridge: Cambridge University Press.

2015. "Aristotle on the Study of Animals and Plants," in K.-D. Fischer and B. Holmes (eds.) *The Frontiers of Ancient Science. Essays in Honour of Heinrich von Staden*. Berlin: De Gruyter, 75–91.

(ed.). 2016 *Brill's Companion to the Reception of Aristotle in Antiquity*. Leiden: Brill.

2018. "Physics I.1," in Quarantotto (ed.), 41–59.

2021a. "Explanatory Strategies in the De incessu animalium," in Falcon and Stavrianeas (eds.), 3–18.

2021b. "The Reception of the De incessu animalium," in Falcon and Stavrianeas (eds.), 19–31.

2021c. "De incessu animalium 1–3: The Theoretical Framework and the Beginning of the Actual Investigation," in Falcon and Stavrianeas (eds.), 101–116.

2021d. "The Reception of Aristotle's Biology in Late Antiquity and Beyond," in Connell (ed.), 246–260.

2024. *The Architecture of the Science of Living Beings: Aristotle and Theophrastus on Animals and Plants*. Cambridge: Cambridge University Press.

Falcon, A., and D. Lefebvre (eds.). 2018. *The Generation of Animals: Cambridge Critical Guide*. Cambridge: Cambridge University Press.

Falcon, A., and S. Stavrianeas (eds.). 2021. *Aristotle on How Animals Move. Aristotle's De incessu animalium: Text, Translation, and Interpretative Essays*. Cambridge: Cambridge University Press.

Fazzo, S. 2004. "Aristotelianism as a Commentary Tradition," in P. Adamson, H. Baltussen, and M. W. F. Stone (eds.) *Philosophy, Science and Exegesis in Greek, Arabic and Latin Commentaries*, vol. I, London: s.n., 1–19.

2008. "Nicolas, l'auteur du Sommaire de la philosophie d'Aristote: doutes sur son identité, sa datation, son origine," *REG* 121.1: 99–126.

2016. "Unmoved Mover as Pure Act or Unmoved Mover in Act? The Mystery of a Subscript Iota," in C. Horn (ed.) *Aristotle's Metaphysics Lambda: New Essays*. Berlin: De Gruyter, 181–205.

Feola, G. 2017. "Alcune considerazioni sull'ordinamento del *corpus* biologico di Aristotele," in M. M. Sassi, E. Coda, and G. Feola (eds.) *La Zoologia di Aristotele e la sua ricezione dall'età ellenistica e romana alle culture medievali*. Pisa: Pisa University Press, 35–57.

Festa, N. 1891. *Iamblichi de communi mathmatica scienta liber ad fidem codicis florentini*. Leipzig: Teubner. Revised by U. Klein, Stuttgart, 1975.

Festugière, A. J. 1956. "Un fragment nouveau du 'Protreptique' d'Aristote," *Revue philosophique de la France et l'Étranger* 146: 117–127.

Fine, K. 1996. "The Problem of Mixture," in F. Lewis and R. Bolton (eds.) *Form, Matter and Mixture*. Malden, MA: Blackwell, 82–182.

Fögen, T. 2007a. "Pliny the Elder's Animals: Some Remarks on the Narrative Structure of *Nat. Hist.* 8–11," *Hermes* 135. Jahrg., H. 2: 184–198.

2007b. "Antike Zeugnisse zu Kommunikationsformen von Tieren," *Antike und Abendland* 53.1: 39–75.

Föllinger, S. 2022. *Aristotle's Generation of Animals: A Comprehensive Approach*. Berlin: De Gruyter.

Frede, M. 1990. "The Definition of Sensible Substances in Met. *Z*," in Devereux and Pellegrin (eds.), 113–129.

1992. "On Aristotle's Conception of the Soul," in Nussbaum and Rorty (eds.), 93–107.

1997. "The Definition of Sensible Substances in Metaphysics *Z*," in Devereux and Pellegrin (eds.), 113–29.

2004. "On Generation and Corruption I.10: Mixture and Mixables," in F. De Haas and J. Mansfeld (eds.) *Aristotle: On Generation and Corruption, Book I, Symposium Aristotelicum*. Oxford: Oxford University Press, 289–314.

Frede, M., and G. Patzig. 1988. *Aristoteles "Metaphysik Z,"* 2 vols. Munich: Verlag C. H. Beck.

Freeland, C. A. 1990. "Scientific Explanation and Empirical Data in Aristotle's 'Meteorology,'" *Oxford Studies in Ancient Philosophy* 8: 67–102.

1994. "Nourishing Speculation: A Feminist Reading of Aristotelian Science," in Bat-Ami Bar-On (ed.) *Engendering Origins: Critical Feminist Readings in Plato and Aristotle*. Albany: State University of New York Press, 145–187.

Freudenthal, G. 1995. *Aristotle's Theory of Material Substance: Heat and Pneuma, Form and Soul*. Oxford: Oxford University Press.

Frey, C. 2007. "Organic Unity and the Matter of Man," *Oxford Studies in Ancient Philosophy* 32: 167–204.

2015a. "From Blood to Flesh: Homonymy, Unity, and Ways of Being in Aristotle," *Ancient Philosophy* 35.2: 375–394.

2015b. "Two Conceptions of Soul in Aristotle," in Ebrey (ed.), 137–160.

2018. "Aristotle on the Intellect and Limits of Natural Science," in J. E. Sisko (ed.) *Philosophy of Mind in Antiquity*. New York: Routledge, 160–174.

2021. "Aristotle on the Soul's Unity," in Cohoe (ed.), 88–103.

Gambra Gutiérrez, J. M. 2012. "The *topoi* from the Greater, the Lesser and the Same Degree: An Essay on the σύγκρισις in Aristotle's *Topics*," *Argumentation* 26: 413–437.

Gasser-Wingate, M. 2019. "Aristotle on the Perception of Universals." *British Journal for the History of Philosophy* 27: 446–467.

2021. *Aristotle's Empiricism*. Oxford: Oxford University Press.

Gelber, J. 2015a. "Are Facts about Matter Primitive?," in Ebrey (ed.), 46–60.

2015b. "Aristotle on Essence and Habitat," *Oxford Studies in Ancient Philosophy* 48: 267–294.

2017. "Uses of Aporiai in *Generation of Animals*," in G. Karamanolis and V. Politis (eds.) *The Aporetic Tradition in Ancient Philosophy*. Cambridge: Cambridge University Press, 155–171.

2018. "Two Ways of Being for an End," *Phronesis* 63.1: 64–86.

2020. "Soul's Tools," in C. King and H. Bartos (eds.) *Heat, Pneuma, and Soul in Ancient Philosophy and Science*. Cambridge: Cambridge University Press, 243–259.

2022. "Aristotle on Seed," in C. Cohoe (ed.) *Cambridge Critical Guide to Aristotle's de Anima*. Cambridge: Cambridge University Press, 104–121.

Gielen, E. 2018. "Physician versus Physician: Comparing the Audience of *On the Constitution of Man* by Meletios and *Epitome on the Nature of Men* by Leo the Physician," in P. Bouras-Vallianatos and S. Xenophontos (eds.) *Greek Medical Literature and its Readers*. London: Routledge, 153–179.

Gill, M. L.. 1989. *Aristotle on Substance: The Paradox of Unity*. Princeton: Princeton University Press.

1993. "Matter against substance," *Synthese* 96.3: 379–397.

2001. "Aristotle's Attack on Universals," *Oxford Studies in Ancient Philosophy* 19: 235–60.

2006. "First Philosophy in Aristotle," in M. L. Gill and P. Pellegrin (eds.) *A Companion to Ancient Philosophy*. Oxford: Blackwell, 347–373.

2010. "Unity of Definition in Metaphysics H.6 and Z.12," in Lennox and Bolton (eds.), 97–121.

2021a. "Method and Nutritive Soul in Aristotle's De Anima II,4," in Lo Presti and Korobili (eds.), 21–42.

2021b. "Aristotle's Hylomorphism Reconceived," *Proceedings of the Aristotelian Society* 121.2: 183–201.

Gottlieb, P., and E. Sober. 2017. "Aristotle on 'Nature Does Nothing in Vain,'" *HOPOS: The Journal of the International Society for the History of Philosophy of Science* 7: 246–271.

Gotthelf, A. (ed.). 1985. *Aristotle On Nature and Living Things*. Pittsburgh: Mathesis.

1985. "Notes toward a Study of Substance and Essence in Aristotle's Parts of Animals II–IV," in Gotthelf (ed.), 27–54. Reprinted in Gotthelf 2012: 217–240.

1987a. "Aristotle's Conception of Final Causality" and "Postscript 1986," in Gotthelf and Lennox (eds.), 204–242. Reprinted in Gotthelf 2012: 3–44.

1987b. "First Principles in Aristotle's Parts of Animals," in Gotthelf and Lennox (eds.), 167–198.

1988. "The Place of the Good in Aristotle's Natural Teleology," in J. J. Cleary and D. C. Shartin (eds.) *Proceedings of the Boston Area Colloquium in Ancient Philosophy*, vol. 4. Lanham, MD: University Press of America, 113–139. Reprinted in Gotthelf 1985: 45–66.

1997a. "Division and Explanation in Aristotle's *Parts of Animals*," in H.-C. Günther and A. Rengakos (eds.) *Beiträge zur antiken Philosophie: Festschrift für Wolfgang Kullmann*. Stuttgart: Steiner-Verlag, 215–230. Reprinted in Gotthelf 2012: 197–214.

1997b. "The Elephant's Nose: Further Reflections on the Axiomatic Structure of Biological Explanation in Aristotle," in Kullmann and Föllinger (eds.), 215–230. Reprinted in Gotthelf 2012: 186–196.

2012. *Teleology, First Principles and Scientific Method in Aristotle's Biology*. Oxford: Oxford University Press.

2012a. "Division and Explanation in Aristotle's *Parts of Animals*," in Gotthelf (ed.), 197–214.

Gotthelf, A., and A. Falcon. 2018. "'One Long Argument'? The unity of Aristotle's *Generation of Animals*," in Falcon and Lefebvre (eds.), 15–34.

Gotthelf, A., and M. Leunissen. 2010. "'What's Teleology Got to Do with It?' A Reinterpretation of Aristotle's *Generation of Animals* Book V," *Phronesis* 55: 325–356.

Graham, D. 1999. *Aristotle, Physics Book VIII*. Oxford: Clarendon.

Gregoric, P. 2005. "Plato's and Aristotle's Explanation of Human Posture," *Rhizai* II.2: 183–196.

2007. *Aristotle on the Common Sense*. Oxford: Oxford University Press.

Gregoric, P., and F. Grgic. 2006. "Aristotle's Notion of Experience," *Archiv für Geschichte der Philosophie* 88: 1–30.

Grene, M. 1985. "About the Division of the Sciences," in Gotthelf (ed.), 9–13.

Grimm-Stadelmann, I. (ed.). 2008. "Theophilos. Der Aufbau des Menschen. Kritische Edition des Textes mit Einleitung, Übersetzung und Kommentar," PhD diss., Ludwig-Maximilians-University.

Güreman, R. 2015. "Merely Living Animals in Aristotle," *Journal of Ancient Philosophy* 9.1: 115–134.

Hadot, I. 1990. "Commentaire," in I. Hadot, Ph. Hoffmann, P. Hadot, and J.-P. Mahé (eds.) *Simplicius. Commentaire sur les "Catégories". Vol. 1: Introduction. Première Partie (p. 1–9,3 KALBFLEISCH)*. Leiden: Brill, 19–168.

Hankinson, R. J. 2016. "Galen's Reception of Aristotle," in A. Falcon (ed.) *Brill's Companion to the Reception of Aristotle in Antiquity*. Leiden: Brill, 238–257.

Hatzimichali, M. 2016. "Andronicus of Rhodes and the Construction of the Aristotelian Corpus," in A. Falcon (ed.) *Brill's Companion to the Reception of Aristotle in Antiquity*. Leiden; Boston: Brill, 81–100.

2021. "The Early Reception of Aristotle's Biology," in Connell (ed.), 228–245.

Hayduck, M. (ed.). 1904. *Michaelis Ephesii in Libros De Partibus Animalium, De Animalium Motione, De Animalium Incessu Commentaria*. Berlin: De Gruyter.
Heath, T. 1949. *Mathematics in Aristotle*. Oxford: Oxford University Press.
Hellmann, O. 2006. "Peripatetic Biology and the Epitome of Aristophanes of Byzantium," in W. W. Fortenbaugh and S. A. White (eds.) *Aristo of Ceos. Text, Translation and Discussion*. London: Routledge, 329–359.
2015. "On the Interface of Philology and Science: The Case of Zoology," in F. Montanari, S. Matthaios, and A. Rengakos (eds.) *Brill's Companion to Ancient Greek Scholarship, Vol. I: History Disciplinary Profiles*. Leiden: Brill, 1235–1266.
Hennig, B. 2020. "Form and Function in Aristotle," *History of Philosophy and Logical Analysis* 23.2: 317–337.
Henry, D. 2007. "How Sexist Is Aristotle's Developmental Biology?," *Phronesis* 52.3: 251–269.
2008. "Organismal Natures," in J. Mouracade (ed.) *Aristotle on Life*. Kelowna, BC: Academic, 47–74.
2011. "Aristotle's Pluralistic Realism," *The Monist* 94.2: 197–220.
2013. "Optimality Reasoning in Aristotle's Natural Teleology," *Oxford Studies in Ancient Philosophy* 45: 225–263.
2014. "The Birds and the Bees: Aristotle on the Biological Concept of ἀνάλογον," *Proceedings of the Boston Area Colloquium in Ancient Philosophy* 29.1: 145–169.
2019. *Aristotle on Matter, Form and Moving Causes: The Hylomorphic Theory of Substantial Generation*. Cambridge: Cambridge University Press.
2021a. "Essence and Definition in Aristotle's Parts of Animals," *British Journal for the History of Philosophy*, 29.5: 763–783.
2021b. "*Parts of Animals* Book I on Methods of Inquiry," in Connell (ed.), 83–96.
Holmes, B., and K.-D. Fischer. 2015. *The Frontiers of Ancient Science: Essays in Honor of Heinrich von Staden*. Berlin: De Gruyter.
Howton, R. 2020. "Why *De Anima* needs III 12–13," in C. Murgier, C. Louguet, and G. Guyomarc'h (eds.) *Aristote et l'Âme Humaine: Lectures de De Anima III Offertes à Michel Crubellier*. Leuven: Peeters, 329–350.
Huffman, C. A. 1993a. "Philolaus' Use of ἀρχαί and the Method of Hypothesis," in Huffman 1993b: 78–92.
1993b. *Philolaus of Croton*. Cambridge: Cambridge University Press.
Hutchinson, D. S., and M. R. Johnson. 2005. "Authenticating Aristotle's *Protrepticus*," *Oxford Studies in Ancient Philosophy* 29: 193–294.
2014. "Protreptic Aspects of Aristotle's *Nicomachean Ethics*," in R. Polansky (ed.) *The Cambridge Companion to Aristotle's Nicomachean Ethics*. Cambridge: Cambridge University Press, 383–409.
2018. "Protreptic and Apotreptic: Aristotle's Dialogue *Protrepticus*," in A. Kotzé and S. van der Meeren (eds.) *When Wisdom Calls: Philosophical Protreptic in Antiquity*. Brepols: Turnhout, 119–162.

Unpublished manuscript. Aristotle's *Protrepticus*: Text, Translation, Notes, and Indices. www.protreptic.info.

Irwin, T. H. 1988. *Aristotle's First Principles*. Oxford: Oxford University Press.

Jazdzewska, K. 2015. "Dialogic Format of Philo of Alexandria's *De animalibus*," *Eos* 102.1: 45–56.

Joachim, H. H. 1922. *Aristotle on Coming to Be and Passing Away: A Revised Text with Introduction and Commentary*. Oxford: Clarendon.

Johansen, T. K. 1997. *Aristotle on the Sense-Organs*. Cambridge: Cambridge University Press.

2012. *The Powers of Aristotle's Soul*. Oxford: Oxford University Press.

2015. "The Two Kinds of End in Aristotle: The View from the *de Anima*," in Ebrey (ed.), 119–136.

Johnson, M. R. 2005. *Aristotle on Teleology*. Oxford: Oxford University Press.

2015. "Aristotle's Architectonic Sciences," in D. Ebrey (ed.), 163–186.

2021. "Biology and Theology in Aristotle's Theoretical and Practical Sciences," in Connell (ed.), 12–29.

2025. "Aristotle's Protreptic to the Study of Animals and Plants in *On the Parts of Animals*," in P. Destree and M. R. Johnson (eds.), *Philosophical Exhortation: Studies in Protreptic Aspects of Aristotle's Works*, 87–114. Louvain; Paris: Peeters.

Jones, C. P. 1972. "Two Friends of Plutarch," *Bulletin de Correspondance Hellénique* 96.1: 263–267.

Kalogeridou, O. 2010. Μιχαήλ Εφεσίου Εις το Αριστοτέλους περί ζώων γενέσεως: από την αρχαία εξηγητική παράδοση στη βυζαντινή ερμηνευτική πρακτική. Thessaloniki: Aristotle University of Thessaloniki Press.

Karbowski, J. 2012. "Slaves, Women, and Aristotle's Natural Teleology," *Ancient Philosophy* 32: 323–350.

2019. "Political Animals and Human Nature in Aristotle's Politics," in Keil and Kreft (eds.), 221–237.

Katz, E. 2022. "De Generatione et Corruptione II.3: Some Puzzles," in Dimas, Falcon, and Kelsey (eds.), 59–78.

Keil, G., and N. Kreft (eds.). 2019. *Aristotle's Anthropology*. Cambridge: Cambridge University Press.

Kelsey, S. (n.d.) "The Cement of Aristotle's Universe." Unpublished manuscript.

King, R. A. H. 2001. *Aristotle on Life and Death*. London: Duckworth.

2021. "Life-Cycles and the Actions of Nutritive Soul in Aristotle," in Connell (ed.), 127–141.

Kirby, J. 1997. "Aristotle on Metaphor," *The American Journal of Philology* 118.4: 517–554.

Koch, L. 2015. "τὸ τῆς λέξεως συνεχές. Michael von Ephesos und die Rezeption der Aristotelischen Schrift De motu animalium in Byzanz," PhD diss., University of Hamburg.

Korobili, G. 2022. *Aristotle. On Youth and Old Age, Life and Death, and Respiration 1–6: With Translation, Introduction and Interpretation*. Cham: Springer.

Koslicki, K. 1997. "Four Eighth Hephaistos: Artifacts and Living Beings in Aristotle," *History of Philosophy Quarterly* 14.1: 77–98.

Form, Matter, Substance. Oxford: Oxford University Press.

Kress, E. 2023. "How the Soul Uses Tools: Flexible Agency in Aristotle's Account in *Generation of Animals*," *Phronesis* 68: 293–325.

Kruk, R. 1979. *The Arabic Version of Aristotle's Parts of Animals. Books XI–XIV of the Kitāb Al-Ḥayawān: A Critical Edition*. Amsterdam and Oxford: North Holland.

Kullmann, W. 1985. "Different Concepts of the Final Cause in Aristotle," in Gotthelf (ed.), 169–175.

1997. *Aristoteles. Über die Teile der Lebewesen*. Berlin: Akademie Verlag.

1999. "Aristoteles' Wissenschaftliche Methode in Seinen Zoologischen Schriften," in *Geschichte der Mathematik und der Naturwissenschaften in der Antike 1: Biologie*: Stuttgart: F. Steiner, 103–123.

2007. *Aristoteles: Über die Teile der Lebewesen. Werke in deutscher Übersetzung*, Band 17, Zoologische Schriften II, Teil I. Berlin: Akademie Verlag.

Kullmann. W., and S. Föllinger (eds.). 1997. *Aristotelische Biologie: Intentionen, Methoden, Ergebnisse*. Stuttgart: Steiner.

Kühn C. G. (ed.). 1821–1833. *Claudii Galeni opera omnia*, 20 vols. Leipzig: C. Cnobloch.

Lambros, S. (ed.). 1885. *Excerptorum Constantini de Natura Animalium Epitome subiunctis Aeliani Timothei aliorumque Eclogis*. Berlin: De Gruyter.

Lefebvre, D. 2018. "Parts and Generation: The Prologue to the *Generation of Animals* and the Structure of the Treatise," in Falcon and Lefebvre (eds.), 35–55.

Lennox, J. G. 1985a. "Are Aristotelian Species Eternal?," in Gotthelf (ed.), 67–94. Reprinted in Lennox 2001: 131–159.

1985b. "Demarcating Ancient Science: A Discussion of GER Lloyd, *Science, Folklore and Ideology*," *Oxford Studies in Ancient Philosophy* 3: 307–324.

1987a. "Divide and Explain: The Posterior Analytics in Practice," in Gotthelf and Lennox (eds.), 90–119. Reprinted in Lennox 2001b: 7–38.

1987b. "Kinds, Forms of Kinds, and the More and the Less in Aristotle's Biology," in Gotthelf and Lennox (eds.), 339–359. Reprinted in Lennox 2001: 160–181.

1991. "Between Data and Demonstration: The Analytics and the *Historia Animalium*," in A. Bowen (ed.) *Science and Philosophy in Classical Greece*. New York: Garland, 2–61. Reprinted in Lennox 2001b: 39–71.

1994. "The Disappearance of Aristotle's Biology: A Hellenistic Mystery," in T. D. Barnes (ed.) *The Sciences in Greco-Roman Society*. Edmonton: Academic, 7–24.

1996a. "Aristotle's Biological Development: The Balme hypothesis," in W. R. Wians (ed.) *Aristotle's Philosophical Development*. Lanham, MD: Rowman & Littlefield Publishers, 229–248.

1996b. "Material and Formal Natures in Aristotle's De Partibus Animalium," in John J. Cleary and W. Wians (eds.) *Proceedings of the Boston Area Colloquium of Ancient Philosophy* 1.11: 217–240. Reprinted in Lennox 2001b: 182–204.

1997. "Nature Does Nothing in Vain . . .," in H. Günther and A. Rengakos (eds.) *Beiträge zur antiken Philosophie: Festschrift für Wolfgang Kullmann.* Stuttgart: Steiner, 199–214. Reprinted in Lennox 2001b: 205–223.

1999. "The Place of Mankind in Aristotle's Zoology," *Philosophical Topics* 27.1: 1–16.

2001a. *Aristotle On the Parts of Animals Book I–IV.* Oxford: Clarendon.

2001b. *Aristotle's Philosophy of Biology: Studies in the Origins of Life Science.* Cambridge: Cambridge University Press.

2001c. "Theophrastus on the Limits of Teleology," in Lennox 2001a: 259–70.

2005a. "The Place of Biology in Aristotle's Natural Philosophy," in R. W. Sharples (ed.) *Philosophy and the Sciences in Antiquity.* Aldershot: Ashgate, 55–71.

2005b. "Getting a Science Going: Aristotle on Entry Level Kinds," in G. Wolters and M. Carrier (eds.) *Homo Sapiens und Homo Faber.* Berlin: de Gruyter, 87–100.

2008. "'As If We Were Investigating Snubness': Aristotle on the Prospects for a Single Science of Nature," *Oxford Studies in Ancient Philosophy* 35: 149–186.

2009. "Aristotle on Mind and the Science of Nature," in M. Rossetto, M. Tsianikas, and G. Palaktsoglou (eds.) *Greek Research in Australia: Proceedings of the Eighth Biennial Conference of Greek Studies.* Adelaide: Flinder's University Press, 1–18.

2010a. "Aristotle's Natural Science: The Many and the One," *Apeiron* 43: 1–23.

2010b. "The Unity and Purpose of On the Parts of Animals I," in Lennox and Bolton (eds.), 56–77.

2010c. "Bios and Explanatory Unity in Aristotle's Biology," in D. Charles (ed.) *Definition in Greek Philosophy.* Oxford: Clarendon Press, 329–355.

2010d. "Aristotle's Natural Science: The Many and the One," in Lesher 2010: 1–23.

2010e. "*De Caelo* 2.2 and Its Debt to the *De Incessu Animalium,*" in A. Bowen and C. Wildberg (eds.) *New Perspectives on Aristotle's De Caelo.* Boston: Brill, 187–214.

2012. "The Complexity of Aristotle's Study of Animals," in C. Shields (ed.) *The Oxford Handbook of Aristotle.* Oxford: Oxford University Press, 287–305.

2014. "Aristotle on the Emergence of Material Complexity: *Meteorology* IV and Aristotle's Biology," *Hopos: The Journal of the International Society for the History of Philosophy of Science* 4.2: 272–305.

2019. "Is Reason Natural? Aristotle's Zoology of Rational Animals," in G. Keil and N. Kreft, eds. Aristotle's Anthropology, Cambridge University Press, 99–117.

2020. "Why Animals Must Keep Their Cool: Aristotle on the Need for Respiration (and Other Forms of Cooling)," in Bartoš and King (eds.), 217–242.

2021. *Aristotle on Inquiry: Erotetic Frameworks and Domain-Specific Norms.* Cambridge: Cambridge University Press.

Lennox, J. G., and R. Bolton. (eds.). 2010. *Being, Nature, and Life in Aristotle.* Cambridge: Cambridge University Press.

Lesher, J. 2010. *From Inquiry to Demonstrative Knowledge: New Essays on Aristotle's Posterior Analytics*. Kelowna, BC: Academic Printing & Publishing.

Leunissen, M. 2010a. *Explanation and Teleology in Aristotle's Science of Nature*. Cambridge: Cambridge University Press

2010b. "Aristotle's Syllogistic Model of Knowledge and the Biological Sciences: Demonstrating Natural Processes," in Lesher 2010: 31–60.

2012. "Crafting Natures: Aristotle on Animal Design," in G. Dicker (ed.) *The Annual Proceedings of the Center for Philosophic Exchange*. Brockport: State University of New York Press, 28–51.

2014. "Commentary on Henry," *Proceedings of the Boston Area Colloquium in Ancient Philosophy* 29.1: 170–181.

2017a. "Surrogate Principles and the Natural Order of Exposition in Aristotle's *De Caelo II*," in W. Wians and R. Polansky (eds.) *Reading Aristotle: Argument and Exposition*. Leiden: Brill, 165–180.

2017b. *From Natural Character to Moral Virtue in Aristotle*. Oxford: Oxford University Press

2018. "Order and Method in Aristotle's Generation of Animals 2," in Falcon and Lefebvre (eds.), 56–74.

2021. "Empiricism and Hearsay in Aristotle's Zoological Collection of Facts," in Connell (ed.), 64–82.

Lewis, O., and P. Gregoric. 2015. "The Context of *De Spiritu*," *Early Science and Medicine* 20: 125–149.

Lloyd, G. E. R. 1983. *Science, Folklore and Ideology: Studies in the Life Sciences in Ancient Greece*. Cambridge: Cambridge University Press.

1987. "Empirical Research in Aristotle's Biology," in Gotthelf and Lennox (eds.), 53–63.

1992. "Aspects of the Relationship between Aristotle's Psychology and His Zoology," in M. Nussbaum and A. Rorty (eds.), 147–168.

1996. *Aristotelian Explorations*. Cambridge: Cambridge University Press.

1996a. "The Theories and Practices of Demonstration," in Lloyd 1996: 7–37.

1996b. "The Unity of Analogy," in Lloyd 1996: 138–159.

Lo Presti, R., and G. Korobili (eds.). 2021. *Nutrition and The Nutritive Soul in Aristotle and Aristotelianism*. Berlin: De Gruyter.

Louis, P. 1964. *Aristote. Histoire des animaux, tom. I, livres I–IV. Texte établi et traduit*. Paris: Les Belles Lettres.

2017. *Aristote: Les parties des animaux. 2 (revised and corrected)*, 5th printing. Paris: Les Belles Lettres.

Loux, M. 1979. "Form, Species, and Predication in *Metaphysics* Z, H, Θ," *Mind* 88.349: 1–23.

Malink, M. 2017. "Aristotle on Principles as Elements," *Oxford Studies in Ancient Philosophy* 53: 163–213.

Marchant, E. C. 1979. *Xenophon Memorabilia and Oeconomicus*. Cambridge, MA: Harvard University Press.

May, M. T. 1968. *Galen On the Usefulness of the Parts of the Body: Translated from the Greek with an Introduction and Commentary*, 2 vols. Ithaca, NY: Cornell University Press.

Mayhew, R. 2011. *Problems, Volume II: Books 20–38. Rhetoric to Alexander.* Cambridge, MA: Harvard University Press.

Mayhoff, K. F. T. (ed.). *Pliny the Elder Naturalis Historia (Pliny the Elder, Naturalis Historia, liber xi*, chapter 8, tufts.edu.

Michalewski, A. 2016. "The Reception of Aristotle in Middle Platonism: From Eudorus of Alexandria to Ammonius Saccas," in A. Falcon (ed.) *Brill's Companion to the Reception of Aristotle in Antiquity*. Leiden, Boston: 218–237.

Menn, S. 1992. "Aristotle and Plato on God as *Nous* and as the Good," *Review of Metaphysics* 45.3: 543–573.

1995. *Plato on God as Nous*. Carbondale, IL: Southern Illinois University Press.

2002. "Plato and the Method of Analysis," *Phronesis* 47.3: 193–223.

Merlan, P. 1960. *From Platonism to Neoplatonism*, 2nd ed. The Hague: Martinus Nijhoff.

Meyer, M. F. 2006. "Der Mensch Als Maß und Muster: Anthropozentrische Momente der Aristotelischen Biologie," *Antike Naturwissenschaft und ihre Rezeption* 16: 19–34.

Moraux, P. 1985. "Galen and Aristotle's *De Partibus Animalium*," trans. A. Preus, in A. Gotthelf (ed.) *Aristotle on Nature and Living Things*. Pittsburgh and Bristol: Bristol University Press, 327–344.

Morani, M. 1987 (ed.). *Nemesii Emeseni de Natura Hominis*. Leipzig: Teubner.

Morrison, D. R. 2015. "Alcinous on Methods of Analysis," in Cerami (ed.), 417–428.

Mossman, J. 2005. "Plutarch on Animals: Rhetorical Strategies in *de sollertia animalium*," *Hermathena* 179: 141–163.

Müller, J. 2019. "Spot the Differences! The Hidden Philosophical Anthropology in Aristotle's Biological Writings," in Keil and Kreft (ed.), 118–139.

Newmyer, S. T. 2020. *Plutarch's Three Treatises on Animals: A Translation with Introductions and Commentary*. London: Routledge.

Niemitz, C. 2010. "The Evolution of the Upright Posture and Gait: A Review and a New Synthesis," *Naturwissenschaften* 97.3: 241–263.

Nussbaum, M. C. 1978. *Aristotle's De Motu Animalium: Text with Translation, Commentary, and Interpretive Essays*. Princeton: Princeton University Press.

Nussbaum, M. C., and A. Rorty (eds.). 1992. *Essays on Aristotle's De Anima*. Oxford: Oxford University Press.

Nussbaum, M. C., and H. Putnam. 1992. "Changing Aristotle's Mind," in Nussbaum and Rorty (eds.), 27–56.

Ogle, W. 1882. *Aristotle on The Parts of Animals: Translated with Introduction and Notes*, London: Kegan Paul.

1912. *De Partibus Animalium* (trans.), in J. A. Smith and W. D. Ross (eds.) *The Works of Aristotle*, vol. 5. Oxford: Clarendon Press.

1937. *Parts of Animals*, in J. Barnes (ed.) *The Complete Works of Aristotle: The Revised Oxford Translation*, 2 vols. Princeton: Princeton University Press.

Olson, S. D. 2007. *Athenaeus. The Learned Banqueters, Vol. I: Books 1–3.106e. Edited and Translated*. Cambridge, MA: Harvard University Press.

Osborne, C. 2007. *Dumb Beasts and Dead Philosophers: Humanity and the Humane in Ancient Philosophy and Literature*. Oxford: Oxford University Press.

Peck, A. L. 1937. *Aristotle. Parts of Animals*. Cambridge, MA: Harvard University Press.

1965. *Aristotle. History of Animals*, vol. 1. Cambridge, MA: Harvard University Press.

1968. *Aristotle. Parts of Animals*. Cambridge, MA: Harvard University Press.

Pellegrin, P. 1982. *La Classification des animaux chez Aristote. Statut de la biologie et unité de l'aristotélisme*. Paris: Les Belles Lettres.

1985. "Aristotle: Zoology without Species," trans. A. Preus, in Gotthelf (ed.), 95–116.

1986. *Aristotle's Classification of Animals: Biology and the Conceptual Unity of the Aristotelian Corpus*. Revision of Pellegrin 1982, trans. A. Preus, Berkeley: University of California Press.

1987. "Logical Difference and Biological Difference: The Unity of Aristotle's Thought," in Gotthelf and Lennox (eds.), 313–338.

2011. *Aristote Les Parties des animaux*. Paris: Flammarion.

2025. "Can Aristotle Be Considered a Biologist?," in D. Lefebvre (ed.), *The Science of Life in Aristotle and the Early Peripatos*. Leiden: Brill, 335–353.

Peramatzis M. 2011. *Priority in Aristotle's Metaphysics*. Oxford: Oxford University Press.

(2013–2014), "Matter in Scientific Definitions in Aristotle," in Oxford Handbook Editorial Board (eds.), *Oxford Handbooks Online*. New York: Oxford University Press.

2015. "What Is a Form in Aristotle's Hylomorphism?," *History of Philosophy Quarterly* 32.3, 195–216.

2018. "Aristotle's Hylomorphism: The Causal-Explanatory Model," *Metaphysics*. 1.1: 12–32. http://doi.org/10.5334/met.2.

2019. "*Form, Matter, Substance*, by Kathrin Koslicki," *Mind*. https://doi.org/10.1093/mind/fzz006.

Perfetti, S. 2000. *Aristotle's Zoology and its Renaissance Commentators (1521–1601)*. Leuven: Leuven University Press.

Pfeiffer, R. 1949. *Callimachus Vol. I. Fragmenta*. Oxford: Oxford University Press.

Pickard-Cambridge, W. A. 1928. *Aristotle: Topica and de sophisticis elenchis. The Works of Aristotle*, vol. I. Oxford: Oxford University Press.

Pistelli, H. 1888. *Iamblichi Protrepticus*. Leipzig: Teubner.

Preus, A. 1975. *Science and Philosophy in Aristotle's Biological Works*. New York: Hildesheim.

1990. "Man and Cosmos in Aristotle: *Metaphysics* Λ and the Biological Works," in D. Devereux and P. Pellegrin (eds.) *Biologie, Logique et Metaphysique Chez Aristote: Actes du Seminaire Cr.S.-N.S.F., 28 Juin–3 Juillet 1987*. Paris: Éditions du Centre National de la Recherche Scientifique, 471–491.

Primavesi, O. 2020. "Introduction, Part II," in Ch. Rapp and O. Primavesi (eds.) *Aristotle's De Motu Animalium. Symposium Aristotelicum*. Oxford: Oxford University Press, 67–158.

Quandt, K. 1983. "On the Programmatic Formula ΠΡΩΤΟΝ ΑΠΟ ΤΩΝ ΠΡΩΤΩΝ in Aristotle," *The American Journal of Philology* 104: 358.

Quarantotto, D. 2018. *Aristotle's Physics Book I: A Systematic Exploration.* Cambridge: Cambridge University Press.

2019. "Aristotle on the Differences in Material Organisation between Spoken and Written Language: An Inquiry into Part-Whole Relations," *Elenchos* 40: 333–362.

2022. "Aristotle on the Order of Embryonic Development and the Homonymy Principle," in Föllinger (ed.), 88–103.

Rapp, C. 2019. "The Planetary Nature of Mankind: A Cosmological Perspective on Aristotle's Anthropology," in Keil and Kreft (eds.), 77–96.

Rashed, M. 2004. "Agrégat de parties ou *vinculum substantiale*? Sur une hésitation conceptuelle et textuelle du corpus aristotelicien," in A. Laks and M. Rashed (eds.) *Aristote et le mouvement des animaux*. Villeneuve d'Ascq: Presses universitaires du Septentrion, 185–202.

2021. *Ptolémée al-Gharīb'. Épître à Gallus sur la vie, le testament et les écrits d'Aristote*. Paris: Les Belles Lettres.

Renehan, R. 1969. *Leo the Physician. Epitome On the Nature of Men*, ed. and trans. R. Renehan. Berlin: De Gruyter.

Robinson, R. 1941. *Plato's Earlier Dialectic*. Ithaca: Cornell University Press.

Rocca, J. 2003. *Galen on the Brain: Anatomical Knowledge and Physiological Speculation in the Second Century AD*. Leiden: Brill.

2019. "Brai," in Oxford Classical Dictionary. https://oxfordre.com/classics/view/10.1093/acrefore/9780199381135.001.0001/acrefore-9780199381135-e-8160.

Rosen, J. 2008. *Necessity and Teleology in Aristotle's Physics*. Princeton: Princeton University Press.

Ross, W. D. 1925. *Aristotle: Ethica Nicomachea. The Works of Aristotle*, vol. IX. Oxford: Oxford University Press.

1928. *Aristotle: Metaphysica. The Works of Aristotle*, vol. VIII, 2nd ed. Oxford: Oxford University Press.

Sandbach, F. H. 1982. "Plutarch and Aristotle," *Illinois Classical Studies* 7.2: 207–232.

Sassi, M. M., E. Coda, and G. Feola (eds.). *La zoologia di Aristotele e la sua ricezione dall'età ellenistica e romana alle culture medievali*. Pisa: Pisa University Press.

Scaltsas T. 1994. *Substances and Universals in Aristotle's Metaphysics*. Ithaca; London: Cornell University Press.

Sedley, D. 2007. *Creationism and Its Critics in Antiquity*. Berkeley: University of California Press.

Sefrin-Weis, H. 2002. "Homogeneity in Aristotle's Metaphysics," PhD dissertation, University of Pittsburgh.

Segev, M. 2012. "The Teleological Significance of Dreaming in Aristotle," *Oxford Studies in Ancient Philosophy* 43: 107–141.

Shields, C. 2016. *Aristotle: De Anima*. Oxford: Oxford University Press.

Spittler, J. 2016. "Apollonios, Amazing Stories (1672)," in S. Schorn and G. Schepens (eds.). http://dx.doi.org/10.1163/1873-5363_jciv_a1672.

Stavrianeas, S. 2018. "The Good, the Bad, and the Ugly: Natural Teleology and Its Failures in Aristotle," in P. Kontos (ed.) *Evil in Aristotle*. Cambridge: Cambridge University Press, 51–72.

2021. "De incessu animalium 7–8 – Number and Distribution of Feet in Animal Progression," in Falcon and Stavrianeas (ed.), 165–193.

Stein, N. 2016. "Explanation and Hypothetical Necessity in Aristotle," *Ancient Philosophy* 36.2: 353–382.

2018. "Definition and the Epistemology of Natural Kinds in Aristotle," *Metaphysics* 1: 33–51.

Stocks, J. L. 1930. *Aristotle: On the Heavens. The Works of Aristotle*, vol. II. Oxford: Oxford University Press

Stoyles, B. J. 2013. "Μέγιστα Γένη and Division in Aristotle's *Generation of Animals*," *Apeiron* 46.1: 1–25.

Taylor, A. E. 1928. *A Commentary on Plato's Timaeus*. Oxford: Oxford University Press.

Thein, K. 2014. "Aristotle on Why Study Lower Animals (*De Partibus Animalium* I,5, 644b22–645a36)," *Eirene* 50: 208–229.

Thompson, D'Arcy W. 1910. *Aristotle Historia Animalium*. Oxford: Clarendon.

Tieleman, T. 2021. "The Use of Aristotle's Biology in Nemesius," *On Human Nature, Journal of Ancient Philosophy* 15.2: 170–189.

Tipton, J. A. 2013. *Philosophical Biology in Aristotle's Part of Animals*. New York: Springer.

Trizio, M. 2007. "Byzantine Philosophy as a Contemporary Historiographical Project," *Recherches de Theologie et Philosophie médiévales* 74.1: 247–294.

Tutrone, F. 2012, *Filosofi e animali in Roma antica. Modelli di animalità e umanità in Lucrezio e Seneca*. Pisa: Edizioni ETS.

2013. "Libraries and Intellectual Debate in the Late Republic: The Case of the Aristotelian Corpus," in J. König, K. Oikonomopoulou, and G. Woolf (eds.) *Ancient Libraries*. Cambridge: Cambridge University Press, 152–166.

van der Eijk, P. 1997. "The Matter of Mind: Aristotle on the Biology of Psychic Processes and the Bodily Aspects of Thinking," in Kullmann and Föllinger (eds.), 231–258. Reprinted in van der Eijk, *Medicine and Philosophy in Classical Antiquity: Doctors and Philosophers on Nature, Soul, Health and Disease*. Cambridge: Cambridge University Press, 206–237.

2013. "Galen and the Scientific Treatise: A Case Study of Mixtures," in M. Asper (ed.) *Writing Science: Medical and Mathematical Authorship in Ancient Greece*. Berlin: De Gruyter, 145–176.

van Oppenraaij, A. M. I. (ed.). 1998. *Aristotle De Animalibus. Michael Scot's Arabic-Latin Translation. Part Two: Books XI–XIV: Parts of Animals*. Leiden: Brill.

von Fritz, K. 1974. "*Nous, Noein*, and Their derivatives in Pre-Socratic philosophy," in A. P. D. Mourelatos (ed.) *The Presocratics: A Collection of Critical*

Essays. Garden City, NY: Anchor Press. Reprinted (excluding Anaxagoras) from *Classical Philology* 40 (1945) and 41 (1946).

von Staden, H. 1989. *Herophilus. The Art of Medicine in Early Alexandria. Edition, Translation and Essays*. Cambridge: Cambridge University Press.

Wedin, M. V. 1988. *Mind and Imagination in Aristotle*. New Haven: Yale University Press

2000. *Aristotle's Theory of Substance: The Categories and Metaphysics Zeta*. Oxford: Oxford University Press.

Wein, A. n.d. "Lingering Problems in *Physics* I.1." Unpublished paper.

Whiting, J. 1992. "Living Bodies," in Nussbaum and Rorty (eds.), 75–92.

1991. "Metasubstance: Critical Notice of Frede-Patzig and Furth," *The Philosophical Review*, 100.4: 607–639.

1986. "Form and Individuation in Aristotle," *History of Philosophy Quarterly*, 3.4: 359–377.

Wilberding, J. 2017. *Forms, Souls, and Embryos: Neoplatonists on Human Reproduction*. London: Routledge.

Williams, C. J. F. 1982. *Aristotle, De Generatione et Corruptione*. Oxford: Clarendon.

Wilson, M. 1997. "Analogy in Aristotle's Biology," *Ancient Philosophy* 17: 335–351.

2000. *Aristotle's Theory of the Unity of Science*. Phoenix Supplementary volume. Toronto: University of Toronto Press.

2009. "A Somewhat Disorderly Nature: Unity in Aristotle's *Meteorologica* I–III," *Apeiron* 42: 63–88.

2013. *Structure and Method in Aristotle's Meteorologica: A More Disorderly Nature*. Cambridge: Cambridge University Press.

Witt, C. 2012. "Aristotle on Deformed Animal Kinds," *Oxford Studies in Ancient Philosophy* 43: 83–106.

Wolters, G., and M. Carrier. 2005. *Homo Sapiens und Homo Faber: Epistemische und Technische Rationalität in Antike und Gegenwart*. Berlin: De Gruyter.

Zalta, E. N. 2006. "Essence and modality," *Mind* 115: 659–693.

General Index

CAMBRIDGE CRITICAL GUIDES

Titles published in this series (continued):

Descartes' *Meditations*
EDITED BY KAREN DETLEFSEN
Augustine's *City of God*
EDITED BY JAMES WETZEL
Kant's *Observations and Remarks*
EDITED BY RICHARD VELKLEY AND SUSAN SHELL
Nietzsche's *On the Genealogy of Morality*
EDITED BY SIMON MAY
Aristotle's *Nicomachean Ethics*
EDITED BY JON MILLER
Kant's *Metaphysics of Morals*
EDITED BY LARA DENIS
Spinoza's *Theological-Political Treatise*
EDITED BY YITZHAK Y. MELAMED AND MICHAEL A. ROSENTHAL
Plato's *Laws*
EDITED BY CHRISTOPHER BOBONICH
Plato's *Republic*
EDITED BY MARK L. MCPHERRAN
Kierkegaard's *Concluding Unscientific Postscript*
EDITED BY RICK ANTHONY FURTAK
Wittgenstein's *Philosophical Investigations*
EDITED BY ARIF AHMED
Kant's *Critique of Practical Reason*
EDITED BY ANDREWS REATH AND JENS TIMMERMANN
Kant's *Groundwork of the Metaphysics of Morals*
EDITED BY JENS TIMMERMANN
Kant's *Idea for a Universal History with a Cosmopolitan Aim*
EDITED BY AMÉLIE OKSENBERG RORTY AND JAMES SCHMIDT
Mill's *On Liberty*
EDITED BY C. L. TEN
Hegel's *Phenomenology of Spirit*
EDITED BY DEAN MOYAR AND MICHAEL QUANTE

www.ingramcontent.com/pod-product-compliance
Lightning Source LLC
LaVergne TN
LVHW011704200926
848484LV00004B/1062

* 9 7 8 1 3 1 6 5 1 6 1 1 9 *